INTRODUCTION TO NUMBER THEORY

For BA, BSc [Hons and General], and PG students of all leading Indian universities

INTRODUCTION TO NUMBER THEORY

Dr Ajay Kr Chaudhuri

Ex-Reader and Head of the
Department of Mathematics [UG & PG]
J K College, Purulia [WB]

New Central Book Agency (P) Ltd

LONDON

HYDERABAD ERNAKULAM BHUBANESWAR

NEW DELHI KOLKATA PUNE GUWAHATI

NCBA

REGD OFFICE
8/1 Chintamoni Das Lane, Kolkata 700 009, India
email: ncbapvtltd@eth.net

OVERSEAS
NCBA (UK) Ltd, 149 Park Avenue North, Northampton, NN3 2HY, UK
email: ncbauk@yahoo.com

EXPORT
NCBA Exports Pvt. Ltd, 212 Shahpur Jat, New Delhi 110 049
email: ncbaexports@ncbapvtltd.com

BRANCHES

2/27 Ansari Road, 1st Floor
Daryaganj, **New Delhi** 110 002
email: ncbadel@ncbapvtltd.com

Shop Nos. 3 & 4, Vinayak Towers
681/B Budhwar Peth, Appa Balwant Chowk
Pune 411 002
email: ncbapune@ncbapvtltd.com

Shop No. 15, Bharati Towers, Ground Floor
Forest Park
Bhubaneswar 751 001
email: ncbabub@ncbapvtltd.com

House No. 3-1-315, 1st Floor
Nimboliadda, Kachiguda, **Hyderabad** 500 027
email: ncbahydb@ncbapvtltd.com

GSS Shopping Complex, 1st Floor
Opposite College Post Office, Convent Road
Ernakulam 682 035
0484-4021054 / 53

Radhamayee Complex (Second Floor)
Opp. SBI, Panbazar, Jaswanta Road
Guwahati 781 001
email: ncbaguwahati@ncbapvtltd.com

INTRODUCTION TO NUMBER THEORY • Dr Ajay Kr Chaudhuri

First Published: January 2009
Reprinted: July 2012
Thoroughly Revised and Updated Second Edition: May 2018

PUBLISHER
New Central Book Agency (P) Ltd
8/1 Chintamoni Das Lane, Kolkata 700 009

TYPESETTER
Anin, Kolkata

PRINTER
New Central Book Agency (P) Ltd
Web-Offset Division, Dhulagarh, Sankrail, Howrah

TECHNICAL EDITOR
Dipan Roy

PROJECT TEAM
Somenath Ganguly and Pradip Biswas

ISBN: 978 81 7381 586 7

Price ₹ 250.00

Dedicated to my mother
Late Sabitri Chaudhuri

CONTENTS

PREFACE TO THE SECOND EDITION

'Introduction to Number Theory' was first published in January 2009. In undergraduate and postgraduate courses in Mathematics of almost all universities Number Theory has already been introduced. So feeling the need of a comprehensive text on the topic and considering the suggestions of teachers and students from all corners, the book has been revised thoroughly by correcting the printing errors and changing statements and proofs of theorems and problems in some cases for better presentation before the users. In particular the most important part 'Fibonacci Sequence' in number theory has been explained through theorems and applications elaborately in Ch-10 and more than 25 newly solved problems are inserted in Ch-11 (solved problems). Hope, the second edition of this volume will be more acceptable to the users. I remain grateful to all the users of the book for their constructive suggestions towards the improvement of its quality.

My sincere thanks are due to New Central Book Agency (P) Ltd, for their excellent co-operation in bringing out the second edition of this volume.

19 February 2018

Ajay Kumar Chaudhuri

PREFACE TO THE FIRST EDITION

Number Theory is that branch of Pure Mathematics which deals with numbers in general, and integers in particular. The intricate Number Theory is subdivided into several sectors—according to the methods used and the type of problems investigated. It can be stated in elementary number theoretical terms, but may also require, to solve, deep consideration, and new approaches outside the realm of elementary number theory.

Gabriel Lame (1795-1875), though a man of physics, made large contributions to Number Theory. There are other mathematical scientists and philosophers who have—in the history of mathematics through the last 2500 years—developed this theory in general.

Pythagoras; Euclid; Diophantus; Kronecker; Dirichlet; Fermat; Gauss; Euler; Eisentien; Edward Waring; Lagrange; Leibnitz; Wilson; Srinivasa Ramanujan; G. H. Hardy.

Our undergraduate text accomodates all the students of Mathematics and Engineering, as well as Researchers. It mainly covers the syllabi of Maths Honours courses of all leading Indian Universities.

After each chapter the solutions of a large number of problems are given. The book also offers an outstanding introduction to partitions plus chapters on multiplicativity—divisibility, congruences, linear and non-linear Diophantine Equations and much more.

My sincere thanks are due to New Central Book Agency (P) Ltd, for their excellent cooperation in bringing out this volume.

The author would thankfully welcome notices for errors and suggestions for the improvement of the book in the subsequent editions.

14 April 2008 **A. K. Chaudhuri**

Chapter 1

Divisibility

The set of numbers 1, 2, 3, ... is called natural numbers and is denoted by $N = \{1, 2, 3, \ldots\}$. N is closed with respect to addition and multiplication but not closed w.r.t. subtraction and division. To make it close w.r.t. subtraction we extended N by the set of numbers ..., -2, -1, 0 and finally N becomes $\{0, \pm 1, \pm 2, \ldots\}$ which is denoted by Z, called as the set of all integers. $N = \{1, 2, 3, \ldots\}$ is further called as the set of (+)ve integers. Again, to make N closed w.r.t. division we extended N by $Q = \left\{\frac{a}{b},\ a, b \in z,\ b \neq 0\right\}$ called the set of rational numbers. It follows that $N \subset Z$ and $Z \subset Q \Rightarrow N \subset Q$. In number theory we deal with the properties of integers only, i.e., the properties of Z and N.

$S = \{\ldots -2, -1, 1, 2, \ldots\}$ is called as the set of non-zero integers. That is, $S = Z - \{0\}$. $Z^+ =$ set of all non-negative integers $= \{0, 1, 2, \ldots\}$.

Definition (1.1): *Let $a \in Z$, and x is any arbitrary member of Z. Then ax is called multiple of a.*

Example: $3 \times 7 = 21$. Then 21 is called a multiple of 3. Also it is called a multiple of 7.

Example: The number 0 is a multiple of every member of Z, since $a \cdot 0 = 0\ \forall a \in Z$.

Example: There exist infinitely many elements which are multiples of $a \in Z$.

Divisibility: The most important and fundamental concept in number theory is divisibility.

Definition (1.2): *$a \in Z$ is said to be divisible by $b(\neq 0) \in Z$, if $a = bq$ for some $q \in Z$.*

The statement can be stated as 'b divides a' and is denoted by $b|a$.

Again, $b|a \Rightarrow a = b \cdot q$ for some $q \in Z \Rightarrow$ 'a is a multiple of b'.

Example: $9|63$ as $63 = 9 \cdot 7,\ 7 \in Z$.

That is, 63 is a multiple of 9.

Example: -81 is divisible by 3 as $-81 = 3 \cdot (-27),\ -27 \in Z$.

Example: 0 is divisible by every $x \in Z$.

That is $x|0\ \forall x \in Z$, as $0 = x \cdot 0\ \ \forall x \in Z$.

Theorem (1.1): *If $b|a$, then every divisor of b divides a.*

Proof: $b|a \Rightarrow a = bq$ for some $q \in Z$.

Let x be any divisor of b. Then $b = xp$ for some $p \in Z$.

Thus, $a = xpq \Rightarrow x|a,\ pq \in Z$.

Theorem (1.2): *Divisibility has the following properties:*

(i) $x|x\ \forall x \in Z$, (*Reflexive Property*)

(ii) $a|b$, $b|c \Rightarrow a|c$ (*Transitivity Property*)

(iii) $a|b,\ a|c \Rightarrow a|(bx+cy)$ (*Linearity Property*)

(iv) $x|y \Rightarrow ax|ay$ for every $a \in Z$. (*Multiplicative Property*)

(v) $ax|ay \Rightarrow x|y$, when $a \neq 0$ (*Cancellation Law*)

(vi) $1|x\ \forall x \in Z$ (1 *divides every integer*)

(vii) $x|y$ and $y \neq 0 \Rightarrow |x| \leq y$ (*Comparison Property*)

(viii) $x|y$ *and* $y|x \Rightarrow |x| = |y|$

(ix) $x|y$ *and* $x \neq 0 \Rightarrow (y/x)|y$ (y/x *is called the divisor conjugate of* x)

(x) $x|y \Rightarrow |x| \not> y$, *when* y *is a* (+)*ve integer.*

Proof: (i) $x = x \cdot 1,\ 1 \in Z$.

Then, x is called a multiple of $x \Rightarrow x|x$.

(ii) $a|b \Rightarrow b = aq$ for some $q \in Z$

$b|c \Rightarrow c = bp$ for some $p \in Z$.

Thus, $c = apq,\ pq \in Z \Rightarrow a|c$.

(iii) $a|b$ and $a|c \Rightarrow b = aq$ and $c = ap$ for some $q, p \in Z$.

Therefore, $bx = axq = ax',\ x' = xq \in Z$

$cy = ayp = ay',\ y' = yp \in Z$.

Hence, $bx + cy = a(x'+y'),\ x'+y' \in Z$

$\Rightarrow$ $a|(bx+cy)$.

(iv) $x|y \Rightarrow y = px$ for some $p \in Z$

$\Rightarrow$ $ay = pax = axp$ (multiplying by $a \in Z$)

$\Rightarrow$ $ax|ay$.

(v) $ax|ay \Rightarrow ay = (ax)p$ for some $p \in Z$

$\Rightarrow$ $y = xp$, when $a \neq 0$

$\Rightarrow$ $x|y$.

(vi) $x = 1 \cdot x\quad \forall x \in Z \Rightarrow 1|x$.

(vii) $x|y \Rightarrow y = x \cdot p$ for some $p \in Z$

$\Rightarrow$ $|y| = |xp| = |x||p|$

$\Rightarrow$ $|x| = \dfrac{|y|}{|p|} \leq |y|$, as $|p| \geq 1,\ y \neq 0$.

(viii) $x|y$ and $y|x$

$\Rightarrow$ $y = xp$ for some $p \in Z$

and $x = yq$ for some $q \in Z$

$\Rightarrow$ $x = xpq \Rightarrow pq = 1 \Rightarrow p = 1,\ q = 1$ or $p = -1,\ q = -1$

$\Rightarrow$ $x = y$ or $x = -y$ [alternatively, $y = x,\ y = -x$]

$\Rightarrow$ $|x| = |y|$.

(ix) $x|y \Rightarrow y = xp \Rightarrow y/x = p$, an integer.

Now $y = (y/x)x,\ x \in Z \Rightarrow (y/x)|y$.

(x) $x|y \Rightarrow y = xp$ for some $p \in Z$

$\Rightarrow \quad |y| = y = |xp| = |x||p|, |p| \geq 1$

$\Rightarrow \quad y \geq |x|$.

That is $|x| \leq y$, i.e., $|x| \not> y$.

Definition (1.3): ***Well-ordering Principle:*** *Every non-empty set S of non-negative integers has a least element.*

That is, $\exists$ an element $a \in S$ such that $a \leq b \ \ \forall b \in S$.

Division Algorithm

Theorem (1.3): *If a and b be two integers with $b > 0$, then $\exists$ two unique integers q and r such that $a = bq + r, \ 0 \leq r < b$.*

In other words, if a is divided by b, $b > 0$, then $\exists$ two unique integers q and r such that $a = bq + r, \ 0 \leq r < b$, where q is called the quotient and r is called the remainder.

Proof: We consider the infinite sequence of multiples of b,

$$\ldots, -qb, \ldots, -2b, -b, 0, b, 2b, \ldots, qb, \ldots$$

Then for any integer a, either $a = bq$ or a lies between two consecutive multiples, say, bq and $b(q+1)$.

That is, $bq \leq a < b(q+1) \Rightarrow 0 \leq a - bq < b$.

Let $a - bq = r$. Then $a = bq + r, \ 0 \leq r < b$. Thus we prove the main part of the theorem. Now we shall show that q and r are unique. If not, then $\exists$ two other integers q', r' such that $a = bq' + r', \ 0 \leq r' < b$.

Now $\quad a = bq + r \quad$ and $\quad a = bq' + r', \ 0 \leq r, \ r' < b$

$\Rightarrow \quad r' - r = b(q - q')$ and also $|r' - r| < b$

$\Rightarrow \quad b|(r' - r)$.

Now if $r' - r \neq 0$, then $b|(r' - r) \Rightarrow b \leq |r' - r|$, which contradicts that $|r' - r| < b$.

Therefore, $r' - r = 0$; i.e., $r' = r$.

Also $\quad r' = r \Rightarrow q' = q$.

Thus, q and r are unique.

Alternative Proof: Let $S = \{a - bt | a, b, t \in Z\}$.

Then the elements of S are of the forms:

$$a - b(-1), \ a, \ a - b, \ a - 2b, \ldots$$

Here $\quad a \in S \Rightarrow a \geq 0$ or, $a < 0$.

If $a < 0$, then $a - ba = a(1 - b) = 0$, if $b = 1$

> 0, if $b > 1$

< 0, if $b < 1$.

Thus, if $b \geq 1$, $a(1 - b) \geq 0 \Rightarrow S$ has a non-negative element.

$\Rightarrow$ S has a least non-negative element $(a - bq)$, say, for some $q \in Z$.

$\Rightarrow \quad 0 \leq a - bq$. (1)

Again, $\quad x \in S, \ x < a - bq \Rightarrow x < 0$ (2)

[$\because$ $(a - bq)$ is least non-negative number of S]

Now, $(a-bq)-b<a-bq \quad (\because\ b>1)$

$\Rightarrow \quad a-(q+1)b<0$ by (2).

$\Rightarrow \quad a-bq<b \quad (3)$

So, from (1) & (3), we get $0 \le a-bq<b$.

Taking $a-bq=r$, we get $0 \le r<b$

and $\quad a=bq=bq+(a-bq)$

$\quad = bq+r,\ 0 \le r<b.$

Hence the proof.

Example: Let $a=5$ and $b=2$.

We consider the infinite sequence $-6,-4,-2,0,2,4,6,\ldots$

Then 5 lies between 4 and 6.

That is $4<5<6 \Rightarrow 0<5-4<2$.

Now $\quad 5-4=1 \Rightarrow 5=4+1=2\cdot 2+1.$

Thus $\quad 5(=a)=2\cdot 2+1.$

Theorem (1.4): *If a and b be two integers, $b>0$, then $\exists$ two integers Q and R such that*

$$a=bQ \pm R,\ 0 \le R<b/2.$$

***Proof*: Case(1):** For any two integers a and b, $b>0$, it has been shown in Theorem (1.3) that $a=bq+r$.

$$0 \le r<b. \quad (1.1)$$

Let $r<b/2$.

Taking $q=Q$ and $r=R$ in (1.1), we have

$$a=bQ+R,\ 0 \le R<b/2.$$

Case(2): Let $r>b/2$.

Now $a=bq+r$ can be put in the form:

$a=b(q+1)+r-b=b(q+1)-(b-r).$

Taking $q+1=Q$ and $b-r=R$, we have

$a=bQ-R,\ 0 \le R<b/2,\ R=b-r<b-b/2=b/2.$

Thus, combining the case (1) and case (2), we have

$a=bQ \pm R,\ 0 \le R<b/2.$

Case(3): Let $r=b/2$.

Then $a=bq+r$ can be put in the form: $a=bQ+R$, taking $q=Q$,

$r=R=b/2. \quad$ (i)

Again, $a=bq+r=b(q+1)+r-b=b(q+1)-(b-r)=bQ+R,$

where $\quad q+1=Q$ and $-(b-r)=R,$

i.e., $\quad R=-(b-b/2)=-b/2. \quad$ (ii)

Thus, it follows that for $r=b/2$, Q and R are not unique.

In this case, R is called the *minimal remainder*, i.e., the absolutely least remainder of a w.r.t. b.

Corollary (1.1): For any two integers a and b, $b > 0$, $\exists\ Q, R \in Z$ such that $a = bQ + R$, $0 \le |R| \le b/2$ for $R = b/2$, R is the minimal remainder. Q and R are unique except when $R = b/2$.

Example: For $a = 50,\ b = 4,$

$$50 = 4 \cdot 12 + 2 \text{ (of the form: } a = bQ + R) \quad \text{(i)}$$

Also $$50 = 4 \cdot 13 - 2 \text{ (of the form: } a = bQ - R). \quad \text{(ii)}$$

In (i), $Q = 12,\ R = 2$

and in (ii), $Q = 13,\ R = -2.$

So Q and R are not unique as $R = \dfrac{b}{2} = \dfrac{4}{2} = 2.$

Theorem (1.5): *Every integer is of the form:*

(i) $3k$ *or,* $3k \pm 1$

(ii) $5k$ *or,* $5k \pm 1$ *or,* $5k \pm 2$

(iii) $6k$ *or,* $6k \pm 1$ *or,* $6k \pm 2$ *or,* $6k \pm 3.$

***Proof*:** (i) Any integer a is of the form:

$$a = bk \pm R,\ 0 \le |R| \le b/2$$

$$b = 3,\ 0 \le |R| \le \frac{3}{2} = 1.5.$$

$\therefore \quad R = 0, \pm 1.$

Therefore, $a = 3k,\ 3k \pm 1.$

(ii) $a = bk \pm R,\ 0 \le |R| \le b/2$

$$b = 5,\ 0 \le |R| \le \frac{5}{2} = 2.5.$$

$\therefore \quad R = 0, \pm 1, \pm 2.$

Therefore, $a = 5k,\ 5k \pm 1,\ 5k \pm 2.$

(iii) $b = 6,\ 0 \le |R| \le \dfrac{6}{2} = 3.$

$\therefore \quad R = 0,\ \pm 1,\ \pm 2,\ \pm 3.$

Therefore, $a = 6k,\ 6k \pm 1,\ 6k \pm 2,\ 6k \pm 3.$

Theorem (1.6): *Every square number is of the form:* $5k,\ 5k \pm 1.$

***Proof*:** Every integer is of the form:

$5m,\ 5m \pm 1,\ 5m \pm 2$ [by (ii) of Th.(1.5)]

Square of these numbers are of the forms:

$(5m)^2 = 5 \cdot 5m^2 = 5k$

$(5m \pm 1)^2 = 25m^2 \pm 10m + 1 = 5k + 1$

$(5m \pm 2)^2 = 25m^2 \pm 20m + 4 = 5(5m^2 \pm 4m + 1) - 1$

$= 5k - 1.$

Hence, every square number is of the form: $5k,\ 5k \pm 1.$

Theorem (1.7): *Cube of any integer is of the form: $9k, 9k+1, 9k+8$ (or $9k, 9k\pm 1$).*

***Proof*:** Every integer is of the form

$3m,\ 3m \pm 1$ [by (i) of Th.(1.5)]

Cubes of these numbers are of the form:

$$(3m)^3 = 9 \cdot 3m^3 = 9k$$

$$(3m+1)^3 = 27m^3 + 27m^2 + 9m + 1 = 9k + 1$$

$$(3m-1)^3 = 27m^3 - 27m^2 + 9m - 1 = 27m^3 - 27m^2 + 9m - 9 + 8$$

$$= 9k + 8 \text{ (or, } 9k - 1).$$

Hence, cubes of any integer is of the form: $9k$, $9k+1$, or, $9k+8$ (or, $9k$ or, $9k \pm 1$).

Problem (1.1): If x and y be two odd integers, then one of the two numbers $\frac{x+y}{2}$ and $\frac{x-y}{2}$ is odd and the other even.

Solution: Let $x = 2m+1$ and $y = 2n+1$, $m, n \in Z$.

Then, $\frac{x+y}{2} = m+n+1$ and $\frac{x-y}{2} = m-n$.

Now if

(i) m, n are both odd, then $m+n$ is even.

Consequently, $(m+n)+1$ is odd while $(m-n)$ is even.

Thus, $\frac{x+y}{2}$ is odd and $\frac{x-y}{2}$ is even.

(ii) m is odd, n is even or m is even, n is odd.

$\Rightarrow$ $(m+n)$ is odd $\Rightarrow$ $(m+n)+1$ is even $\Rightarrow$ $\frac{x+y}{2}$ is even.

While $m-n$ is odd $\Rightarrow$ $\frac{x-y}{2}$ is odd.

(iii) m and n are both even

$\Rightarrow$ $m+n$ is even $\Rightarrow$ $(m+n+1)$ is odd $\Rightarrow$ $\frac{x+y}{2}$ is odd.

Again, $m-n$ is even $\Rightarrow$ $\frac{x-y}{2}$ is even.

Problem (1.2): Show that $a(a^4+4)/5$ is an integer for any integer a.

Solution: Any integer a is of the form: $5k$, $5k \pm 1$ or, $5k \pm 2$.

When $a = 5k$, $\frac{a(a^4+4)}{5} = \frac{5k(625k^4+4)}{5} = k(625k^4+4)$, which is an integer.

$a = 5k \pm 1$, $\frac{a(a^4+4)}{5} = \frac{(5k\pm 1)(5m)}{5}$,

$m \in Z$ as the constant term is $1^4 + 4 = 5$

$= (5k \pm 1)m$, which is an integer.

$a = 5k \pm 2$, $\frac{a(a^4+4)}{5} = \frac{(5k\pm 2)5n}{5}$,

$n \in Z$ as the the constant term is $2^4 + 4 = 20$, i.e., multiple of 5

$= (5k \pm 2)n$ which is an integer.

Problem (1.3): Square of any odd integer is of the form: $8k+1$.

Solution: $2n+1$ is an odd integer $\forall n \in Z$.

Then $(2n+1)^2 = 4n(n+1)+1$, for every value of n, $n(n+1)$ is multiple of 2. So $n(n+1) = 2k,\ k \in Z$.

Therefore, $(2n+1)^2 = 8k+1$.

Theorem (1.8): *One of every three consecutive integers is divisible by* 3.

***Proof*:** Let a, $a+1$, $a+2$ be any three consecutive integers.

If $a = 3q$, theorem holds.

If $a = 3q+1$, then $a+2 = 3q+3 = 3(q+1) = 3k,\ k \in Z$. So $a+2$ is divisible by 3.

If $a = 3q+2$, then $a+1 = 3q+3 = 3(q+1) = 3k,\ k \in Z$. So $a+1$ is divisible by 3.

Hence the theorem.

Theorem (1.9): *The product of any three consecutive integers is divisible by* 3!.

***Proof*:** Let a be any integer. Then we have to prove that $a(a+1)(a+2)$, product of three consecutive integers, is divisible by 3!. For $a = 1$, $a(a+1)(a+2) = 1\cdot 2\cdot 3 = 3!$ which is divisible by 3!.

Let the result be true for $a = k$. Then $k(k+1)(k+2)$ is divisible by 3!

We shall show that the result also holds for $a = k+1$.

Now for $a = k+1$,

$$\begin{aligned} a(a+1)(a+2) &= (k+1)(k+2)(k+3) \\ &= k(k+1)(k+2) + 3(k+1)(k+2). \end{aligned}$$

1st part is divisible by 3! by our assumption. In the 2nd part, $(k+1)(k+2)$ is multiple of 2 $\forall k \in Z$ and consequently, $3(k+1)(k+2)$ is divisible by 3!.

Hence, the result holds $a = k+1$.

Therefore, the theorem holds $\forall a \in Z$.

Greatest common divisor

Let $a, b \in Z$.

Then $d \in Z$ is called a common divisor of a and b if $d|a$ and $d|b$.

$1|x\ \forall x \in Z$.

Then 1 is a common divisor of a and b.

Again, when $a = b = 0$, then every integer $x|0$.

Thus, the set of (+)ve common divisors of a and b is infinite. So, if at least one of a and $b \neq 0$, then the set of (+)ve common divisors is finite.

Every finite set has the largest number. So the set of (+)ve common divisors of a and b, where at least one of a and b is not zero, has the largest number which is called the greatest common divisor (G.C.D.) of a and b. We follow the following definition.

Definition (1.4): *Let $a, b \in Z$ such that at least one of them is different from zero. Then the (+)ve integer d is called the greatest common divisor of a and b, and denoted by $d = (a, b)$, if it satisfies the following conditions:*

(i) *$d|a$ and $d|b$.*

(ii) *If $c|a$, $c|b$, then $c|d$.*

Problem (1.4): The (+)ve divisors of -16 are 1, 2, 4, 8, 16 and the (+)ve divisors of 40 are 1, 2, 4, 5, 8, 10, 20, 40. Find the G.C.D.

Solution: The set of common (+)ve divisors of -16 and 40 are 1,2,4,8.

$S = \{1, 2, 4, 8\}$, a finite set, has the largest no. 8.

So 8 is called the G.C.D. of -16 and 40.

That is, $(-16, 40) = 8$.

From definition, (i) $8|-16, 8|40$ (ii) $4|-16, 4|40$, then $4|8$.

$\therefore$ $(-16, 40) = 8$.

Euclid's Algorithm

Theorem (1.10): *Apply the division algorithm repeatedly for any two (+)ve integers a and b, a > b to obtain a set of remainders, then the last non-zero remainder in the process of division is the G.C.D. of a and b.*

***Proof*:** By divisional algorithm, we have

$$a = bq + r, \ 0 \leq r < b. \tag{1.1}$$

Now putting $a = r_0$, $b = r_1$, $q = q_1$ and $r = r_2$,

(1) assumes the form:

$$r_0 = r_1 q_1 + r_2, \ 0 < r_2 < r_1. \tag{1.2}$$

Again, dividing r_1 by r_2, we get

$$r_1 = r_2 q_2 + r_3, \ 0 < r_3 < r_2. \tag{1.3}$$

Continuing the process, we have

$$r_{n-2} = r_{n-1} q_{n-1} + r_n, \ 0 < r_n < r_{n-1} \qquad (n-1)$$

$$r_{n-1} = r_n q_n + r_{n+1}, \ r_{n+1} = 0. \qquad (n)$$

Since r_i are decreasing and non-negative numbers, there is a stage (n) at which $r_{n+1} = 0$.

Then from the last stage,

$$r_{n-1} = r_n q_n \ \Rightarrow \ r_n | r_{n-1}.$$

Then $r_{n-2} = r_n q_n q_{n-1} + r_n = r_n(q_n q_{n-1} + 1) \ \Rightarrow \ r_n | r_{n-2}$.

Finally we have $r_n | r_1$, i.e., $r_n | b$.

Thus we have from Eq. (2) $r_n | r_0$, i.e., $r_n | a$, since

$$r_n | r_1, \ r_n | r_2 \Rightarrow \ r_0 = k r_n q_1 + k' r_n = r_n(k q_1 + k')$$
$$\Rightarrow r_n | r_0, \text{ i.e., } r_n | a.$$

Therefore, (i) $r_n | a$, $r_n | b \ \Rightarrow \ r_n$ is the common divisor of a and b.

Now let d be any other common divisor of a and b. Then $d|a$ and $d|b$.

$$d|a \ \Rightarrow \ d|r_0$$
$$d|b \Rightarrow \ d|r_1 \ \Rightarrow \ d|r_2 \text{ [by Eq.(2)]} \ \Rightarrow \ d|r_3$$

and hence $d | r_n$.

Thus the two conditions (i) $r_n | a$ and $r_n | b$ and (ii) $d|a$, $d|b \ \Rightarrow \ d|r_n$ are satisfied.

Therefore r_n is the G.C.D. of a and b. That is $(a, b) = r_n$.

Theorem (1.11): *The greatest common divisor of a and b can be expressed as the linear combination of a and b.*

That is, $\exists$ *two integers* $x, y \in Z$ *such that* $(a, b) = ax + by$.

Proof: By divisional algorithm for any two integers a and b the greatest common divisor is obtained as described in Theorem(1.10).

Since $(|a|, |b|) = (a, b)$, there is no harm in assuming $a \geq b > 0$.

The results obtained from Th(1.10)

are $\quad r_0 = r_1 q_1 + r_2, \; 0 < r_2 < r_1 \qquad (1)$

$r_1 = r_2 q_2 + r_3, \; 0 < r_3 < r_2 \qquad (2)$

$r_2 = r_3 q_3 + r_4, \; 0 < r_4 < r_3 \qquad (3)$

$\vdots$

$r_{n-2} = r_{n-1} q_{n-1} + r_n, \qquad (n-1)$

$r_{n-1} = r_n q_n + r_{n+1}, \; r_{n+1} = 0, \qquad (n)$

From Eq. (1), $r_2 = r_0 - r_1 q_1 = a - bq_1 \; [r_0 = a, \; b = r_1, \; r = r_2]$

$$\Rightarrow \quad r_3 = r_1 - r_2 q_2 = b - q_2(a - bq_1)$$
$$= a(-q_2) + b(1 + q_2 q_1).$$
$$\Rightarrow \quad r_4 = r_2 - r_3 q_3 = (a - bq_1) - q_3[a(-q_2) + b(1 + q_1 q_2)]$$
$$= a(1 + q_2 q_3) + b[(-q_1) + (-q_3)(1 + q_1 q_2)].$$

Finally, $\Rightarrow \quad r_n = ax + by,$

where $x, y \in Z$ are obtained in terms of $q_1, q_2, \cdots, q_n$.

Therefore, $\exists$ two integers x, y such that $r_n = ax + by$,

$\Rightarrow$ G.C.D. of a and b is a linear combination of a and b. [by Th(1.10)]

That is, $\quad (a, b) = ax + by$, for some $x, y \in Z$.

Corollary (1.2): For any (+) ve integer k, $(ka, kb) = k(a, b)$.

Proof: We have by divisional algorithm,

$$a = q_1 b + r_1, \; 0 \leq r_1 < b$$
$$b = q_2 r_1 + r_2, \; 0 \leq r_2 < r_1$$
$$r_1 = q_3 r_2 + r_3, \; 0 \leq r_3 < r_2$$
$$r_2 = q_4 r_3 + r_4, \; 0 \leq r_4 < r_3$$
$$\cdots\cdots\cdots\cdots\cdots\cdots$$
$$r_{n-2} = q_n r_{n-1} + r_n, \; 0 \leq r_n < r_{n-1}$$
$$r_{n-1} = q_{n+1} r_n + r_{n+1}, \; r_{n+1} = 0$$

since r_2 are decreasing set of non-negative integer, there is a stage at which $r_{n+1} = 0$.

Hence, the last non-zero remainder r_n is the G.C.D. of a and b.

That is, $r_n = (a, b)$ [by Th. 1.10]

Since $k > 0$, multiplying both sides of the above results by k, we get

$$ak = q_1(bk) + r_1 k, \quad 0 \leq r_1 k < bk$$
$$bk = q_2(r_1 k) + r_2 k, \quad 0 \leq r_2 k < r_1 k$$

$r_1k = q_3(r_2k) + r_3k, 0 \le r_3k < r_2k$

$r_2k = q_4(r_3k) + r_4k, 0 \le r_4k < r_3k$

....

$r_{n-2}k = q_n(r_{n-1}k) + r_nk,\ 0 \le r_nk < r_{n-1}k$

$r_{n-1}k = q_{n+1}(r_nk) + 0$

$\Rightarrow\ (ak, bk) = r_nk = kr_n = k(a, b).$

Problem (1.5): Find the G.C.D. of 723 and 24 by Euclidean algorithm.

Solution: $723 = 24 \times 30 + 3\ [r_1 = 3]$

$24 = 3 \times 8 + 0\ [r_2 = 0].$

The last non-zero remainder is 3.

$\therefore\quad (723, 24) = 3$

Problem (1.6): Find (426,246).

Solution: $426 = 246 \times 1 + 180$

$246 = 180 \times 1 + 66$

$180 = 66 \times 2 + 48$

$66 = 48 \times 1 + 18$

$48 = 18 \times 2 + 12$

$18 = 12 \times 1 + 6$

$12 = 6 \times 2 + 0$

The last non-zero remainder is 6.

$\therefore\quad (426, 246) = 6.$

Definition (1.5): *The greatest common divisor of any n number of given integers, $a_1,\ a_2,\ \ldots,\ a_n,\ a_i \neq 0\ \forall i = 1, 2, \ldots, n$, denoted by $(a_1, a_2, \ldots, a_n) = d$, a positive integer, satisfies the following properties*

(i) $d|a_i\ \forall i = 1, 2, \ldots, n.$

(ii) *If* $c|a_i,\ i = 1, 2, \ldots, n$, *then* $c|d$.

Some Fundamental Laws

(i) Commutative law: $(a, b) = (b, a)\ \forall a, b \in z.$

(ii) Associative law: $((x, y), z) = (x, (y, z))\ \forall x, y, z \in Z.$

(iii) Distributive law: $(ac, bc) = |c|(a, b)$ for any $c \in Z.$

(iv) $(a, 1) = (1, a) = 1\ \forall a \in Z$

(v) $(a, 0) = (0, a) = |a|$

We prove (ii) and (iii).

Let $(x, y) = d_1$.

***Proof of* (ii):** Then $((x, y), z) = (d_i, z) = d_2$ (say).

$$\left.\begin{aligned} &\text{Now}\quad (d_1, z) = d_2 \ \Rightarrow\ d_2|d_1,\ d_2|z \\ &\text{and}\quad (x, y) = d_1 \ \Rightarrow\ d_1|x,\ d_1|y \end{aligned}\right\} \Rightarrow d_2|x,\ d_2|y,\ d_2|z.$$

Thus, we have (i) $d_2|x,\ d_2|y,\ d_2|z$.

Next, if $c|d_1,\ c|z$, then $c|d_2$ [as $(d_1, z) = d_2$]

Thus, $c|x,\ c|y$ and $c|z \Rightarrow c|d_2$.

So from the conditions (i) and (ii) of Def. [1.5], we have

$$(x, y, z) = d_2 = (d_1, z) = ((x, y), z). \quad (1.1)$$

Similarly, we can show that

$$(x, (y, z)) = (x, y, z). \quad (1.2)$$

Hence, from Eqs. (1.1) & (1.2),

$$((x, y), z) = (x, (y, z)).$$

***Proof* (iii):** Let $(ac, bc) = d$ and let $(a, b) = p$.

Now $\quad (a, b) = p \Rightarrow \exists\ x, y \in Z$ such that $p = ax + by$

$\Rightarrow \quad pc = acx + bcy \quad$ (i)

$\Rightarrow \quad pc|d$ as $d|ac$ and $d|bc$.

Again, $d|ac$ and $d|bc \Rightarrow d|(acx + bcy) \Rightarrow d|pc$ [by(i)]

So from $pc|d$ and $d|pc$,

$$d = pc \Rightarrow |d| = |pc| = |c||p|$$

$\Rightarrow \quad d = |c|p$, as $d, p > 0$.

Hence, $(ac, bc) = |c|(a, b)$.

Euclid's Lemma (1.11A): If $a|bc$ and $(a, b) = 1$, then $a|c$.

***Proof*:** $(a, b) = 1$

$\Rightarrow \quad \exists\ x, y \in Z$ such that $ax + by = 1$. By Th. (1.11)

$\Rightarrow \quad acx + bcy = c. \quad$ (i)

But $a|ac$ and $a|bc$ (given)

$\Rightarrow \quad a|acx + bcy \Rightarrow a|c$.

Theorem (1.12): *The G.C.D. is unique.*

***Proof*:** Let d_1 and d_2 be the two G.C.D. of $a_1, a_2, \ldots a_n$.

i.e., $\quad (a_1, a_2, \ldots, a_n) = d_1$ and $(a_1, a_2, \ldots, a_n) = d_2$.

Then $\ d_1|a_i\ \forall\ i = 1, 2, \ldots n$.

Again, $\ d_2|a_i\ \forall\ i = 1, 2, \ldots n$.

$\Rightarrow \quad d_2|d_1$ [by condition (ii) of definition of G.C.D.]

Conversely, it can be shown that $d_1|d_2$.

$\therefore \quad d_1 = d_2$.

Hence the theorem.

Theorem (1.13): *Let $a_1,\ a_2,\ \ldots,\ a_n$ be any non-zero integers and $d = (a_1,\ a_2, \ldots,$ $a_n)$. Then $\exists\ x_1,\ x_2,\ \ldots,\ x_n \in Z$ such that $d = a_1x_1 + a_2x_2 + \cdots + a_nx_n$.*

***Proof*:** Consider the set S defined by

$$S = \{a_1u_1 + a_2u_2 + \cdots + a_nu_n,\ u_i \in Z\}.$$

If a_1 is a (+)ve integer, then

$$a_1 = a_1 \cdot 1 + a_2 \cdot 0 + \cdots + a_n \cdot 0 \in S \quad (1)$$

Again, if a_1 is a (−)ve integer, then

$$-a_1 = a_1 \cdot (-1) + a_2 \cdot 0 + \cdots + a_n \cdot 0 \in S. \quad (2)$$

From Eqs. (1) & (2) it follows that $\exists$ (+)ve integer in S and consequently S has a least (+)ve number. Let t be the least (+)ve number in S.

Then $t = a_1x_1 + a_2x_2 + \cdots + a_nx_n,\ x_i \in Z$.

We shall show that $t = d$.

Now dividing a_1 by t, we have

$$a_1 = tq + r,\ 0 \le r < t,\ q, r \in Z.$$

$$\Rightarrow \quad r = a_1 - tq = a_1 - (a_1x_1 + \cdots + a_nx_n)q$$
$$= a_1(1 - x_1q) + a_2(-x_2q) + \cdots + a_n(-x_nq)$$
$$\Rightarrow \quad r \in S \Rightarrow r = 0.$$

For, if $r > 0$, then r being a member of S is less than t, a contradiction. So $r = 0$.

Therefore, $a_1 = tq \Rightarrow t|a_1$.

Similarly, we can show that $t|a_2, \ldots, t|a_n$.

Thus $\quad t|a_i\ \forall i = 1, 2, \ldots, n$

$$\Rightarrow t|d. \quad (3)$$

Also $\quad (a_1, a_2, \ldots, a_n) = d \Rightarrow d|a_i\ \forall i = 1, 2, \ldots, n$

$$\Rightarrow \quad d|a_1x_1 + \cdots + a_nx_n \Rightarrow d|t. \quad (4)$$

So from (3) & (4), $t = d$.

Thus, the G.C.D. of $(a_1, a_2, \ldots, a_n)$ is the least (+)ve integer in the set $S = \{a_1u_1 + \cdots + a_nu_n,\ u_0 \in Z\}$.

Corollary (1.2): If $(a, b) = d$, then $\exists\ x, y \in Z$ such that $d = ax + by$.

It is the consequence of the above theorem when $n = 2$, $a_1 = a$, $a_2 = b$, $x_1 = x$, $x_2 = y$.

Definition (1.6): *If $(a, b) = 1$, then a and b are said to be relatively prime. a, b both are not zero.*

Theorem (1.14): *Two integers a and b, both not zero, are relatively prime, if and only if $1 = ax + by$, for some $x, y \in Z$.*

Proof: Let a and b be relatively prime. Then $(a, b) = 1$. Therefore, $\exists$ two integers x and y such that $1 = ax + by$ [by Th.(1.11)].

Next let $I = ax + by$ and let $(a, b) = d$.

Then $\quad (a, b) = d \Rightarrow d|a$ and $d|b$

$$\Rightarrow \quad d|ax + by,\ x, y \in Z$$

$\Rightarrow \quad d|1 \Rightarrow d = 1$ as d is a (+)ve integer.

Therefore, $(a, b) = 1$.

Hence, a and b are relatively prime.

Corollary (1.3): If $(a, b) = d$, then $\exists\ x, y \in Z$ such that $d = ax + by$ but the converse is not true.

Let $a = 7$, $b = 35$. Then $(7, 35) = 7 = 7 \cdot (-4) + 1 \cdot 35$.

Now let $x = 7$, $y = (-1)$.

Then $ax + by = 7 \cdot 7 + (-1)35 = 14 \neq (7, 35)$.

Definition (1.7): *A set of integers $a_1, a_2, \ldots, a_n$, not all zero, are said to be relatively prime, if $(a_i, a_j) = 1 \; \forall \; i, j = 1, 2, \ldots, n$.*

Example: The set of integers 21, 92, 143, 85 are relatively prime, since $(21, 92) = 1$, $(21, 143) = 1$, $(21, 85) = 1$, $(92, 143) = 1$, $(92, 85) = 1$, $(143, 85) = 1$.

It is to be noted that three integers to be relatively prime as a triple, that is $(a, b, c) = 1$, though the integers in pairs are not relatively prime, which follows from the example: $(10, 12, 15) = 1$.

But $(10, 12) = 2$, $(10, 15) = 5$, $(12, 15) = 3$, that is, they are not prime in pairs.

Theorem (1.15): *If $(a, b) = 1$, then for any integer c, $(ac, b) = (c, b)$.*

Proof: Let $(ac, b) = d$ and $(c, b) = d_1$.

Since $(a, b) = 1$, $\exists \; x, y \in Z$ such that $ax + by = 1$

$$\Rightarrow \quad acx + bcy = c. \tag{1}$$

Now $(ac, b) = d \;\Rightarrow\; d|ac,\; d|b \;\Rightarrow\; d|ac,\; d|bc$ [as $b|bc$]

$\Rightarrow \quad d|(acx + bcy) \;\Rightarrow\; d|c$ [by Eq. (1)].

Then $\quad d|c,\; d|b \;\Rightarrow\; d|d_1$

as $(c, b) = d_1$. (2)

Also $(c, b) = d_1 \;\Rightarrow\; d_1|c,\; d_1|b \;\Rightarrow\; d_1|ac,\; d_1|b$ [as $c|ac$]

$\Rightarrow \quad d_1|d,$ (3)

since $(ac, b) = d$.

So from (2) & (3), it follows

$$d = d_1 \;\Rightarrow\; (ac, b) = (c, b).$$

Example: $(2, 5) = 1$. $c = 10$, $(20, 5) = 5$. Also $(10, 5) = 5$.

$\therefore \quad (2 \cdot 10, 5) = (10, 5)$.

Theorem (1.16): *If $(a, b) = 1$, then for any integer q, $(a + bq, b) = 1$.*

Proof: Let $(a + bq, b) = k$, where $k \geq 1$.

If $k = 1$, theorem holds.

Let the theorem be true for $k > 1$.

Then $\quad (a + bq, b) = k \; (> 1)$

$\Rightarrow \quad k|a + bq,\; k|b \;\Rightarrow\; k|(a + bq) \cdot 1 + b(-q)$

$\Rightarrow \quad k|a$

Therefore, $k|a,\; k|b \;\Rightarrow\; (a, b) \neq 1$, a contradiction.

Hence, the theorem is true for $k = 1$ only.

Therefore, $\quad (a + bq, b) = 1$.

Corollary (1.4): (i) If $(a+bq, b) = 1$, then $(a, b) = 1$.

(ii) If $(a, b) = 1$, then $(a, b + aq) = 1$.

Theorem (1.17): *If* $(a_i, b) = 1 \ \forall\ i = 1, 2, \ldots, n$, *then* $(a_1 a_2 \cdots a_n, b) = 1$.

***Proof*:** Using the Theorem (1.15),

$$(a, b) = 1 \ \Rightarrow\ (ac, b) = (c, b) \text{ for any integer } c.$$

We have $(a_1, b) = 1$.

$\Rightarrow \quad (a_1 a_2, b) = (a_2, b) = 1$

$\Rightarrow \quad (a_1 a_2 a_3, b) = (a_3, b) = 1$

$\ldots \quad \ldots \quad \ldots$

$\Rightarrow \quad (a_1 a_2 \cdots a_n, b) = (a_n, b) = 1.$

Corollary (1.5): If $(a, b) = 1$, then $(a^n, b) = 1$.

The result follows from above theorem by taking $a_1 = a_2 = \cdots = a_n = a$.

Corollary (1.6): If $(a, b) = 1$, then $(a, b^n) = 1$.

Corollary (1.7): If $(a, b) = 1$, then $(a^n, b^n) = 1$.

Theorem (1.18): *If* $(a, b) = d$, *then* $(a + xb,\ b) = d$, *where* x *is any integer.*

***Proof*:** Let $(a + xb,\ b) = d_1$.

Then $d_1 | a + xb,\ d_1 | b$.

$\Rightarrow \quad d_1 | a$

Thus, $d_1 | a,\ d_1 | b \Rightarrow d_1 | d$ as $(a, b) = d$. (1)

Again, $(a, b) = d \Rightarrow d|a,\ d|b \Rightarrow d | a + xb$.

Thus, $d | a + xb,\ d|b \Rightarrow d | d_1$ as $(a + xb, b) = d_1$. (2)

So from (1) and (2) it follows that $d_1 = d$. Hence the result.

Problem (1.7): If $(a, b) = 1$, then $(a + b, a^2 - ab + b^2) = 1$ or 3.

Solution: Let $(a + b, a^2 - ab + b^2) = d$.

Then $\quad (a^2 - ab + b^2, a + b) = d \Rightarrow (-2ab + b^2, a + b) = d$ [by Th.(1.18)]

$\Rightarrow \quad (-3ab, a + b) = d \Rightarrow (3a^2, a + b) = d \Rightarrow d | 3a^2$.

Also $\quad (3b^2, a + b) = d \Rightarrow d | 3b^2$.

Therefore, $d | 3a^2 \Rightarrow d = 3$ or $d | a^2$ and $d | 3b^2 \Rightarrow d = 3$ or $d | b^2$.

So $d = 3$ or $d | a^2$ and $d | b^2$

$\Rightarrow \quad d = 1$, since $(a, b) = 1$

$\Rightarrow \quad (a^2, b^2) = 1$ by Cor.(1.7) of Th.(1.17) for $n = 2$.

Thus we have $d = 3$ or, $d = 1$.

Lemma (1.17A): If $(x, a) = 1$ and $(x, b) = 1$, then $(x, ab) = 1$.

***Proof*:** $(x, a) = 1$.

$\Rightarrow \quad \exists\ p, q \in Z$ such that $px + qa = 1$. (i)

Also $(x, b) = 1 \Rightarrow \exists\ r, s \in Z$ such that $rx + sb = 1$. (ii)

Now (i) $\times\ b$ gives $pxb + qab = b$.

Then from (ii), $rx + s(pxb + qab) = 1$

$\Rightarrow \quad xp' + abq' = 1,\quad$ where $p' = r + spb$ and $q' = qs,\ p', q' \in Z$.

$\Rightarrow \quad (x, ab) = 1.$

Problem (1.8): If $(m_1, m_2) = 1,\ (a, m_1) = 1,\ (b, m_2) = 1$, then show that $(am_2 + bm_1, m_1m_2) = 1$.

Solution: $(a, m_1) = 1 \Rightarrow (am_2, m_1) = (m_2, m_1) = 1$, by Th.(1.15)

$\Rightarrow \quad (am_2 + bm_1, m_1) = 1$, by Th.(1.16) (1)

Again, $(b, m_2) = 1 \Rightarrow (bm_1, m_2) = (m_1, m_2) = 1$

$\Rightarrow \quad (am_2 + bm_1, m_2) = 1.$ (2)

So Eqs. (1) and (2) show that

$(am_2 + bm_1, m_1m_2) = 1$ [by Lemma 1.17A].

Example: Let $m_1 = 7,\ m_2 = 5,\ a = 2,\ b = 9.$

Then $\quad (m_1, m_2) = 1,\ (a, m_1) = 1,\ (b, m_2) = 1.$

Therefore, $(am_2 + bm_1, m_1m_2) = (2\cdot 5 + 9 \cdot 7, 7 \cdot 5) = (73, 35) = 1.$

Least Common Multiple (L.C.M.)

Definition (1.8): *Let $a_1, a_2, \ldots, a_n$ be given non-zero integers. Then a (+)ve integer b satisfying the following properties:*

(i) $a_i | b\ \forall i = 1, 2, \ldots, n.$

(ii) *If $\exists$ an integer c such that $a_i | c$, then $b | c$*

is called the least common multiple of $a_1, a_2, \ldots, a_n$ and denoted by $[a_1, a_2, \ldots, a_n]$.

Example: Find the L.C.M. of 4,10,12,20,24,30.

Multiple of 4 = ⋯ 120, 124, 128, 132, 136, 140, 144, ⋯
" " 10 = ⋯ 120, 130, 140, 150, 160, ⋯ ⋯ ⋯
" " 12 = ⋯ 120, 132, 144, 156, 168, ⋯ ⋯
" " 20 = ⋯ 120, 140, 160, 180, 200, ⋯ ⋯ ⋯
" " 24 = ⋯ 120, 144, 168, 192, ⋯ ⋯ ⋯ ⋯
" " 30 = ⋯ 120, 150, 180, 210, ⋯ ⋯ ⋯

Common multiples are 120, 240, 360, ⋯

Least common multiple = 120.

As per definition:

(i) 4|120, 10|120, 12|120, 20|120, 24|120, 30|120.

(ii) $\exists$ an integer 240 such that 4|240, 10|240, 20|240, 24|240, 30|240 and 120|240, therefore, L.C.M. of 4,10,12,20,24,30 is 120.

Theorem (1.18): *The L.C.M. of any set of non-zero integers is unique.*

Proof: If possible, let l and l' be the L.C.M. of the set of non-zero integers $a_1, a_2, \ldots, a_n$. Then $a_i | l$ and l' being the L.C.M., by properties (ii), $l' | l$ (1)

Again, $a_i | l'$ and l being the L.C.M., by properties (ii), $l | l'$. (2)

So from (1) & (2), $l = l'$.

Theorem (1.19): *If $(a, b) = d$, then $[a, b] = \frac{ab}{d}$.*

Proof: Let $\frac{ab}{d} = l$.

Now $(a, b) = d \Rightarrow d|a$ and $d|b$.

Then $d|a \Rightarrow a = dx$ for some $x \in Z$ and $d|b \Rightarrow b = dy$ for some $y \in Z$.

$$l = \frac{ab}{d} = \frac{dxb}{d} = bx \Rightarrow b|l.$$

Also $l = \frac{ab}{d} = \frac{a \cdot dy}{d} = ay \Rightarrow a|l.$

Thus the condition (i) of Def. [1.8] $a|l$, $b|l$ is satisfied.

Again, let $\exists\ c \in Z$ such that $a|c$ and $b|c$.

Then $c = ap$ and $c = bq$ for some $p, q \in Z$. (1)

Since $d = (a, b)$, $\exists\ r, s \in Z$ such that $d = ar + bs$.

Then $\frac{c}{l} = \frac{cd}{ab} = \frac{c(ar + bs)}{ab} = \left(\frac{c}{b}\right) r + \left(\frac{c}{a}\right) s = qr + ps$ [by (1)]

$= t$ (say)

$\therefore$ $c = lt$ for some $t \in Z$.

$\Rightarrow l|c$, condition (ii) of Def. [1.8] is satisfied.

Hence, $[a, b] = l = \frac{ab}{d}$.

Problem (1.9): If $(a, b) = 1$, then show that $[a, b] = ab$.

Result follows immediately from the above theorem by putting $d = 1$.

Alternatively: Let $ab = l$.

Now $a|ab$ and $b|ab \Rightarrow a|l$, $b|l$.

So condition (1) of Def.(1.8) is satisfied.

Let $\exists\ c \in Z$ such that $a|c$ and $b|c$.

Then $c = ap$ and $c = bq$ for some $p, q \in Z$.

Since $(a, b) = 1$, $\exists\ x, y \in Z$ such that $1 = ax + by$.

$$\frac{c}{l} = \frac{c}{ab} = \frac{c(ax + by)}{ab} = \left(\frac{c}{b}\right) x + \left(\frac{c}{a}\right) y = qx + py = t \text{ (say)}.$$

$\therefore$ $c = lt$, for some $t \in Z$.

$\therefore$ $l|c$, condition (ii) of Def.(1.8) is satisfied.

Hence, $[a, b] = l = ab$.

Theorem (1.20): *If $[a_1, a_2] = M_2$, $[M_2, a_3] = M_3$, $[M_3, a_4] = M_4 \cdots [M_{n-1}, a_n] = M_n$, then $[a_1, a_2, \ldots, a_n] = M_n$.*

Proof: $[M_{n-1}, a_n] = M_n \Rightarrow M_{n-1}|M_n$, $a_n|M_n$.

Again, $[M_{n-2}, a_{n-1}] = M_{n-1} \Rightarrow M_{n-2}|M_{n-1}$, $a_{n-1}|M_{n-1}$.

Then $M_{n-1}|M_n$, $a_{n-1}|M_{n-1} \Rightarrow a_{n-1}|M_n$.

Continuing the process we can show that

(i) $a_i|M_n\ \forall i = 1, 2, \ldots, n$.

Let $\exists$ an integer c such that $a_i|c\ \forall i$.

Then $[a_1, a_2] = M_2 \Rightarrow M_2|c$, by property (ii) of definition (1.8).

Also $\qquad [M_2, a_3] = M_3 \Rightarrow M_3|c$, since $M_2|c$, $a_3|c$.

Finally, we have

(ii) $M_n|c$.

So conditions (i) & (ii) are satisfied.

Therefore, $[a_1, a_2, \ldots, a_n] = M_n$.

Problem (1.10): Find L.C.M. of 4,10,12,20,24,30.

Solution: $[4, 10] = \dfrac{4 \times 10}{(4, 10)} = \dfrac{40}{2} = 20$

$$[20, 12] = \frac{20 \times 12}{(20, 12)} = \frac{240}{4} = 60$$

$$[60, 20] = \frac{60 \times 20}{(60, 20)} = \frac{1200}{20} = 60$$

$$[60, 24] = \frac{60 \times 24}{(60, 24)} = \frac{1440}{12} = 120$$

$$[120, 30] = \frac{120 \times 30}{(120, 30)} = \frac{3600}{30} = 120.$$

$\therefore$ L.C.M. of $4, 10, 12, 20, 24, 30 = 120$.

Problem (1.11): If a is an odd integer, then $32|(a^2 + 3)(a^2 + 7)$.

Solution: Let $a = 2n + 1$.

Then $a^2 + 3 = (2n + 1)^2 + 3 = 4(n^2 + n + 1) = 4p,\ p \in Z$

and $\qquad a^2 + 7 = 4n^2 + 4n + 8 = 4(n^2 + n + 2)$

$\qquad = 8q$, since $n^2 + n + 2 = 2q\ \forall n \in Z,\ q \in Z$.

Therefore, $(a^2 + 3)(a^2 + 7) = 4p \cdot 8q = 32pq,\ pq \in Z$.

$\Rightarrow \qquad 32|(a^2 + 3)(a^2 + 7)$.

Problem (1.12): If $d|n$, then $2^d - 1|2^n - 1$. $d, n \geq 1$.

Solution: $2^d - 1 = (2 - 1)(2^{d-1} + 2^{d-2} + \cdots + 1) = 2^{d-1} + 2^{d-2} + \cdots + 1$

$$2^n - 1 = 2^{n-1} + 2^{n-2} + \cdots + 1.$$

Now $d|n \Rightarrow n = dk$, for some $k \in Z$.

$$2^n - 1 = 2^{dk-1} + 2^{dk-2} + \cdots + 1 \qquad (1)$$

Now $2^{dk} - 1 = (2 - 1)(2^{dk-1} + 2^{dk-2} + \cdots + 1)$

$\qquad = 2^{dk-1} + 2^{dk-2} + \cdots + 1$

$\Rightarrow \qquad 2^{dk-1} + 2^{dk-2} + \cdots + 1 = (2^d)^k - 1 = (2^d - 1)[(2^d)^{k-1} + (2^d)^{k-2} + \cdots + 1]$

[since $x^k - 1 = (x - 1)(x^{k-1} + \cdots + 1)$, where $x = 2^d$]

$\qquad = (2^d - 1)p$, where $p = (2^d)^{k-1} + \cdots + 1 \in Z$.

From Eq. (1), $2^{n-1} = (2^d - 1)p$, for some $p \in Z$

$\Rightarrow \qquad 2^d - 1|2^n - 1$.

Problem (1.13): If a is an integer, not divisible by 2 or 3, then $24|(a^2+23)$.

Solution: $24|(a^2+23) \Rightarrow a^2+23 = 24k$ for some $k \in Z$

$\Rightarrow \quad a^2 = 24k - 23 = 24(k-1)+1$

$\Rightarrow \quad a^2 = 2m+1 \quad \text{or} \quad 3n+1$

$\Rightarrow \quad a \neq 2m \quad \text{or} \quad 3n$

$\Rightarrow \quad 2 \nmid a \quad \text{or} \quad 3 \nmid a.$

Theorem (1.21): Show that $(a+b, [a,b]) = (a,b)$ for any two non-zero integers a and b.

[Go through the examples.]

Example: (i) Let $a = 14,\ b = 49$.

Then $(a,b) = 7$ (a prime number)

Now $a+b = 63,\ [a,b] = 98$.

Therefore, $(a+b,[a,b]) = (63,98) = 7 = (14,49)$.

(ii) Let $a = 12,\ b = 30,\ (a,b) = (12,30) = 6 =$ composite number, $[a,b] = [12,30] = 60$.

Then $(a+b,[a,b]) = (42,60)$.

Now $6|42,\ 6|60$. Also $6 = 3 \cdot 2$.

So $\exists$ a (+)ve number 3 (or 2) such that $3|42$, $3|60$ and $3|6$.

$\therefore \quad (42,60) = 6.$

So $\quad (a+b,[a,b]) = (a,b)$.

Proof of the main theorem: Let $d = (a,b)$ and $l = [a,b]$.

Then $(a+b,[a,b]) = (a+b,l)$

$d = (a,b) \Rightarrow d|a,\ d|b.$

Also $a|l,\ b|l.$

$$\left.\begin{array}{l}\text{Then } d|a,\ a|l \Rightarrow d|l. \\ \text{Also } d|a,\ d|b \Rightarrow d|(a+b)\end{array}\right\} \quad (1)$$

If d is prime, then $(a+b,l) = d = (a,b)$.

Now, let d be not prime.

That is, d is a composite number.

Then $d = d_1 \cdot d_2$ (say), where $(d_1,d_2) = 1,\ d_1,d_2 < d$.

Thus, $\exists$ a (+)ve number d_1 (or d_2) such that $d_1|a+b$, $d_1|l$ and $d_1|d$.

So $(a+b,l) = d$. That is, $(a+b,[a,b]) = (a,b)$.

Theorem (1.22): *If $a > 1$, then $(a^m-1,\ a^n-1) = a^{(m,n)}-1$.*

Proof: Let $(m,n) = d$.

Again, $d|m$ and $d|n \Rightarrow m = dx$,

$n = dy$ for some x and $y \in Z$ such that $(x,y) = 1$

$a^m - 1 = a^{dx} - 1 = (a^d)^x - 1$ has a factor $a^d - 1\ \forall x$.

Also $\quad a^n - 1 = a^{dy} - 1 = (a^d)^y - 1$ has a factor $a^d - 1\ \forall y$.

Thus, $\quad a^d - 1 | a^m - 1,\ a^d - 1 | a^n - 1$. (1)

Again, $(a-1)|a^m - 1, (a-1)|a^n - 1,\ a > 1$

such that $(a-1)|a^d - 1$. (2)

From (1) & (2) it follows that

$$(a^m - 1, a^n - 1) = a^d - 1 = a^{(m,n)} - 1.$$

Note: (1) When $a = 2$, (1) $\Rightarrow (a^m - 1, a^n - 1) = (a^d - 1) = (a^{(m,n)} - 1)$.

(2) When $a = 2$ and m, n are both even, d is even, then (2) stands as $a^2 - 1|a^m - 1$, $a^2 - 1|a^n - 1$ such that $a^2 - 1|a^d - 1$.

Exercise 1

1. If $(a, b) = 1$, and if $d|(a+b)$, then $(a, d) = (b, d) = 1$.
2. If n is an odd integer, show that $n^4 + 4n^2 + 11$ is of the form 16k.
3. Prove that $3a^2 - 1$ is never a perfect square.
4. Prove that, if both a and b are odd integers, $a^4 + b^4 - 2$ is divisible by 16.
5. For any integer a, show that

   (a) $(2a+1,\ 9a+4) = 1$

   (b) $(5a+2,\ 7a+3) = 1$.
6. Show that $2^n + 1$ or $2^n - 1$ is divisible by 3, according as n is odd or even.
7. If $u_n = (3+\sqrt{5})^n + (3-\sqrt{5})^n$, show that u_n is an integer and that $u_{n+1} = 6u_n - 4u_{n-1}$. Hence, prove that the integer next greater than $(3+\sqrt{5})^n$ is divisible by 2^n.
8. Verify that any integer n can be expressed as $n = 2^k m$, where $k \geq 0$ and m is an odd integer.
9. Give a proof or exhibit a counter example:

   (a) If $b^2|n$ and $a^2|n$ and $a^2 \leq b^2$, then $a|b$.

   (b) If b^2 is the largest square divisor of n, then $a^2|n \Rightarrow a|b$.
10. If $(a, b) = 1$ and one of a and b is even and other is odd, then $(a+b, a-b) = 1$.
11. Find $(48, 72, 30, 27)$.
12. Prove that $n^4 + 4$ is composite, if $n > 1$.
13. If $(a, b) = 1$, then prove that $24|a^2 + 23$.

Chapter 2

Prime Numbers

Definition (2.1): *An integer $p > 1$ is said to be a prime number if p has no other (+)ve divisor except* 1 *and itself.*

$2, 3, 5, 7, \ldots$ *are all prime numbers.*

2 *is the only even prime number.*

Composite Number

Any integer $p > 1$, which is not a prime number, is called as a composite number.

Theorem (2.1): Euclid : *There exists an infinte number of primes.*

Proof: Let $p_1 = 2,\ p_2 = 3,\ p_3 = 5,\ p_4 = 7, \ldots$ be the primes in ascending order and suppose that p_n is the last prime in the given order of primes.

We consider a (+)ve integer

$$P = p_1 p_2 \ldots p_n + 1. \tag{1}$$

$P > 1 \Rightarrow P$ is divisible by some prime p (say).

That is, $p|P$. (2)

But since $p_1, p_2, \ldots, p_n$ are the only primes, p must be equal to one of them.

Thus, $p|p_1 p_2 \ldots p_n$. (3)

So from (2) and (3) it follows that $p|P - p_1 p_2 \ldots p_n$.

That is, $p|1$ which contradicts that $p > 1$.

Hence, $\exists$ infinite number of primes.

Theorem (2.2): *Every composite number N has at least one prime divisor.*

Proof: Since N is not prime, it has a divisor n different from 1 and N. We suppose that n is not greater than any other divisor of N.

Then n must be prime. For if n be not prime, then it would have a divisor k such that $1 < k < n$. Thus $k|N$, which contradicts our assumption that n is not greater than other divisor N.

Hence, N has at least one prime divisor.

Example: $N = 24$.

N is a composite and it has divisors 2, 3, 4, 6, 8, 12 and 24.

Then N has at least one divisor 2 which is prime.

Theorem (2.3): *Fundamental Theorem on Arithmetic. Every positive integer is either* 1 *or a prime or can be expressed uniquely as the product of prime factors.*

Proof: Let n be a (+)ve integer.

Then either $n = 1$ or $n > 1$.

If n being greater than 1 is a prime number, then nothing to prove.

So if $n(> 1)$ be a composite number, then it is sufficient to show that n can be expressed uniquely as the product of prime factors.

Let n be a composite number. Then by theorem (2.2) n has at least a prime factor p_1 (say), $p_1 < n$.

We have $n = p_1 a$, where $1 < a < n$.

If a is prime, then proof is complete. Now if a is not prime, then it has a prime factor p_2 (say) such that $a = p_2 b, 1 < b < a$.

Then $n = p_1 p_2 b,\ 1 < a < n,\ 1 < b < a$, that is $n > a > b > \cdots$

Thus, the sequence $\{n, a, b, \ldots\}$ being a decreasing sequence of prime numbers must end in a prime number, p_m (say).

Thus, $n = p_1, p_2, p_3, \ldots, p_m$.

Now to prove the uniqueness, let us assume that $n = p_1 p_2 p_3 \cdots p_m = q_1 q_2 q_3 \cdots q_k$,

since $\quad p_1|n, p_1|q_1 q_2 \cdots q_k$.

Then $\quad p_1|q_r$ for some r, $1 \le r \le k$.

$\Rightarrow \quad p_1 = q_r,$

since q_r, being a prime, is not divisible by any (+)ve integer other than 1 or q_r, and as $p_1 > 1$, so $p_1 = q_r$.

Therefore, cancelling the common factor, we have

$$p_2 p_3 \cdots p_m = q_1 q_2, \ldots, q_{r-1} q_{r+1}, \cdots q_k.$$

The process is continued.

If $m < k$, then the left-hand side is reduced to 1 while the right-hand side contains the product of $(k - m)$ numbers of prime numbers, which is not possible.

Therefore, $m \ge k$. (1)

Again, $\quad q_1|n \Rightarrow q_1|p_1 p_2 \cdots p_m$

$\Rightarrow \quad q_1|p_s$ for some $s, 1 \le s \le m$.

$\Rightarrow \quad q_1 = p_s,$

since p_s, being a prime number, is not divisible by any (+)ve integer other than 1 or p_s and as $q_1 > 1$, so $q_1 = p_s$.

Then by cancelling the common factor, we have

$$p_1 p_2, \cdots p_{s-1} p_{s+1}, \ldots, p_m = q_2 q_3 \cdots q_k.$$

The process is continued.

If $k < m$, then the right-hand side is reduced to 1 while the left-hand side contains the product of $(m - k)$ numbers of prime numbers, which is absurd.

Therefore, $k \ge m$. (2)

Hence, from (1) and (2) it follows that $m = k$

and $p_i = q_j \ \forall\ i, j,\ 1 \le i, j \le m, i = j$.

Thus, n can be expressed as the product of prime factors uniquely.

Theorem (2.4): *Every positive integer $N > 1$ can be expressed uniquely in the form $2^{m_1} 3^{m_2} 5^{m_3} \ldots p_n^{m_n}$, where p_n is the nth prime and $2, 3, 5, \ldots, p_n$ are increasing order of prime numbers.*

Proof: We first divide N by 2 if $2|N$. If $2 \nmid N$, then we divide it by 3 provided $3|N$. If not, then by the next prime 5 and so on.

Let $2|N$ and let $N = 2^{m_1}N_1$, when N is divisible by 2, m_1 times and N_1 is not divisible by 2. Thus, m_1 is a (+)ve integer or zero, if $2 \nmid N$.

Next let $3|N_1$ and let $N_1 = 3^{m_2}N_2$, when N_1 is divisible by 3, m_2 times and N_2 is not divisible by 3. Thus, m_2 is a (+)ve integer or zero, if $3 \nmid N_1$.

Proceeding in the above way we get $N = 2^{m_1} \cdot 3^{m_2} \cdot 5^{m_3} \ldots N_n$.

Let p_n be the last prime which divides N_n and let $N_n = P_n^{m_n}$, where m_n is a (+)ve integer and $\neq 0$.

Thus, finally $N = 2^{m_1} \cdot 3^{m_2} \ldots p_n^{m_n}$ where $m_r (1 \le r \le n)$ is a (+)ve integer and zero but $m_n \neq 0$.

Now to prove the uniqueness, let

$$N = 2^{m_1} \cdot 3^{m_2} \cdots p_n^{m_n} = 2^{q_1} \cdot 3^{q_2} \ldots P_n^{q_n}$$

$$2^{m_1}|N \Rightarrow 2^{m_1}|2^{q_1} \cdot 3^{q_2} \cdots p_n^{q_n}$$

$$\Rightarrow \quad 2^{m_1}|2^{q_1} \quad \text{since} \quad (2^{m_1}, 3^{q_2}, 5^{q_3} \ldots q_n^{q_n}) = 1$$

$$\Rightarrow \quad m_1 \le q_1. \tag{1}$$

Similarly, we can show that $q_1 \le m_1$ as $2^{q_1}|2^{m_1} \cdot 3^{m_2} \cdots p_n^{m_n} \Rightarrow 2^{q_1}|2^{m_1}$.

In general, we have $m_r \le q_r$ and $q_r \le m_r$ for all values of r, $1 \le r \le n$.

Thus, $m_r = q_r$, $r = 1, 2, \ldots, n$.

Hence, $N = 2^{m_1}3^{m_2} \cdots q_n^{m_n}$ is the unique expression of N.

Example: $N = 2700$

$2|N$ and $N = 2^2 \times 675$, where 675 is not divisible by 2.

Now $3|675$ and $675 = 3^3 \cdot 25$, where 25 is not divisible by 3.

Finally, 5 is the last prime which divides 25, that is, $25 = 5^2$.

Hence, $N = 2^2 \cdot 3^3 \cdot 5^2$.

Note: Every integer $N(> 1)$ is of the form $N = p_1^{a_1} p_2^{a_2} \ldots P_m^{a_m}$, where p_i are primes in ascending order and a_i are (+)ve integers or zero $\forall\ i = 1, 2, \ldots, m$.

That is, $N = \prod_{i=1}^{m} p_i^{a_i}$.

Theorem (2.5): *If* $a = \prod_{i=1}^{m} p_i^{a_i}$ *and* $b = \prod_{i=1}^{m'} p_i^{b_i}$, *then* $[a, b] = \prod_{i=1}^{m} p_i^{c_i}$, *where* $c_i = \max(a_i, b_i)$ *and* p_i *are primes.*

Proof: Let $M = \prod_{i=1}^{m} p_i^{c_i}, c_i = \max(a_i, b_i), a_i \le \max(a_i, b_i)$ and $b_i \le \max(a_i, b_i)$, it follows that $\prod_{i=1}^{m} p_i^{a_i}$ and $\prod p_i^{b_i}$, that is, a and b both divide M. So condition (i) $a|M$ and $b|M$ holds. (ii) Let $\prod p_i^{a_i}$ and $\prod p_i^{b_i}$ both divide an integer c.

Then $\quad c = K \prod p_i^{\max(a_i, b_i)}, K \in z, K \ge 1$

$\quad\quad = K \prod p_i^{c_i} = K \cdot M.$

Then $\quad M|c.$

So conditions (i) and (ii) of definition of L.C.M. (1.8) are satisfied.

Hence, $M = \prod_{i=1}^{m} p_i^{\max(a_i,b_i)} = [a, b]$.

Example: Find [72, 150].

$72 = 2^3 \cdot 3^2 \cdot 5^0$

$150 = 2^1 \cdot 3^1 \cdot 5^2$

$[72,\ 150] = 2^{\max(3,1)} \cdot 3^{\max(2,1)} \cdot 5^{\max(0,2)} = 2^3 \cdot 3^2 \cdot 5^2 = 1800.$

Theorem (2.6): *Let* $a = \prod_{i=1}^{m} p_i^{a_i}$ *and* $b = \prod_{i=1}^{m} p_i^{b_i}$, *where each* p_i *are primes. Then* $(a, b) = \prod_{i=1}^{m} p_i^{c_i}$ *where* $c_i = \min(a_i, b_i)$.

[Proof is left]

Example: (72, 150)

$(72,\ 150) = 2^{\min(3.1)} \cdot 3^{\min(2.1)} \cdot 5^{\min(0.2)} = 2^1 \cdot 3^1 \cdot 5^0 = 6.$

Theorem (2.7): *If* $n = \prod_{i=1}^{m} p_i^{a_i}$, *where* p_i *are primes, then the set of positive divisors of* n *is the set of the numbers of the form* $\prod_{i=1}^{m} p_i^{r_i}$, *where* $0 \le r_i \le c_i$ *for* $i = 1, 2, \ldots, m$.

[Go through the example]

Example: Let $n = 2^3\ 3^4\ 5^7$.

Divisors of 2^3 are $1, 2, 2^2, 2^3$

Divisors of 3^4 are $1, 3, 3^2, 3^3, 3^4$

Divisors of 5^7 are $1, 5, 5^2, \ldots, 5^7$.

$S = \{2^0\ 3^0\ 5^0,\ 2^1\ 3^0\ 5^1,\ 2^1\ 3^1\ 5^0, \ldots\}$

is the set of positive divisors of $n = 2^3\ 3^4\ 5^7$. Thus the set of positive divisors of n is the set of numbers of the form $\prod 2^{a_i}\ 3^{b_i}\ 5^{c_i}$,

where $0 \le a_i \le 3,\ 0 \le b_i \le 4,\ 0 \le c_i \le 7$.

Proof: If $n = \prod_{i=1}^{m} p_i^{a_i} = p_1^{a_1}\ p_2^{a_2} \ldots p_m^{a_m}$, then

divisors of $p_1^{a_i}$ are $p_1^0, p_1^{b_1}, p_1^{b_2}, \ldots, p_1^{b_i}$, $0 < b_i \le a_1$,

divisors of $p_2^{a_2}$ are $p_2^0, p_2^{c_1} p_2^{c_2}, \ldots, p_2^{c_i}$, $0 \le c_i \le a_2$,

divisors of $p_m^{a_m}$ are $p_m^0, p_m^{t_1}, p_m^{t_2}, \ldots, p_m^{t_i}$, $0 \le t_i \le a_m$.

So divisors of n is the set S,

$S = \{p_1^0\ p_2^0 \cdots p_m^0,\ p_1^0\ p_2^1 \cdots p_m^0, \ldots\}$,

where the members of the set is of the form:

$\prod_{i=1}^{m} p_i^{r_1}$, $0 \le r_i \le a_i\ \forall\ i = 1, 2, \ldots, m$.

Theorem (2.8): *Let* $N = p_1^{a_1} p_2^{a_2} \dots p_m^{a_m} = \prod_{i=1}^{m} p_i^{a_i}$, *where* p_i *are primes. Let* n *denote the number of positive divisors of* N *and* s *denote the sum of the positive divisors, then*

$$n = (a_1 + 1)(a_2 + 1) \cdots (a_n + 1)$$

and $$s = \frac{p_1^{a_1+1} - 1}{p_1 - 1} \cdot \frac{p_2^{a_2+1} - 1}{p_2 - 1} \cdots \frac{p_m^{a_m+1} - 1}{p_m - 1}.$$

[Go through the example]

Example: Let $N = 2^2\, 3^3 = 108$.

Divisors of N are of the forms,

$$2^0 3^0,\ 2^0 3^1,\ 2^0 3^2,\ 2^0 3^3,\ 2^1 3^0,\ 2^1 3^1,\ 2^1 3^2,\ 2^1 3^3,\ 2^2 3^0,\ 2^2 3^1,\ 2^2 3^2,\ 2^2 3^2.$$

$n =$ total no. of divisors $= 12 = (2+1)(3+1)$

$s =$ sum of the divisors

$$= 2^0 3^0 + 2^0 3^1 + 2^0 3^2 + 2^0 3^3 + 2^1 3^0 + 2^1 3^1 + 2^1 3^2 + 2^1 3^3 + 2^2 3^0 + 2^2 3^1 + 2^2 3^2 + 2^2 3^3$$

$$= 1 + 3 + 9 + 27 + 2 + 6 + 18 + 54 + 4 + 12 + 36 + 108 = 280.$$

Now, $(1 + 2 + 2^2) \cdot (1 + 3 + 3^2 + 3^3) = \dfrac{2^3 - 1}{2 - 1} \cdot \dfrac{3^4 - 1}{3 - 1} = 7 \cdot 40 = 280.$

$\therefore \quad s = (1 + 2 + 2^2) \cdot (1 + 3 + 3^2 + 3^3).$

Proof: Given, $N = p_1^{a_1} \cdot p_2^{a_2} \cdots p_m^{a_m}$, p_i are primes.

We consider the product

$$(1 + p_1 + p_1^2 + \cdots + p_1^{a_1})(1 + p_2 + p_2^2 + \cdots + p_2^{a_2}) \cdots (1 + p_m + p_m^2 + \cdots + p_m^{a_m}) \quad (2.1)$$

(i) Each term of the product is of the form $p_1^{r_1} p_2^{r_2} \cdots p_m^{r_m}$, where $0 \le r_i \le a_i, i = 1, 2, \dots, m$.

(ii) Each term of the product is distinct.

(iii) Each term of the product is the divisor of N.

Then the total no. of divisors is given by

$n =$ (total no. of terms in the 1st expression) × (total no. of terms in the 2nd expression) × (total no. of terms in the last expression) of the product in (2.1)

$= (a_1 + 1)(a_2 + 1) \cdots (a_n + 1).$

$s =$ sum of the divisors of N is given by the product of the sums in each brackets in (1)

$$= \frac{p_1^{a_1+1} - 1}{p_1 - 1} \cdot \frac{p_2^{a_2+1}}{p_2 - 1} \cdots \frac{p^{a_m+1} - 1}{p_m - 1}.$$

Theorem (2.9): *Let* $f(n) = a^n - 1$, $n > 1$. *Then* $f(n)$ *is prime only if* $a = 2$ *and* n *is prime.*

Proof: Let $f(n)$ be a prime number. Then $f(n)$ has two factors 1 and $f(n)$ itself.

Now $f(n) = a^n - 1$ has a factor $(a - 1)$, which is 1, when $a = 2$ and the other is itself, when n is prime.

If n is not prime, then n is a composite number and let $n = n_1 \cdot n_2$, n_1, n_2 be primes.

In this case, $a^n - 1 = a^{n_1 n_2} - 1 = (a^{n_1})^{n_2} - 1$ has a factor $a^{n_1} - 1$, which is not 1, when $a = 2$, which contradicts that $f(n)$ is prime.

Thus, $f(n)$ is prime only when $a = 2$ and n is prime.

Example: $a^5 - 1$ is prime, when $a = 2$.

Here, $n = 5$, a prime number.

$$a^5 - 1 = (a - 1)(a^4 + a^3 + a^2 + a + 1)$$

For $a = 2$, $a^5 - 1 = 31$, $a - 1 = 1$, $a^4 + a^3 + a^2 + a + 1 = 31$.

Thus, $a^5 - 1$ has only two factors 1 and 31, when $a = 2$.

So $a^5 - 1$ is prime.

Example: $a^6 - 1$ is not prime, when $a = 2$.

Here $n = 6 = 2 \cdot 3$ (a composite number)

$$a^6 - 1 = (a^2)^3 - 1 = (a^2 - 1)(a^4 + a^2 + 1).$$

For $a = 2$, $a^6 - 1 = 63$, $a^2 - 1 = 3$, $a^4 + a^2 + 1 = 21$.

So $a^6 - 1$ has two factors 3 and 21 and hence $a^6 - 1$ is not prime.

Theorem (2.10): *$2^n + 1$ is prime when n is power of 2.*

Proof: Let $2^n + 1$ be prime and n has an odd multiple, say, $n = n_1 h$, where h is odd.

Then $2^n + 1 = 2^{n_1 h} + 1 = (2^{n_1})^h + 1$ has a factor $2^{n_1} + 1$, and $2^{n_1} + 1 \neq 1$ which contradicts that $2^n + 1$ is prime.

If n_1 is also odd, then $2^n + 1$ has a factor $2^h + 1 \neq 1$, a contradiction.

So n_1, h both should be even numbers.

Hence, n = power of 2, i.e., $n = 2^m, m \in N$.

Example: $2^6 + 1$, when $N = 6 = 2.3$, has an odd multiple 3.

$2^6 + 1 = 64 + 1 = 65$ is not prime, while $2^4 + 1$, $2^8 + 1$ are primes.

Corollary (2.1): (i) $2^n - 1$ is prime when n is prime

(ii) If $a^c + 1$ be prime, $a > 1$, then a is even and $c = 2^n \ \forall \ n \in N$.

Theorem (2.11): *If p is prime and $p|ab$, then $p|a$ or $p|b$.*

Proof: If $p|a$, then proof is complete.

Let $p \not| a$. Then $(p, a) = 1$, since p is prime, the divisors of p are 1 and p and $p \not| a$.

Therefore $(p, a) = 1$.

$\Rightarrow$ $\exists \ x, y \in Z$ such that $1 = px + ay$

$\Rightarrow$ $b = pbx + aby$. (1)

Now $p|pb$ and $p|ab$ (given) $\Rightarrow$ $(p|(pbx + aby)$ $\Rightarrow$ $p|b$ [by (1)].

Similarly, if $p \not| b$, then we can prove $p|a$, proceeding as above.

Theorem (2.12): *If p is prime and $p|a_1 a_2 \cdots a_n$, then $p|a_k$ for some k, $1 \leq k \leq n$.*

Proof: If $p|a_1a_2\cdots a_n \Rightarrow p|a_1$, then proof is complete. But if $p\not|a_1$, then by above theorem $p|a_2a_3\ldots a_n$. Now if $p|a_2$, then the result holds. But if $p\not|a_2$, then $p|a_3\ldots a_n$. So after finite no. of steps $\exists\ k,\ 1\le k\le n$ such that $p|a_k$.

Theorem (2.13): *If $d=(a,b)$, then $\frac{a}{d}$ and $\frac{b}{d}$ are two integers prime to each other.*

Proof: $d=(a,b) \Rightarrow d|a$ and $d|b$.

Also $d|a \Rightarrow a=dx$ for some $x\in Z$

and $d|b \Rightarrow b=dy$ for some $y\in Z$

such that $(x,y)=1$.

Now $x=\dfrac{a}{d},\ y=\dfrac{b}{d} \Rightarrow \dfrac{a}{d},\dfrac{b}{d}$ are integers and $(x,y)=1 \Rightarrow \left(\dfrac{a}{d},\dfrac{b}{d}\right)=1$.

Therefore, $\dfrac{a}{d},\dfrac{b}{d}$ are prime to each other.

Theorem (2.14): *$\sqrt{p}$ is irrational for any prime p.*

Proof: If possible, let $\sqrt{p}$ be a rational number.

Then $\sqrt{p}=\dfrac{a}{b}$ where $a,b\in Z,\ b\neq 0,\ (a,b)=1$

$$\Rightarrow \quad p=\frac{a^2}{b^2} \Rightarrow a^2=b^2p=b(bp) \Rightarrow b|a^2.$$

If $b>1$, then by fundamental theorem of Arithmetic [Th. 2.3], $\exists$ a prime p such that $p|b$

So $p|b$ and $b|a^2 \Rightarrow p|a^2 \Rightarrow p|a$.

Then $(a,b)\ge p>1$, a contradiction arises unless $b=1$.

When $b=1$, we have $a^2=p$ which is not possible, as the square of any integer cannot be prime.

Hence, $\sqrt{p}$ is rational.

Therefore, $\sqrt{p}$ is irrational for any prime p.

Example: $\sqrt{2},\sqrt{3},\sqrt{5}$ are all irrationals.

Theorem (2.15): *Any prime of the form $3n+1$ is also of the form $6n+1$.*

Proof: $3n+1$ is prime for some even integer n,

for if $n=2m-1$, then $3n+1=3(2m-1)+1$

$=6m-2=2(3m-1)$, which is not prime for $m\in N$.

Thus, $3n+1$ is prime for $n=2m$ and consequently it assumes the form $3(2m)+1=6m+1$ [i.e., of the form $(6n+1)$]. Hence the theorem.

Theorem (2.16): *Any integer of the form 8^n+1, $n\ge 1$ is composite.*

Proof: $8^n+1=(2^3)^n+1=(2^n)^3+1=(2^n+1)(2^{2n}-2^n+1)$.

Then $2^n+1|8^n+1$, where $2^n+1\ge 3\ \forall\ n\ge 1$.

Thus, 8^n+1 is a composite number as it has a factor ≥ 3 for $n\ge 1$.

Definition (2.2): Fermat's Number : *The number of the form $2^{2^n}+1$ is known as a Fermat's number and it has been also found that the Fermat's number $2^{2^n}+1$ is prime $\forall\ n\ge 0$ and it is denoted by F_n.*

Theorem (2.17): *Fermat's numbers are relatively prime to each other.*

Proof: Let F_n and F_{n+k} be two Fermat's numbers and if possible let $r(>1)$ be their common divisor.

Since F_n and F_{n+k} being odd integers cannot have any even integer as a common divisor. Therefore, r is odd.

Again, $F_n = 2^{2^n} + 1, F_{n+k} = 2^{2^{n+k}} + 1 = (2^{2^n})^{2^k} + 1.$

Then $F_{n+k} - 2 = (2^{2^n})^{2^k} - 1$ has a factor $2^{2^n} + 1 = F_n$.

That is $F_n | F_{n+k} - 2$. Also $r | F_n$.

Therefore, $r | F_{n+k} - 2$.

Thus $r | F_{n+k}$ and $r | F_{n+k} - 2 \Rightarrow r | F_{n+k} - F_{n+k} + 2 \Rightarrow r | 2,$

which is not possible since r is odd.

Hence, $(F_n,\ F_{n+k}) = 1.$

Therefore, Fermat's numbers are prime to each other.

Theorem (2.18): *The product of the first n Fermat's numbers is* $2^{2^n} - 1$.

Proof: $F_n = 2^{2^n} + 1$

$F_0 = 3,\ F_1 = 5,\ F_2 = 17.$

$F_0 F_1 = 3.5 = 15 = 2^4 - 1 = 2^{2^2} - 1$

$F_0 \cdot F_1 \cdot F_2 = 15 \times 17 = 255 = 256 - 1 = 2^{2^3} - 1.$

Result holds for the 1st two products and first three products. Let the result be true for the first k products.

Then $F_0 \cdot F_1 \cdot F_2 \cdots F_{k-1} = 2^{2^k} - 1.$

Therefore, the first $(k+1)$ products $F_0 \cdot F_1 \cdots F_k$

$$= (2^{2^k} - 1) \cdot (2^{2^k} + 1) = (2^{2^k})^2 - 1 = 2^{2^{k+1}} - 1.$$

Thus the result is also true for $n = k + 1$.

Hence the theorem is true $\forall\ n$, by method of induction.

Problem (2.1): $a^{2^n} + 1$ divides $a^{2^m} - 1$, where a is a(+)ve integer and $m > n$.

Solution: Let $m = n + k,\ k \in N$.

Then for any (+)ve integer a,

$a^{2^m} - 1 = a^{2^{n+k}} - 1 = (a^{2^n})^{2^k} - 1$ has a factor $a^{2^n} + 1$ and so $a^{2^n} + 1 | a^{2^m - 1}$.

Definition (2.3): *Symbol* $I\left[\dfrac{x}{y}\right]$

If a be a rational or an irrational number, the symbol [a] denotes the integral part of a which is the greatest integer, not exceeding a.

If $x = qy + r,\ 0 \le r < y,$ then $\dfrac{x}{y} = q + \dfrac{r}{y}.$

So $I\left[\dfrac{x}{y}\right] = q.$

Example: $I[20] = 20$

$I[2.5] = 2$

$I[-7.15] = I[-8 + 0.85] = -8$

$I[0.2] = 0$

$I[\sqrt{3}] = 1.$

Theorem (2.19): *If $x_1, x_2, \ldots, x_n$ be any integers and s is their sum and a is any number < x_i, then* $I\left[\frac{s}{a}\right] \geq I\left[\frac{x_1}{a}\right] + I\left[\frac{x_2}{a}\right] + \cdots + I\left[\frac{x_n}{a}\right]$.

Proof: Let $x_1 = ay_1 + r_1, x_2 = ay_2 + r_2, \ldots, x_n = ay_n + r_n$, where $0 \leq r_1 < a,\ y_i \in Z$ and $I\left[\frac{x_i}{a}\right] = y_i, i = 1, 2, \ldots, n.$

$$s = x_1 + x_i + \cdots + x_n.$$

$$\frac{s}{a} = \frac{x_1}{a} + \frac{x_2}{a} + \cdots + \frac{x_n}{a}$$

$$I\left[\frac{s}{a}\right] = I\left[\frac{x_1}{a} + \frac{x_2}{a} + \cdots + \frac{x_n}{a}\right] = 1\left[\frac{ay_1 + r_1}{a} + \cdots + \frac{a_n + r_n}{a}\right]$$

$$= I\left[y_1 + y_2 + \cdots + y_n + \sum_{i=1}^{n} r_i/a\right].$$

Then $\quad I[s/a] > y_1 + y_2 + \cdots + y_n,$ when $\sum_{i=1}^{n} r_i/a \geq 1$

and $\quad I[s/a] = y_1 + y_2 + \cdots + y_n,$ when $\sum r_i/a < 1.$

$\therefore \quad I\left[\frac{s}{a}\right] \geq y_1 + y_2 + \cdots + y_n = I\left[\frac{x_1}{a}\right] + I\left[\frac{x_2}{a}\right] + \cdots + I\left[\frac{x_n}{a}\right].$

That is, $I\left[\frac{\Sigma x_i}{a}\right] \geq \sum_{i=1}^{n} I\left[\frac{x_i}{a}\right].$

Example: Let $x_1 = 7,\ x_2 = 11,\ x_3 = 4$ and $a = 3$,
such that $x_1 = 3 \cdot 2 + 1,\ x_2 = 3 \cdot 3 + 2,\ x_3 = 3 \cdot 1 + 1.$

$$s = x_1 + x_2 + x_3 = 22$$

$$\frac{s}{a} = \frac{22}{3} = 7.33$$

$$I\left[\frac{s}{a}\right] = 7. \tag{1}$$

Again, $I\left[\frac{x_1}{a}\right] = I\left[\frac{7}{3}\right] = 2$

$$I\left[\frac{x_2}{a}\right] = I\left[\frac{11}{3}\right] = 3$$

$$I\left[\frac{x_3}{a}\right] = I\left[\frac{4}{3}\right] = 1.$$

$$I\left[\frac{x_1}{a}\right] + I\left[\frac{x_2}{a}\right] + I\left[\frac{x_3}{a}\right] = 6. \tag{2}$$

From Eqs. (1) and (2) it follows that $I\left[\frac{s}{a}\right] > I\left[\frac{x_1}{a}\right] + I\left[\frac{x_2}{a}\right] + I\left[\frac{x_3}{a}\right].$

Here to be noted that $\sum r_i/a = \frac{1+2+1}{3} > 1.$

Example: Let $x_1 = 6,\ x_2 = 9,\ x_3 = 12,\ a = 2.$

$$x_1 = 2 \cdot 3 + 0 \quad x_2 = 2 \cdot 4 + 1 \quad x_3 = 2 \cdot 6 + 0$$

$$\frac{s}{a} = \frac{x_1 + x_2 + x_3}{a} = \frac{27}{2} = 13.5$$

$$I\left[\frac{s}{a}\right] = 13. \tag{1}$$

Again, $I\left[\frac{x_1}{a}\right] = I\left[\frac{6}{2}\right] = 3$

$I\left[\frac{x_2}{a}\right] = I\left[\frac{9}{2}\right] = 4$

$I\left[\frac{x_3}{a}\right] = I\left[\frac{12}{2}\right] = 6.$

So $\quad I\left[\frac{x_1}{a}\right] + I\left[\frac{x_2}{a}\right] + I\left[\frac{x_3}{a}\right] = 13. \qquad (2)$

In this case, it follows from Eqs. (1) & (2)

that $\quad I\left[\frac{s}{a}\right] = I\left[\frac{x_1}{a}\right] + I\left[\frac{x_2}{a}\right] + I\left[\frac{x_3}{a}\right].$

Here to be noted that $\sum r_i/a = \dfrac{0+1+0}{2} < 1.$

That is, equality sign may hold.

Theorem (2.20): *The highest power of a prime p contained in* $\underline{|n}$ *is equal to* $I\left[\frac{n}{p}\right] + I\left[\frac{n}{p^2}\right] + \cdots + I\left[\frac{n}{p^r}\right]$, *where* $I\left[\frac{n}{p^r}\right] < p$.

Proof: Let $k_1 = I\left[\frac{n}{p}\right]$

$k_2 = I\left[\frac{k_1}{p}\right]$

$k_3 = I\left[\frac{k_2}{p}\right]$

$\cdots\cdots$

$k_r = I\left[\frac{k_{r-1}}{p}\right].$

Now $\quad I\left[\frac{n}{p}\right] = k_1$

$I\left[\frac{n}{p^2}\right] = I\left[\frac{1}{p}\cdot I\left[\frac{n}{p}\right]\right] = I\left[\frac{k_1}{p}\right] = k_2$

$I\left[\frac{n}{p^3}\right] = I\left[\frac{1}{p}\cdot I\left[\frac{n}{p^2}\right]\right] = I\left[\frac{k_2}{p}\right] = k_3$

$\cdots\cdots\cdots\cdots\cdots\cdots\cdots\cdots\cdots$

$I\left[\frac{n}{p^r}\right] = I\left[\frac{1}{p}\cdot I\left[\frac{n}{p^{r-1}}\right]\right] = I\left[\frac{k_{r-1}}{p}\right] = k_r.$

The process in continued till $I\left[\frac{n}{p^n}\right] < p$. Therefore, the highest power of prime p contained in $\underline{|n}$ is the sum of $k_1, k_2, \ldots, k_r$.

Therefore, $k_1 + k_2 + \cdots + k_r = I\left[\frac{n}{p}\right] + \cdots + I\left[\frac{n}{p^r}\right].$

Hence p occurs in $\underline{|n}$ in the form $p^{\sum_{i=1}^{r} k_i}$.

Example: (i) The highest power of a prime 5 in $\lfloor\underline{213}$ is 51.

$$I\left[\frac{213}{5}\right] = 42$$

$$I\left[\frac{213}{5^2}\right] = I\left[\frac{42}{5}\right] = 8$$

$$I\left[\frac{213}{5^3}\right] = I\left[\frac{8}{5}\right] = 1\ (< 5).$$

Therefore, the highest power of 5 in $\lfloor\underline{213}$ is $42 + 8 + 1 = 51$.

That is 5^{51} is a factor of $\lfloor\underline{213}$.

(ii) The highest power of a prime 2 in $\lfloor\underline{213}$ is 208.

$$I\left[\frac{213}{2}\right] = 106$$

$$I\left[\frac{213}{2^2}\right] = I\left[\frac{106}{2}\right] = 53$$

$$I\left[\frac{213}{2^3}\right] = I\left[\frac{53}{2}\right] = 26$$

$$I\left[\frac{213}{2^4}\right] = I\left[\frac{26}{2}\right] = 13$$

$$I\left[\frac{213}{2^5}\right] = I\left[\frac{13}{2}\right] = 6$$

$$I\left[\frac{213}{2^6}\right] = I\left[\frac{6}{2}\right] = 3$$

$$I\left[\frac{213}{2^7}\right] = I\left[\frac{3}{2}\right] = 1\ (< 2).$$

Therefore, the highest power of 2 in $\lfloor\underline{213}$ is $106 + 53 + 26 + 13 + 6 + 3 + 1 = 208$.

That is, 2^{208} is a factor of $\lfloor\underline{213}$.

Theorem (2.21): *The product of n consecutive integers is divisible by* $\lfloor\underline{n}$.

[Go through the example.]

We consider the product of 6 consecutive integers, starting from the integer 8 (say). Then the 6 consecutive integers are 8, 9, 10, 11, 12, 13, which are of the form $(7+1)(7+2)(7+3)(7+4)(7+5)(7+6)$. Their product is $8 \cdot 9 \cdot 10 \cdot 11 \cdot 12 \cdot 13 = \dfrac{\lfloor\underline{13}}{\lfloor\underline{7}}$.

We now consider the number $\dfrac{\lfloor\underline{13}}{\lfloor\underline{7} \cdot \lfloor\underline{6}}$.

Now the primes contained in $\lfloor\underline{13}$ are 2, 3, 5, 7, 11, 13. The highest power of 2, 3, 5, 7, 13 are respectively 10, 5, 2, 1, 1, 1, proceeding as in Example (i) & (ii).

So $\lfloor\underline{13} = 2^{10} \cdot 3^5 \cdot 5^2 \cdot 7 \cdot 11 \cdot 13$.

Now primes contained in $\lfloor\underline{7}$ are 2, 3, 5, 7 and their highest powers are 4, 2, 1, 1 and hence, $\lfloor\underline{7} = 2^4 \cdot 3^2 \cdot 5 \cdot 7$.

Again, the primes contained in $\lfloor\underline{6}$ are 2, 3, 5 and their highest powers are 4, 1, 1 and hence, $\lfloor\underline{6} = 2^4 \cdot 3 \cdot 5$.

Therefore, $\dfrac{\lfloor 13}{\lfloor 7 \cdot \lfloor 6} = \dfrac{2^{10} \cdot 3^5 \cdot 5^2 \cdot 7 \cdot 11 \cdot 13}{2^4 \cdot 3^2 \cdot 5 \cdot 7 \cdot 2^4 \cdot 3 \cdot 5} = \dfrac{2^{10} \cdot 3^5 \cdot 5^2 \cdot 7 \cdot 11 \cdot 13}{2^8 \cdot 3^3 \cdot 5^2 \cdot 7} = 2^2 \cdot 3^2 \cdot 11 \cdot 13.$

Hence, $\dfrac{\lfloor 13}{\lfloor 7 \cdot \lfloor 6} = 2^2 \cdot 3 \cdot 11 \cdot 13 = \text{integer}.$

Therefore, $\dfrac{8 \cdot 9 \cdot 10 \cdot 11 \cdot 12 \cdot 13}{\lfloor 6} = \text{integer}.$

Hence, $8 \cdot 9 \cdot 10 \cdot 11 \cdot 12 \cdot 13$ is divisible by $\lfloor 6$. That is product of 6 consecutive integers is divisible by $\lfloor 6$.

***Proof*:** Let us consider the product of n consecutive integers $(m+1)(m+2)-(m+n)$, starting from the $(m+1)$th integer.

Now $\quad (m+1)(m+2)\cdots(m+n) = \dfrac{\lfloor m+n}{\lfloor m}$

We consider the term $\dfrac{\lfloor m+n}{\lfloor m \cdot \lfloor n}$.

If p be any prime occurring in $\lfloor m+n$ and also in $\lfloor m$ and $\lfloor n$, then we have

$$I\left[\frac{m+n}{p^r}\right] \geq I\left[\frac{m}{p^r}\right] + I\left[\frac{n}{p^r}\right] \quad \text{for } r = 1, 2, 3 \cdots.$$

Therefore, the sum of the highest powers of any prime p occurring in $\lfloor m$ and $\lfloor n$ is always less than or at most equal to the highest power of the prime p occurring in $\lfloor m+n$.

Hence, $\quad \dfrac{\lfloor m+n}{\lfloor m \cdot \lfloor n}$ is of the form $\dfrac{p_1^{a_1} p_2^{a_2} \cdots p_n^{a_n}}{p_1^{r_1} p_2^{r_2} \cdots p_n^{r_n}}$,

where $0 \leq r_1 \leq a_i$, $i = 1, 2, \ldots, n$; p_i are primes occurring in $\lfloor m+n$, $\lfloor m$ and $\lfloor n$.

Then $\quad \dfrac{\lfloor m+n}{\lfloor m \cdot \lfloor n} = \prod_{i=1}^{n} p_i^{a_i - r_i} = \text{integer}, \; a_i - r_i \geq 0.$

Therefore, $\dfrac{(m+1)(m+2)\cdots(m+n)}{\lfloor n} = k, \; k \in N,$

which shows that $(m+1)(m+2)\cdots(m+n)$ is divisible by $\lfloor n$.

Problem (2.2): If n be prime, ^nC_r is divisible by n.

Solution: $^nC_r = \dfrac{\lfloor n}{\lfloor r \cdot \lfloor n-r} = \dfrac{n(n-1)(n-2)\cdots(n-r+1)}{\lfloor r}$,

where $\quad r < n$.

Then by Theorem (2.21), $n(n-1)(n-2)\cdots(n-r+1)$ is divisible by $\lfloor r$.

Since n is prime and $r < n$, $(\lfloor r, n) = 1$ and hence $\lfloor r \mid (n-1)(n-2)\cdots(n-r+1)$.

Therefore, $(n-1)(n-2)\cdots(n-r+1) = \lfloor r \cdot k$ for same (+)ve integer k.

So $\quad n(n-1)(n-2)\cdots(n-r+1) = n\lfloor r \cdot k,$

Hence, $\dfrac{n(n-1)(n-2)\cdots(n-r+1)}{\lfloor r}$ is divisible by n. That is, ^nC_r is divisible by n.

Problem (2.3): If n is prime to 5, then $n^5 - n$ is divisible by 30.

Solution: $n^5 - n = n(n^2-1)(n^2+1)$

Every square number is of the form: $5m$ or $5m \pm 1$.

When $n^2 = 5m + 1$, $n^2 - 1$ is divisible by 5 and when $n^2 = 5m - 1$, $n^2 + 1$ is divisible 5.

So either $n^2 - 1$ or $n^2 + 1$ in the product $n(n^2 - 1)(n^2 + 1)$ is divisible by 5.

Also $n(n^2-1)(n^2+1) = n(n-1)(n+1)(n^2+1)$ is divisible by 6, since $n(n-1)(n+1)$ being the product of 3 consecutive integers is divisible by 3!, i.e., by 6.

Hence, $n^5 - n$ is divisible by $5 \cdot 6 = 30$.

Mathematical Induction

Peano's axioms: *Let N be the set of natural numbers. Then the following axioms are known as the Peano's axioms.*

A(1) $I \in N$.

A(2) For every $n \in N$, there corresponds a natural number n^*, called the successor of n. That is $n^* = n + 1$.

A(3) If $n^* = m^*\ \forall\ n, m \in N$, then $n = m$.

A(4) $I \in N$ is not the successor of any natural number n.

A(5) If (i) $I \in N$ and (ii) $n \in N$, then $n + 1 \in N$, which is known as the *principle of finite inductions.*

Method of Induction

Theorem (2.22): *Let $S \subset N$ with the properties that (i) $I \in S$ and (ii) $m \in S \Rightarrow m + 1 \in S$.*

Then $S = N$.

Proof: Let $T \subset N$ such that $S^c = T$.

To prove the theorem it is sufficient to show that $T = \phi$.

Let T be a non-empty subset of N.

So $T \subset N$, by well ordering property T has a least element, say m. Also since $I \in S$ and $m > 1$, we have $0 < m - 1 < m$. Again, m being the least element of T,

$$m - 1 \notin T.$$

Now $m - 1 \notin T \Rightarrow m - 1 \in S$

$\Rightarrow$ $(m - 1) + 1 = m \in S$ [by (ii)]

$\Rightarrow$ $m \notin S^c = T$, which contradicts that m is the least element of T.

Therefore, $T = \phi$,

which shows that $S^c = \phi$

$\Rightarrow$ $N - S = \phi \Rightarrow S = N$.

It immediately follows from the above theorem [Theorem (2.22)]:

If P_n be any statement for any natural number n such that (i) P_1 is true and (ii) P_{n+1} is true when P_n is true, then P_n is true $\forall\ n \in N$.

Problem (2.4): Show that $7^{2n} + 16n - 1$ is divisible by 64.

Solution: Let $P_n = 7^{2n} + 16n - 1$.

For $n = 1$, $P_1 = 49 + 16 - 1 = 64$ is divisible by 64.

Let the result be true for $n = m$.

That is $P_m = 7^{2m} + 16m - 1$ is divisible by 64.

Now $\quad P_{m+1} = 7^{2(m+1)} + 16(m+1) - 1 = 7^{2m} \cdot 49 + 16m + 15.$

Then $\quad P_{m+1} + 15P_m = 7^{2m}(49+15) + (16+240)m = 7^{2m} \cdot 64 + 256m$

$$= 64[7^{2m} + 4m],$$

which is divisible by 64. P_m is also divisible by 64.

Hence, P_{m+1} is divisible by 64.

Thus, the result is true for $n = m + 1$.

Therefore, P_n is divisible by 64 $\forall\ n \in N$.

Problem (2.5): Show that for any $n \in N$, $\dfrac{1}{n} + \dfrac{1}{n+1} + \dfrac{1}{n+2} + \cdots + \dfrac{1}{2n-1} = 1 - \dfrac{1}{2} + \dfrac{1}{3} - \dfrac{1}{4} + \cdots + \dfrac{1}{2n-1}$ holds.

Solution: Let $P_n = \dfrac{1}{n} + \dfrac{1}{n+1} + \dfrac{1}{n+2} + \cdots + \dfrac{1}{2n-1}$ (1)

Then $\quad P_{n+1} - P_n = -\dfrac{1}{n} + \dfrac{1}{2n} + \dfrac{1}{2n+1} = -\dfrac{1}{2n} + \dfrac{1}{2n+1}.$

Again, if $\quad R_n = 1 - \dfrac{1}{2} + \dfrac{1}{3} - \dfrac{1}{4} + \cdots + \dfrac{1}{2n-1}.$ (2)

Then $\quad R_{n+1} - R_n = -\dfrac{1}{2n} + \dfrac{1}{2n+1} = P_{n+1} - P_n.$ (3)

From Eqs. (1) & (2) it follows that $P_1 = R_1$ is true.

So if $P_n = R_n$, then $P_{n+1} = R_{n+1}$ by Eq. (3).

Hence, it is true $\forall\ n \in N$.

Therefore, $P_n = R_n\ \forall\ n \in N$.

Problem (2.6): $2^{2n+1} - 9n^2 + 3n - 2$ is divisible by 54 $\forall\ n \in N$.

Solution: Let $P_n = 2^{2n+1} - 9n^2 + 3n - 2$.

Then $\quad P_1 = 8 - 9 + 3 - 2 = 0$ is divisible by 54.

Let $\quad P_m = 2^{2m+1} - 9m^2 + 3m - 2$ be divisible by 54.

Then $\quad P_{m+1} = 2^{2m+3} - 9(m+1)^2 + 3(m+1) - 2$

$$= 2^3 \cdot 2^{2m} - 9m^2 - 15m - 8.$$

Now $\quad P_{m+1} - 4P_m = 27m^2 - 27m = 27m(m-1).$

Since $m(m-1)$ is divisible by 2 $\forall\ m \in N$,

$$P_{m+1} - 4P_m = 27 \times 2k = 54k,\ k \in N,$$

which shows that P_{m+1} is divisible by 54.

Hence the result.

Problem (2.7): $1\underline{|1} + 2\underline{|2} + \cdots + n\underline{|n} = \underline{|n+1} - 1\ \forall\ n \geq 1$.

Solution: Let $P_n = 1\underline{|1} + 2\underline{|2} + \cdots + n\underline{|n}$.

$$P_1 = 1\underline{|1} = 1 = 2 - 1 = \underline{|1+1} - 1.$$

Let P_n be true for $n = m \in N$.

That is, $P_m = \underline{|m+1} - 1$.

Now $P_{m+1} = 1\lfloor 1 + \cdots + m\lfloor m + (m+1)\lfloor m+1 = (\lfloor m+1 - 1) + (m+1)\lfloor m+1$

$= \lfloor m+1(1+m+1) - 1 = \lfloor m+1(m+2) - 1$

$= \lfloor m+2 - 1 = \lfloor (m+1)+1 - 1.$

So it is true for $n = m+1$.

Hence, it is true $\forall\, n \geq 1$.

Problem (2.8): $\frac{1}{1^2} + \frac{1}{2^2} + \frac{1}{3^2} + \cdots + \frac{1}{n^2} \leq 2 - \frac{1}{n}$. $\forall\, n \geq 1$.

Solution: Let $P_n = \frac{1}{1^2} + \frac{1}{2^2} + \frac{1}{3^2} + \cdots + \frac{1}{n^2}$

$P_1 = \frac{1}{1^2} = 1 = 2 - \frac{1}{1}$, P_n is true for $n = 1$.

Let P_n be true for $N = m$.

That is $P_m < 2 - \frac{1}{m}$.

Then $P_{m+1} = \frac{1}{1^2} + \frac{1}{2^2} + \cdots + \frac{1}{m^2} + \frac{1}{(m+1)^2}$

$\leq 2 - \frac{1}{m} + \frac{1}{(m+1)^2} = 2 - \frac{1}{m+1} + \frac{1}{m+1} - \frac{1}{m} + \frac{1}{(m+1)^2}$

$= \left(2 - \frac{1}{m+1}\right) - \frac{1}{m(m+1)} + \frac{1}{(m+1)^2}$

$= 2 - \frac{1}{m+1} - \left\{\frac{m+1-m}{m(m+1)^2}\right\} = 2 - \frac{1}{m+1} - \frac{1}{m(m+1)^2}.$

$\therefore \quad P_{m+1} \leq \left(2 - \frac{1}{m+1}\right) - \frac{1}{m(m+1)^2} < 2 - \frac{1}{m+1}$

Therefore, P_n is true for $n = m+1$.

Hence, it is true $\forall\, n \geq 1$.

$\therefore \quad P_n \leq 2 - \frac{1}{n}$.

Problem (2.9): If the numbers a_n are defined by $a_1 = 11$, $a_2 = 21$ and $a_n = 3a_{n-1} - 2a_{n-2}\ \forall\, n \geq 3$, then $a_n = 5 \cdot 2^n + 1\ \forall\, n \geq 1$.

Solution: For $n = 3$, $a_3 = 3a_2 - 2a_1 = 3 \cdot 21 - 2 \cdot 11 = 63 - 22 = 41 = 5 \cdot 2^3 + 1$.

Result is true for $n = 3$.

Let the result be true up to $n = m$.

Then $a_{m-1} = 5 \cdot 2^{m-1} + 1$, $a_m = 5 \cdot 2^m + 1$.

Therefore, $a_{m+1} = 3a_m - 2a_{m-1} = 3[5 \cdot 2^m + 1] - 2[5 \cdot 2^{m-1} + 1]$

$= 3 \cdot 5 \cdot 2^m + 3 - 5 \cdot 2^m - 2 = 5 \cdot 2^m \cdot 2 + 1 = 5 \cdot 2^{m+1} + 1.$

So the result is true for $n = m+1$.

Also $a_1 = 11 = 10 + 1 = 5 \cdot 2^1 + 1$

$a_2 = 21 = 20 + 1 = 5 \cdot 2^2 + 1.$

Hence, the result is true $\forall\, n \geq 1$.

Therefore, $a_n = 5 \cdot 2^n + 1\ \forall\, n \geq 1$.

Exercise 2

1. Show that $\sqrt{5}$ is irrational.

2. Show that if $2^n - 1$ is prime, then n is prime.

3. Show that $2^n + 1$ is prime if n is a power of 2.

4. Show that the highest power of 5 contained in $\underline{|158}$ is 38.

5. If p is prime and $p = a^2 - b^2$, then $a = \frac{1}{2}(p+1)$, $b = \frac{1}{2}(p-1)$.

6. Prove that $\underline{|2n}$ is divisible by $\underline{|n}\,\underline{|n+1}$.

7. The only prime p for which $3p + 1$ is a perfect square is $p = 5$.

8. Prove that if $n > 2$, then there exists a prime p such that $n < p < n!$.

9. Prove that any prime of the form $3n + 1$ is also of the form $6m + 1$.

10. Show that any integer of the form $8^n + 1$, $n \geq 1$ is composite.

11. If $(a, b) = 1$, then show that $(a + b,\ ab) = 1$ and $(a^2 + b^2,\ a^2b^2) = 1$.

12. Prove, the following by mathematical induction,

$$\frac{1}{2} + \frac{2}{2^2} + \frac{3}{2^3} + \cdots + \frac{n}{2^n} = 2 - \frac{n+2}{2^n},\ \forall\, n \geq 1.$$

13. If $1 + a > 0$, then show that $(1 + a)^n \geq 1 + na$.

Chapter 3

Congruences

The great German Mathematician K. Gauss introduced the concept of congruences and created a new branch of mathematics in the theory of numbers and also used a remarkable notation which plays an important role in divisibility of integers.

Definition (3.1): *For any two given integers a and b, there exists a (+)ve integer m such that $m|(a-b)$, then we write it $a \equiv b \pmod{m}$, which is read as 'a is congruent to b modulo m'. b is called the residue of a modulo m. Thus, the relation $a \equiv b \pmod{m}$ is equivalent to the divisibility relation $m|(a-b)$.*

If $m \nmid (a-b)$, then we write $a \not\equiv b \pmod{m}$.

Example: (i) For two integers 3 and −7, ∃ an integer 2 such that $2|3-(-7)$, i.e., $2|10$.

Therefore, we write it,

$$3 \equiv -7 \pmod{2}, \quad \text{but} \quad 3 \not\equiv -7 \pmod{3}.$$

(ii) n is even iff $n \equiv 0 \pmod{2}$.

(iii) n is odd iff $n \equiv 1 \pmod{2}$.

(iv) $a \equiv b \pmod{1}\ \forall\ a, b \in Z$.

Theorem (3.1): *If $a \equiv b \pmod{m}$, then $a \equiv b \pmod{d}$ when $d|m$, $d, m > 0$.*

Proof: $d|m \Rightarrow m = dk$ for some $k \in Z$. (1)

Again, $a \equiv b \pmod{m}$

$\Rightarrow$ $m|a-b \Rightarrow a-b = mt$ for some $t \in Z$ (2)

$\Rightarrow$ $a-b = dkt$ [by (1)]

$= dr,\ r = kt \in Z.$

$\Rightarrow$ $d|a-b \Rightarrow a \equiv b \pmod{d}$.

Theorem (3.2): *Congruence is an equivalence relation in Z.*

Proof: We define a relation ρ for any two elements $a, b \in Z$ such that $a\rho b \iff a \equiv$ (mod m), $m(>0) \in Z$. Then

(i) $a \equiv a \pmod{m}\ \forall\ a \in z$

$\Rightarrow$ $a\rho a\ \forall\ a \in Z$

$\Rightarrow$ ρ is reflexive.

(ii) Let $a\rho b$, $a, b \in Z$.

Then $a \equiv b \pmod{m}$

$\Rightarrow$ $m|(a-b) \Rightarrow m|(b-a) \Rightarrow b \equiv a \pmod{m}$

$\Rightarrow$ $b\rho a \Rightarrow \rho$ is symmetric.

(iii) Let $a\rho b$ and $b\rho c\ \forall\ a, b, c \in Z$.

Then $a\rho b \Rightarrow m|a-b$ and $b\rho c \Rightarrow m|b-c$.

$\therefore$ $m|(a-b)$ and $m|b-c \Rightarrow m|(a-b)+(b-c) \Rightarrow m|a-c$

$\Rightarrow$ $a \equiv c(\text{mod } m)$

$\Rightarrow$ ρ is transitive.

So from (i), (ii) and (iii) it follows that ρ is an equivalence relation.

Basic properties of congruences

Theorem (3.3): (1) *If* $a \equiv b(\text{mod } m)$ *and* $x \equiv y(\text{mod } m)$*, then*

(i) $ap + xq \equiv (bp + yq)(\text{mod } m),\ p, q \in Z$

(ii) $ax \equiv by(\text{mod } m)$

(iii) $a + x \equiv b + y(\text{mod } m)$.

(2) *If* $a \equiv b(\text{mod } m)$*, then*

(i) $a^n \equiv b^n(\text{mod } m)$ *for every +ve integer n.*

(ii) $ac \equiv bc(\text{mod } m)$ *for any* $c \in Z$

(3) *If* $a \equiv b(\text{mod } m)$ *and* $b \equiv c(\text{mod } m)$*, then* $a \equiv c(\text{mod } m)$

***Proof*:** (1) (i) We have $a - b = mk_1$ for some $k_1 \in Z$

and $x - y = mk_2$ for some $k_2 \in Z$.

Then $ap - bp = mk_1p$ and $xq - yq = mk_2q$.

Therefore, $(ap - bp) + (xq - yq) = m(k_1p + k_2q) = mk,\ k = k_1p + k_2q \in Z$

$\Rightarrow$ $m|[(ap + xq) - (bp + yq)] \Rightarrow ap + xq \equiv bp + yq(\text{mod } m)$.

(ii) $a - b = mk_1$

$x - y = mk_2$

Then $ax - bx = mk_1x$ and $bx - by = mk_2b$.

Adding, we get $ax - by = m(k_1x + k_2b) = mk,\ k \in Z$

$\Rightarrow$ $m|ax - by \Rightarrow ax \equiv by(\text{mod } m)$.

(iii) $a \equiv b(\text{mod } m) \Rightarrow a - b = k_1m,\ k_1 \in Z$

$x \equiv y(\text{mod } m) \Rightarrow x - y = k_2m,\ k_2 \in Z$

Thus adding, we get $(a + x) - (b + y) = (k_1 + k_2)m$

$\Rightarrow$ $a + x \equiv b + y(\text{mod } m)$.

(2) (i) For $n = 1$, $a \equiv b(\text{mod } m)$. (1)

So the result holds for $n = 1$.

Let the result hold for $n = p$.

Then $a^p \equiv b^p(\text{mod } m)$. (2)

From (1) & (2), we have $a \cdot a^p \equiv b\dot{b}^p(\text{mod } m)$ [by ii]

or, $a^{p+1} \equiv b^{p+1}(\text{mod } m)$.

So the result holds for $n = p + 1$.

Then it is true for all values of n.

Hence, $a^n \equiv b^n \pmod{m}$.

(ii) $a - b = mk_1$

$\Rightarrow$ $ac - bc = mk_1c = mck_1 = mr$ where $r = ck_1 \in Z$ for any $c \in Z$

$\Rightarrow$ $m|ac - bc \Rightarrow ac \equiv bc \pmod{m}$.

(3) $a \equiv b \pmod{m} \Rightarrow a - b = k_1m,\ k_1 \in Z$

$b \equiv c \pmod{m} \Rightarrow b - c = k_2m,\ k_2 \in Z$.

Adding $(a-b) + (b-c) = (k_1 + k_2)m \Rightarrow a - c = (k_1 + k_2)m$

$\Rightarrow$ $a - c = k'm,\ k' = k_1 + k_2 \in Z$

$\Rightarrow$ $a \equiv c \pmod{m}$.

Theorem (3.4): *If* $a_1 \equiv b_1 \pmod{m},\ a_2 \equiv b_2 \pmod{m}, \ldots, a_n \equiv b_n \pmod{m}$, *then* $a_1a_2a_3, \ldots, a_n \equiv b_1b_2 \ldots b_n \pmod{m}$.

Proof: Follows from the part (ii) of Theorem (3.3).

Note: Proof of part (iii) of Theorem (3.3) may be deduced from [Th. 3.4] by taking $a_1 = a_2 = \cdots = a_n = a$ and $b_1 = b_2 = \cdots = b_n = b$.

Theorem (3.5): *If* $ac \equiv bc \pmod{m}$ *and* $(c, m) = d$, *then* $a \equiv b \left(\text{mod } \dfrac{m}{d}\right)$.

Proof: $(c, m) = d$

$\Rightarrow$ $d|c$ and $d|m$

$\Rightarrow$ $c = dk_1$ and $m = dk_2$ for some $k_1, k_2 \in Z$, where $(k_1, k_2) = 1$.

Now $ac \equiv bd \pmod{m}$

$\Rightarrow$ $m|ac - bc \Rightarrow ac - bc = mt$, for some $t \in Z$.

$\Rightarrow$ $(a-b)c = mt \Rightarrow (a-b) = \dfrac{m}{c}t = \dfrac{dk_2}{dk_1} = \dfrac{k_2}{k_1}t$

$\Rightarrow$ $(a-b)k_1 = k_2t \Rightarrow k_2|(a-b)k_1$

$\Rightarrow$ $k_2|(a-b)$ as $(k_1, k_2) = 1$.

$\Rightarrow$ $a \equiv b \pmod{k_2}$.

That is $a \equiv b \left(\text{mod } \dfrac{m}{d}\right)$, where $k_2 = \dfrac{m}{d} \in Z$.

Theorem (3.6): *If* $ac \equiv bc \pmod{m}$ *and* $(c, m) = 1$, *then* $a \equiv b \pmod{m}$, *that is the cancellation law holds when c is prime to m.*

Proof: $ac \equiv bc \pmod{m}$

$\Rightarrow$ $m|(ac - bc) = m|(a-b)c \Rightarrow m|(a-b)$ as $(m, c) = 1$

$\Rightarrow$ $a \equiv b \pmod{m}$.

Theorem (3.7): *If* $a \equiv b \pmod{m}$, *then* $(a, m) = (b, m)$. *That is, congruent numbers have the same G.C.D. with m.*

Proof: Let $(a, m) = d$ and $(b, m) = c$.

Then $d|a,\ d|m$ and $c|b,\ c|m$.

Therefore, $a = dr,\ m = ds$ for some $r, s \in Z$.

Now $a \equiv b \pmod{m}$

$\Rightarrow$ $a - b = mp$, for some $p \in Z$.

or, $dr - b = m.p = dsp$ or, $b = d(r - sp) = dt,\ t = r - sp \in Z$.

Then $d|b$. Also $d|m$.

Therefore, $d|c$ as $(b, m) = c$.

Similarly, with the same arguments as above we say that $c|d$.

So $d|c$ and $c|d \Rightarrow d = c$.

Therefore, $(a, m) = (b, m)$.

Theorem (3.8): *If $f(x, y, z)$ be a polynomial in x, y, z with integral coefficient and $x \equiv x'(\text{mod } m)$, $y \equiv y'(\text{mod } m)$, $z \equiv z'(\text{mod } m)$, then $f(x, y, z) \equiv f(x', y', z')(\text{mod } m)$.*

Proof: Each term of $f(x, y, z)$ is of the form $x^a y^b z^c$, $a, b, c \geq 0$. Also let $R(a, b, c)$ denote the integral coefficient of the corresponding term $x^a y^b z^c$. That is, if the coefficient of $z^2y\ z^3$ be 5, then $R(a, b, c) = R(2, 1, 3) = 5$, then the coefficient of the corresponding term $x'^a y'^b z'^c$ in $f(x', y', z')$ is also $R(a, b, c)$.

Now $\quad x \equiv x'(\text{mod } m) \Rightarrow x^a \equiv x'^a(\text{mod } m)$

$\quad y \equiv y'(\text{mod } m) \Rightarrow y^b \equiv y'^b(\text{mod } m)$

$\quad z \equiv z'(\text{mod } m) \Rightarrow z^c \equiv z'^c(\text{mod } m)$

and hence $x^a y^b z^c \equiv x'^a y'^b z'^c \cdots (\text{mod } m)$

$\Rightarrow \quad R(a, b, c)x^a y^b z^c \equiv R(a, b, c)x'^a y'^b z'^c(\text{mod } m)$

$\Rightarrow \quad \sum R(a, b, c)x^a y^b z^c \equiv \sum R(a, b, c)x'^a y'^b z'^c(\text{mod } m)$

$\Rightarrow \quad f(x, y, z) \equiv f(x', y', z')(\text{mod } m)$

Note: In general, $f(x, y, z, \ldots) \equiv f(x', y', z', \ldots)(\text{mod } m)$.

Example: Let $f(x, y, z) = 2 + 5xyz + 3x^2yz + 6yz^3 + x + y + z + 5x^2 + 6y^2 + 3z^2$. (1)

Then $f(x', y', z') = 2 + 5x'y'z' + 3x'^2y'z' + 6y'z'^3 + x' + y' + z' + 5x'^2 + 6y'^2 + 3z'^2$ (2)

which follows that $R(a, b, c)$, the integral coefficient of $z^a y^b z^c$, is same as that of $x'^a y'^b z'^c$, where $0 \leq a \leq 2$, $0 \leq b \leq 2$, $0 \leq c \leq 3$.

Now $\quad x^a y^b z^c \equiv x'^a y'^b z'^c(\text{mod } m)$

$1 \equiv 1(\text{mod } m) \Rightarrow 2 \equiv 2(\text{mod } m)$

$xyz \equiv x'y'z'(\text{mod } m) \Rightarrow 5xyz = 5x'y'z'(\text{mod } m)$

$x^2yz \equiv x'^2y'z'(\text{mod } m \Rightarrow 3x^2yz \equiv 3x'^2y'z'(\text{mod } m)$

....

$z^2 \equiv z'^2(\text{mod } m) \Rightarrow 3z^2 = 3z'^2(\text{mod } m)$.

Adding the results on the R.H.S, we have

$$2 + 5xyz + 3x^2yz + 6yz^3 + x + y + z + 5x^2 + 6y^2 + 3z^2 \equiv 2 + 5x'y'z' + 3x'^2y'z' + 6y'z'^3 + x' + y' + z' + 5x'^2 + 6y'^2 + 3z'^2(\text{mod } m)$$

$\Rightarrow \quad f(x, y, z) \equiv f(x', y', z')(\text{mod } m)$.

Corollary (3.1): If $f(x) = a_0 + a_1x + a_2x^2 + \cdots + a_nx^n$ be a polynomial in x with integral coefficient a_i, $i = 0, 1, \ldots, n$ and if $p \equiv q \pmod m$, then $f(p) \equiv f(q) \pmod m$.

Proof: $p \equiv q \pmod m$

$\Rightarrow\ p^k \equiv q^k \pmod m,\ k > 0$ [by 2(i) of Th. [3.3]]

$\Rightarrow\ a_ip^k \equiv a_iq^k \pmod m,\ a_i \in Z,\ i = 0, 1, 2, \ldots, n.$

$$\Rightarrow \sum_0^n a_ip^k \equiv \sum_{i=0}^n a_iq^k \pmod m$$

$\Rightarrow\ a_0 + a_1p + a_2p^2 + \cdots + a_np^n \equiv a_0 + a_1q + a_2q^2 + \cdots + a_nq^n \pmod m$

$\Rightarrow\ f(p) \equiv f(q) \pmod m.$

Note: (1) If $f(x) = a_0 + a_1x + a_2x^2 + \cdots + a_nx^n$, $a_i \in Z$, then a is a solution of the congruence equation $f(x) \equiv 0 \pmod m$, if $f(a) \equiv 0 \pmod m$.

(2) If a is a solution of $f(x) \equiv 0 \pmod m$ and $a \equiv b \pmod m$, then b is also a solution of it, i.e., $f(b) \equiv 0 \pmod m$.

Hints: $f(a) \equiv 0 \pmod m,\ a \equiv b \pmod m$

$\Rightarrow\ f(a) \equiv f(b) \pmod m$ [by Cor. 3.1]

$\Rightarrow\ f(b) \equiv f(a) \pmod m.$

Hence, $f(b) \equiv 0 \pmod m$.

Divisibility test with the congruence relation

Every integer N can be expressed in the following forms :

(i) $a_m10^m + a_{m-1}10^{m-1} + \cdots + a_210^2 + a_110 + a_0$

(ii) $a_m(100)^m + a_{m-1}(100)^{m-1} + \cdots + a_2100^2 + a_1100 + a_0$

(iii) $a_m(1000)^m + a_{m-1}(1000)^{m-1} + \cdots + a_21000^2 + a_11000 + a_0$,

where $a_m, a_{m-1}, \ldots, a_2, a_1, a_0$ are all integral coefficients.

Example: $N = 205769$

Then (1) $N = 2 \cdot (10)^5 + 0 \cdot 10^4 + 5 \cdot 10^3 + 7 \cdot 10^2 + 6 \cdot 10 + 9$

(2) $N = 20(100)^2 + 57(100) + 69$

(3) $N = 205(1000) + 769.$

Theorem (3.9): *Let* $N = a_m10^m + a_{m-1}10^{m-1} + \cdots + a_110 + a_0$ *be decimal expansion of the (+)ve integer* N, $0 \leq a_r < 10$

and (i) *if* $S = a_0 + a_1 + \cdots + a_n$, *then* $9|N$, *iff* $9|S$

(ii) *if* $T = a_0 - a_1 + a_2 + \cdots (-1)^m a_m$, *then* $11|N$, *iff* $11|T$.

Proof: (i) We consider a polynomial $f(x) = a_mx^m + a_{m-1}x^{m-1} + \cdots + a_1x + a_0$, where $a_r \in Z$ for $r = 0, 1, 2, \ldots, m$.

Now $10 \equiv 1 \pmod 9 \Rightarrow f(10) \equiv f(1) \pmod 9$

$\Rightarrow\ a_m10^m + a_{m-1}10^{m-1} + \cdots + a_110 + a_0 \equiv a_m + a_{m-1} + \cdots + a_1 + a_0 \pmod 9$

$\Rightarrow\ N \equiv S \pmod 9$

$\Rightarrow\ N - S$ is divisible by 9, i.e., $9|N - S$

$\Rightarrow\ 9|N$, iff $9|S$, i.e., $N \equiv 0 \pmod 9$, iff $S \equiv 0 \pmod 9$.

(ii) $10 \equiv -1 \pmod{11}$

$\Rightarrow \quad f(10) \equiv f(-1) \pmod{11}$

$\Rightarrow \quad N \equiv T \pmod{11} \quad [\because\ f(-1) = T]$

$\Rightarrow \quad N - T$ is divisible by 11

$\Rightarrow \quad 11|N - T \Rightarrow 11|N$, iff $11|T$

Theorem (3.10): *Let* $N = a_m 10^m + a_{m-1}10^{m-1} + \cdots + a_8 10^8 + \cdots + a_2 10^2 + a_1 10 + a_0$. *Then* $7, 11, 13$ *will divide* N, *if and only if* $7, 11, 13$ *divides*

$$M = (100a_2 + 10a_1 + a_0) - (100a_5 + 10a_4 + a_3) + (100a_8 + 10a_7 + a_6) - \cdots$$

Proof: $1001 = 7 \cdot 11 \cdot 13 = [7, 11, 13]$.

Now, $a \equiv b \pmod 7$, $a \equiv b \pmod{11}$, $a \equiv b \pmod{13}$

$\Rightarrow \quad a \equiv b \pmod{[7, 11, 13]}$ i.e., $a \equiv b \pmod{1001}$

$\Rightarrow \quad N$ is divisible by $7, 11, 13$.

We have, when n is even, $10^{3n} \equiv 1 \pmod{1001}$

$$10^{3n+1} \equiv 10 \pmod{1001}$$
$$10^{3n+2} \equiv 100 \pmod{1001}$$

and when n is odd, $10^{3n} \equiv -1 \pmod{1001}$

$$10^{3n+1} \equiv -10 \pmod{1001}$$
$$10^{3n+2} \equiv -100 \pmod{1001}.$$

So, it follows the followings.

When, $n = 0$, $1 \equiv 1 \pmod{1001} \Rightarrow 10 \equiv 10 \pmod{1001}$, $100 \equiv 100 \pmod{1001}$

$n = 1$, $10^3 \equiv -1 \pmod{1001}$

$\Rightarrow \quad 10^4 \equiv -10 \pmod{1001}$, $10^5 \equiv -100 \pmod{1001}$

$n = 2$, $10^6 \equiv 1 \pmod{1001} \Rightarrow 10^7 \equiv 10 \pmod{1001}$, $10^8 \equiv 100 \pmod{1001}$

$n = 3$, $10^9 \equiv -1 \pmod{1001}$

$\Rightarrow \quad 10^{10} \equiv -10 \pmod{1001}$, $10^{11} \equiv -100 \pmod{1001}$

Thus $N = (a_2 10^2 + a_1 10 + a_0) + (a_5 10^5 + a_4 10^4 + a_3 10^3) + (a_8 10^8 + a_7 10^7 + a_6 10^6) + \cdots + (a_{m-2}10^{m-2} + a_{m-1}10^{m-1} + a_m 10^m)$.

Using above congruent relations, we get

$$N \equiv (100a_2 + 10a_1 + a_0) - (100a_5 + 10a_4 + a_3) + (100a_8 + a_7 10^7 + a_6) \pmod{1001}.$$

Therefore, $N \equiv M \pmod{1001}$, since $M = (100a_2 + 10a_1 + a_0) - (100a_5 + 10a_4 + a_3) + (100a_8 + 10a_7 + a_6) - \cdots$.

$\Rightarrow \quad N - M$ is divisible by 1001

$\Rightarrow \quad N$ is divisible by 7, 11, 13, if and only if m is divisible by 7, 11, 13.

Example: (1) Let $N = 10\,10\,90\,88\,99$.

Then $N = 1 \cdot 10^9 + 0 \cdot 10^8 + 1 \cdot 10^7 + 0 \cdot 10^6 + 9 \cdot 10^5 + 0 \cdot 10^5 + 0 \cdot 10^4 + 8 \cdot 10^3 + 8 \cdot 10^2 + 9 \cdot 10 + 9.$

Applying the results shown in above theorem, we have

$$N \equiv [-1 + 0\cdot 100 + 1\cdot 10 + 0\cdot 1 + 9(-100) + 0(-10) + 8(-1) + 8\cdot 100 + 9\cdot 10 + 9\cdot 1] (\text{mod } 1001)$$

i.e. $N \equiv [-909 + 909] (\text{mod } 1001)$

i.e. $N \equiv 0 (\text{mod } 1001)$

$\Rightarrow$ N is divisible by 7, 11, 13.

Example: (2) Show that $N = 93000\,9970\,89$ is divisible by 7, 11, 13.

Solution: Here $N = 9\cdot 10^{10} + 3\cdot 10^9 + 9\cdot 10^5 + 9\cdot 10^4 + 7\cdot 10^3 + 8\cdot 10 + 9$.

Then $N \equiv [9(-10) + 3(-1) + 9(-100) + 9(-10) + 7(-1) + 8\cdot 10 + 9$ (mod 1001) [by Th. 3.]

i.e., $N \equiv [-90 - 3 - 900 - 90 - 7 + 80 + 9] (\text{mod } 1001)$

i.e., $N \equiv [-1090 + 89] (\text{mod } 1001)$

i.e., $N \equiv (-1001) (\text{mod } 1001)$

i.e., $N \equiv 0 (\text{mod } 1001)$

$\Rightarrow$ N is divisible by 7, 11 & 13.

Theorem (3.11): *If the integer N is represented by*

$$N = a_m b^m + a_{m-1} b^{m-1} + \cdots + a_2 b^2 + a_1 b + a_0, \quad 0 \le a_n \le b$$

with base b, then show that $(b-1)|N$, if and only if $(b-1)|(a_m + a_{m-1} + \cdots + a_2 + a_1 + a_0)$.

Proof: Let $f(x) = a_m x^m + a_{m-1} x^{m-1} + \cdots + a_2 x^2 + a_1 x + a_0$.

Then $f(b) = a_m b^m + a_{m-1} b^{m-1} + \cdots + a_2 b^2 + a_1 b + a_0 = N$

Now, $b \equiv 1 (\text{mod } (b-1))$

$\Rightarrow$ $f(b) \equiv f(1) (\text{mod } (b-1))$

$\Rightarrow$ $N \equiv (a_m + a_{m-1} + \cdots + a_2 + a_1 + a_0) (\text{mod } (b-1))$

$\Rightarrow$ $(b-1)|N$, if and only if $(b-1)|(a_m + a_{m-1} + \cdots + a_2 + a_1 + a_0)$.

Theorem (3.12): *N is divisible by 5 if the last digit is either 0 or 5.*

Proof: Let $N = a_m 10^m + a_{m-1} 10^{m-1} + \cdots + a_1 10 + a_0$, where the last digit is a_0.

Let $f(x) = a_m x^m + a_{m-1} x^{m-1} + \cdots + a_1 x + a_0$

$f(10) = N$ and $f(0) = a_0$.

We have $10 \equiv 0 (\text{mod } 5) \Rightarrow f(10) \equiv f(0) (\text{mod } 5)$

$\Rightarrow$ $N \equiv a_0 (\text{mod } 5) \Rightarrow 5|N - a_0$.

Now $5|N$. Then $5|N - a_0$ iff $5|a_0$.

Therefore, $f|a_0 \Rightarrow$ either $a_0 = 0$ or, $a_0 = 5$.

Hence, $5|N$ if the last digit is either 0 or 5.

Problem (3.1): Show that the integer 23456785 is divisible by 11.

Solution: Let $N = 23456785$.

Then $N = 23(1000)^2 + 456(1000) + 785$.

Let $f(x) = 23x^2 + 456x + 785$.

Then $f(1000) = N$ and $f(-1) = 23 - 456 + 785 = 352$.

Now $1000 = -1(\text{mod } 11) \Rightarrow f(1000) \equiv f(-1)(\text{mod } 11)$

$\Rightarrow$ $N = 352(\text{mod } 11) \Rightarrow 11|N - 352$.

Now $11|N$, if $11|352$ and this is true.

Therefore, $11|N$.

Problem (3.2): Show that the integer 205769 is not divisible by 3.

Solution: Let $N = 205769$.

Then $N = 20(100)^2 + 57(100) + 69$.

Let $f(x) = 20x^2 + 57x + 69$.

Then $f(100) = N$, $f(1) = 20 + 57 + 69 = 146$.

Now $100 \equiv 1(\text{mod } 3) \Rightarrow f(100) \equiv f(1)(\text{mod } 3)$

$\Rightarrow$ $N \equiv 146(\text{mod } 3) \Rightarrow 3|N - 146$.

Now $3|N$, iff $3|146$, but this is not possible.

Therefore, $3 \nmid N$.

Problem (3.3): Show that an integer N is divisible by 3, if and only if the sum of the digits is divisible by 3.

Solution: Let $N = a_m 10^m + a_{m-1}10^{m-1} + \cdots + a_1 10 + a_0$

Let $f(x) = a_m x^m + a_{m-1}x^m + \cdots + a_1 x + a_0$.

Then $f(10) = N$

and $f(1) = a_m + a_{m-1} + \cdots + a_1 + a_0 =$ sum of the digits of $N = S$.

Now $10 = 1(\text{mod } 3) \Rightarrow f(10) \equiv f(1)(\text{mod } 3)$

$\Rightarrow$ $N \equiv S(\text{mod } 3) \Rightarrow 3|N - S$.

Then $3|N$, if and only if $3|S$.

Residues, least residues, complete system of residues

Definition (3.2): *If $a \equiv b(\text{mod } m)$, then b is called the residue of a modulo m.*

Again, $a \equiv b(\text{mod } m) \Rightarrow a \equiv b + mt(\text{mod } m)$, $t \in Z$.

$\Rightarrow$ $b + mt$ is the residue of a modulo m.

$\Rightarrow$ $\exists$ an infinite no. of residues of a modulo m for different values of $t \in Z$.

If an integer a is divided by m, then by divisional algorithm, $\exists$ two unique integers q and r such that

$$a = mq + r,\ 0 \leq r < m,$$

q is called the quotient and r is called the least non-negative remainder.

Then $a - r = mq \Rightarrow a \equiv r(\text{mod } m)$.

r is called the **least residue** of a modulo m.

It follows that every integer is congruent to modulo m to exactly one of the values $0, 1, 2, \ldots, m - 1$.

$S = \{0, 1, 2, \ldots, m - 1\}$ is called the set of least non-negative residues modulo m.

Example: Let $a = 8$, $m = 3$.

Then $8 = 3 \cdot 2 + 2 \Rightarrow 8 - 2 = 3 \cdot 2$

$\Rightarrow \quad 8 \equiv 2 \pmod 3$.

Then 2 is the least residue of 8 modulo 3. We consider the set $S = \{0, 1, 2\}$ and let $a = 7$ be any integer.

Then 7 is congruent to modulo 3, to exactly one of the member S, and that is 1.

If $a = 32$, then 32 is congruent to only $2 \in S$ modulo 3.

Thus, if a be any integer, then it must be congruent to mod 3, to exactly one of the members of S. $S = \{0, 1, 2\}$ is called the set of least non-negative residues of an integer, modulo 3.

Complete system of residues

Definition (3.3): *A set of integers $a_1, a_2, \ldots, a_n$ is said to form a complete system of residues if no two of them are congruent modulo m, that is, $a_i \not\equiv a_j \pmod m$ $\forall\, i, j$ $(i \neq j)$.*

Examples: (i) Let $m = 11$.

$$S = \{0, 1, 2, \ldots, 10\}$$

$$a_i \not\equiv a_j \pmod m \ \forall\, i, j = 0, \ldots, 10,\ i \neq j.$$

S forms a complete system of residues.

(ii) Let $m = 5$.

Then $S = \{-12, -15, 82, -1, 31\}$ forms a complete system of residues modulo 5.

Theorem (3.13): *If $(a_1, a_2, \ldots, a_n)$ be a complete system of residues modulo m and also $b_1, b_2, \ldots, b_n$ any set of integers such that $b_i \equiv a_i \pmod m$, $i = 1, 2, \ldots, n$, then $(b_1, b_2, \ldots, b_n)$ is also a complete system* (mod m).

Proof: Let r_i be the least residues of a_i, modulo m, for $i = 1, 2, \ldots, n$.

Then $\quad a_i \equiv r_i \pmod m$. (1)

Again $\quad b_i \equiv a_i \pmod m$ [given condition]. (2)

Therefore, r_i are also the same residues of $b_i \pmod m$. That is, $b_i \equiv r_i \pmod m$.

Also from (1) it shows that $r_i \not\equiv r_j \pmod m$,

for if $r_i \equiv r_j \pmod m$,

then $r_i - r_j$ is divisible by m,

and so $r_i - r_j = mk$ for some $k \in Z$.

Now from (1), $a_i - r_i = mt_1$,

and $\quad a_j - r_j = mt_2$, for some $t_1, t_2 \in Z$.

Therefore, $(a_i - a_j) - (r_i - r_j) = m(t_1 - t_2)$

$\Rightarrow \quad (a_i - a_j) = m(t_1 - t_2) + mk = mk'$, $k' = t_1 - t_2 + k \in Z$

$\Rightarrow \quad a_i \equiv a_j \pmod m$, a contradiction.

Hence, $\quad r_i \not\equiv r_j \pmod m \ \forall\, i, j, i \neq j$.

Therefore, from (2) it follows that $b_i \not\equiv b_j \pmod m$, $\forall\, i, j$ $(i \neq j)$ and hence $(b_1, b_2, \ldots, b_n)$ forms a complete residue system (mod m).

Corollary (3.2): If $a_1, a_2, \ldots, a_n$ be a complete residue system modulo m, then

(i) $k + a_1,\ k + a_2, \ldots, k + a_n$

and (ii) $ka_1, ka_2, \ldots, ka_n, \quad (k, m) = 1$

also form complete system of residues.

Theorem (3.14): *A set of m integers which are in A.P. with common difference d, $(d, m) = 1$, forms a complete residue system modulo m.*

Proof: Let us consider the A.P. of m terms with common difference d,

$a, a + d, a + 2d, \ldots, a + (m - 1)d$, such that $(d, m) = 1$.

Now $a + id \equiv a + jd \pmod m$, $i, j = 0, 2, \ldots, m - 1$, $i \neq j$

$\Rightarrow$ $id \equiv jd \pmod m \Rightarrow i \equiv j \pmod m$, $(d, m) = 1$,

which contradicts that $i \neq j \pmod m$.

Therefore, $a + id \not\equiv a + jd \pmod m$.

Hence the theorem.

Example: We consider an A.P. series 4, 9, 14, 19, 24, 29 of 6 terms with common difference 5, where $(5, 6) = 1$.

Then no two of them are congruent mod (6).

Problem (3.4): Find the least (+)ve residues of $2^{44} \pmod{89}$.

Solution: We have $2^6 \equiv -25 \pmod{89}$

That is, $2^6 \equiv -5^2 \pmod{89} \Rightarrow (2^6)^2 \equiv (-5^2)^2 \pmod{89}$

or, $2^{12} \equiv 625 \pmod{89}$ or, $2^{12} \equiv 625 - 7 \times 89 \pmod{89}$

or, $2^{12} \equiv 2 \pmod{89}$ or, $2^{11} \equiv 1 \pmod{89}$ [since (2,89)=1]

or, $(2^{11})^4 \equiv 1 \pmod{89}$ or, $2^{44} \equiv 1 \pmod{89}$.

Least (+)ve residue is 1.

Therefore, $89 | 2^{44} - 1$.

Problem (3.5): What is the remainder when the following sum is divided by 4?

$$1^5 + 2^5 + 3^5 + 4^5 + \cdots + 100^5$$

Solution: We have the following results:

$1 \equiv 1 \pmod 4 \Rightarrow 1^5 \equiv 1 \pmod 4$

$3 \equiv -1 \pmod 4 \Rightarrow 3^5 \equiv -1 \pmod 4$

$5 \equiv 1 \pmod 4 \Rightarrow 5^5 \equiv 1 \pmod 4$

$7 \equiv -1 \pmod 4 \Rightarrow 7^5 \equiv -1 \pmod 4$

....

$99 \equiv -1 \pmod 4 \Rightarrow 99^5 \equiv -1 \pmod 4$.

Adding the congruence relations, we get

$1^5 + 3^5 + 5^5 + 7^5 + \cdots + 99^5 \equiv [1 + -1 + 1 - 1 + \cdots + 1 - 1] \pmod 4$

$= 0 \pmod 4$, since the no. of terms in this sum is even, i.e., 50. (1)

Again, $2 \equiv -2 \pmod 4 \Rightarrow 2^5 \equiv -2^5 \pmod 4$

or, $2^5 \equiv -32 + 8 \cdot 4 \pmod 4$ or, $2^5 \equiv 0 \pmod 4$

$4 \equiv 0 \pmod 4 \Rightarrow 4^5 \equiv 0 \pmod 4$

$6 \equiv -2 \pmod 4 \Rightarrow 6^5 \equiv 0 \pmod 4$

$8 \equiv 0 \pmod 4 \Rightarrow 8^5 \equiv 0 \pmod 4$

....

$100 \equiv 0 \pmod 4 \Rightarrow 100^5 \equiv 0 \pmod 4$. (2)

So from (1) and (2) it follows that

$$1^5 + 2^5 + 3^5 + \cdots + 100^5 \equiv 0 \pmod 4.$$

Hence, the remainder is 0, when the sum is divided by 4.

Problem (3.6): Find the remainder when the sum $1! + 2! + 3! + 4! + \cdots + 79! + 80!$ is divided by 8.

Solution: $4! \equiv 0 \pmod 8$

$5! = 5 \cdot 4! \equiv 0 \pmod 8$

Thus, it follows that $k! \equiv 0 \pmod 8\ \forall\ k \geq 4$ in the given sum.

Hence, $4! + 5! + \cdots 80! \equiv 0 \pmod 8$.

Therefore, $1! + 2! + 3! + 4! + \cdots 80! \equiv (1! + 2! + 3!) \pmod 8 \equiv 9 \pmod 8$

$$\equiv 1 \pmod 8.$$

So the remainder is 1

Problem (3.7): Show that $5n^3 + 7n^5 \equiv 0 \pmod{12}\ \forall\ n \geq 1$.

Solution: Let $f(n) = 5n^3 + 7n^5$.

$$f(1) = 5 + 7 = 12 \equiv 0 \pmod{12}.$$

Result is true for $n = 1$.

Let the result be true for $n = m \in N$.

That is, $f(m) \equiv 0 \pmod{12}$.

Now for $n = m + 1$

$$f(m+1) = 5(m+1)^3 + 7(m+1)^5 = (m+1)^3[5 + 7m^2 + 14m + 7]$$

$$= (m+1)^3[7m(m+2) + 12]$$

$$= 7m(m+1)(m+2)(m+1)^2 + 12(m+1)^3.$$

Now $m(m+1)(m+2)(m+1)^2$ has a factor 12 $\forall\ m \in N$.

Then $7m(m+1)(m+2)(m+1)^2 \equiv 0 \pmod{12}$.

Also $12(m+1)^3 \equiv 0 \pmod{12}$.

Hence, $7m(m+1)(m+2)(m+1)^2 + 12(m+1)^3 \equiv 0 \pmod{12}$.

That is, $f(m+1) \equiv 0 \pmod{12}$

So the result is true for $n = m + 1$.

Therefore, it is true $\forall\ n \in N$.

Hence, $f(n) \equiv 0 \pmod{12}$.

Problem (3.8): Show that $3.5^{2n+1} + 2^{3n+1} \equiv 0 \pmod{17}\ \forall\ n \geq 1$.

Solution: $3 \cdot 5^{2n+1} + 2^{3n+1} = 15 \cdot 5^{2n} + 2 \cdot 2^{3n}$

We have $25 \equiv 8 (\text{mod } 17) \Rightarrow 5^2 \equiv 8 (\text{mod } 17)$

$\Rightarrow \quad (5^2)^n \equiv 8^n (\text{mod } 17)$

$\Rightarrow \quad 15 \cdot 5^{2n} \equiv 8^n \cdot 15 (\text{mod } 17)$. (1)

Also $\quad 8 \equiv 8 (\text{mod } 17) \Rightarrow 2^{3n} \equiv 8^n (\text{mod } 17)$

$\Rightarrow \quad 2 \cdot 3^{2n} \equiv 8^n \cdot 2 (\text{mod } 17)$ (2)

Adding (1) & (2), we get

$$15 \cdot 5^{2n} + 2 \cdot 2^{3n} \equiv 8^n (15 + 2) (\text{mod } 17) \equiv 0 (\text{mod } 17).$$

$$\therefore \quad 3 \cdot 5^{2n+1} + 2^{3n+1} \equiv 0 (\text{mod } 17).$$

Linear Congruence

The general form of a linear congruence is $ax \equiv b(\text{mod } m)$, where x is a variable of power 1.

In finding the solution we observe that the set of non-negative integers $S = \{0, 1, 2, \ldots, m - 1\}$ forms a complete residue system modulo m. There exists at least one member of S which satisfies the linear congruence $ax \equiv b(\text{mod } m)$.

In other words, the system has either a single solution or more than one solution which are incongruent to each other with mod (m).

We first go through the following examples.

Example: (1): $4x \equiv 3(\text{mod } 5)$ (1)

$$S = \{0, 1, 2, 3, 4\}$$

Only $2 \in S$ satisfies (1).

So 2 is the only solution of (1).

Here we also observe that $(a, m) = (4, 5) = 1$.

Example (2): $6x \equiv 3(\text{mod } 9)$ (2)

$$S = \{0, 1, 2, \ldots, 8\}$$

$2, 5, 8 \in S$ satisfies (2).

Therefore, the system (2) has more than one solution.

Also $\quad 2 \not\equiv 5(\text{mod } 9),\ 5 \not\equiv 8(\text{mod } 9),\ 2 \not\equiv 8(\text{mod } 9)$.

Hence the solutions are incongruent to each other.

Here we observe that $(a, m) = (6, 9) \neq 1$,

that is, not prime to each other.

We write the solution of the system as $x \equiv 2, 5, 8(\text{mod } 9)$.

Theorem (3.15): *The congruence* $ax \equiv b(\text{mod } m)$ *where* $(a, m) = 1$ *has a unique solution.*

Proof: We consider the set of integers $0 \cdot a,\ 1 \cdot a,\ 2 \cdot a,\ \ldots,\ (m - 1) \cdot a$ which are in A.P. with common difference a. If they are divided by m, the remainders are 0, 1, 2, $\ldots$, $m - 1$, taken in certain order, since $(a, m) = 1$. Then $S = \{0, 1, 2, \ldots, m - 1\}$ forms a complete residue system modulo m.

Hence, one of the products $0 \cdot a,\ 1 \cdot a,\ \ldots,\ (m - 1)a$ is congruent to b modulo m. Thus, there exists only one member of S which satisfies congruence $ax \equiv b(\text{mod } m)$, where $0 \leq x < m$.

In finding the solution when $(a, m) = 1$, let $\frac{a}{m}$ be expressed as a simple continued fraction with an even number of quotients and $\frac{y_0}{x_0}$ be the last convergent but one. Then $ax_0 - my_0 = 1$ so that $ax_0 \equiv 1 (\text{mod } m)$.

$\Rightarrow \quad abx_0 \equiv b(\text{mod } m)$

$\Rightarrow$ bx_0 is the required solution of $ax \equiv b(\text{mod } m)$.

That is, $x \equiv bx_0(\text{mod } m)$.

Problem (3.9): Solve $47x \equiv 11(\text{mod } 249)$.

Solution: Here $(a, m) = (47, 249) = 1$.

We express it as a simple continued fraction with even number of quotients.

0	47	249	5
		235	
3	42	14	2
1	5	10	
	4	4	4
	1	4	
		0	

$$\frac{47}{249} = 0 + \frac{1}{5+}\,\frac{1}{3+}\,\frac{1}{2+}\,\frac{1}{1+}\,\frac{1}{4}$$

Last but one convergent is $\frac{p_5}{q_5} = \frac{10}{53} \left(= \frac{y_0}{x_0} \right)$.

So $\quad x \equiv 11 \times 53(\text{mod } 249)$

or, $\quad x \equiv 583 - 2 \times 249(\text{mod } 249)$

or, $\quad x \equiv 85(\text{mod } 249)$.

Problem (3.10): Solve $37x = 7(\text{mod } 127)$.

Solution: Here $(37, 127) = 1$.

We express $\frac{37}{127}$ as a simple continued fraction with even number of quotients.

0	37	127	3
2	32	111	
5	5	16	3
	5	15	
	0	1	

$$\frac{37}{127} = 0 + \frac{1}{3+}\,\frac{1}{2+}\,\frac{1}{3+}\,\frac{1}{5}$$

$$= 0 + \frac{1}{3+}\,\frac{1}{2+}\,\frac{1}{3+}\,\frac{1}{4+}\,\frac{1}{1}$$

Last but one convergent is $\frac{p_5}{q_5} = \frac{30}{103} \left(= \frac{y_0}{x_0} \right)$.

Now $\quad x \equiv bx_0(\text{mod } m)$

Here $\quad x \equiv 7 \times 103(\text{mod } 127)$ or, $x \equiv 721 - 5 \times 127(\text{mod } 127)$

or, $\quad x \equiv 721 - 635(\text{mod } 127)$ or, $x \equiv 86(\text{mod } 127)$.

Note: If $ax \equiv b(\text{mod } m)$, where $(a, m) = 1$, then the linear congruence $ax \equiv 1(\text{mod } m)$ has a unique solution called the reciprocal of $a(\text{mod } m)$. If a' be the reciprocal of $a(\text{mod } m)$, then ba' is a solution of $ax \equiv b(\text{mod } m)$.

Therefore, $\quad a \cdot ba' \equiv b(\text{mod } m)$.

Theorem (3.16): *If $(a, m) = d(\neq 1)$, then there are d incongruent solutions of $ax \equiv b(\text{mod } m)$ provided $d|b$. If $d \nmid b$, there does not exist any solution.*

Proof: $(a, m) = d$

$\Rightarrow \quad d|a$ and $d|m$

$\Rightarrow \quad a = da'$ and $m = dm'$ for some integers a' and m' and $(a', m') = 1$.

Then from $ax \equiv b(\text{mod } m)$, (1)

we have $da'x \equiv b(\text{mod } dm')$.

We require those values of x for which $da'x - b$ is divisible by dm'. No such value of x is obtained unless $d|b$.

Thus if $dm'|da'x - b$ and $d|b$,

that is, if $b = db'$ for some integer b',

then $dm'|da'x - db'$.

i.e., $m'|a'x - b'$

i.e., $a'x - b'$ is divisible by m'

i.e., $a'x \equiv b'(\text{mod } m')$ where $(a', m') = 1$, (2)

which is the equivalent form of (1).

Now by Theorem (3.12), the congruence (2) has one solution, say $\alpha < m'$, and then d distinct solutions are

$$\alpha;\ \alpha + m', \alpha + 2m', \ldots, \alpha + (d-1)m'. \quad (3)$$

Therefore, $\alpha, \alpha + \dfrac{m}{d}, \alpha + \dfrac{2m}{d}, \ldots, \alpha + (d-1)\dfrac{m}{d}$

$\left(\text{putting } m' = \dfrac{m}{d}\right)$ are d distinct incongruence solution of (1), (4)

for if $\alpha + r\dfrac{m}{d} \equiv \alpha + s\dfrac{m}{d}(\text{mod } m)$, where $0 \le s < r \le d-1$,

then $r\dfrac{m}{d} \equiv s\dfrac{m}{d}(\text{mod } m)$.

Since $\left(\dfrac{m}{d}, m\right) = \dfrac{m}{d}$, we have by Th. [3.5],

$r \equiv s(\text{mod } d)$, $0 \le r - s < d$ and therefore, $d|r - s$.

But this is impossible when $0 < r - s < d$.

Hence, $r - s = 0$.

That is, $r = s$, a contradiction.

So the d distinct solutions are incongruent to each other modulo (m).

Again, if y is a solution of (1),

then $ay \equiv b(\text{mod } m)$

and hence, $ay \equiv a\alpha(\text{mod } m)$,

since $ay - b = km$, (i)

for some $k \in Z$, and α being a solution of (1),

$a\alpha \equiv b(\text{mod } m) \Rightarrow a\alpha - b = mk'$, for some $k' \in Z$. (ii)

Therefore, (i) – (ii) gives $ay - a\alpha = (k - k')m$,

which implies that $ay \equiv a\alpha(\text{mod } m)$,

Therefore, $ay \equiv a\alpha(\text{mod } m) \Rightarrow da'y \equiv da'\alpha(\text{mod } dm')$

$\Rightarrow a'y \equiv a'\alpha(\text{mod } m') \Rightarrow y \equiv \alpha(\text{mod } m')$.

That is, $y \equiv \alpha\left(\text{mod } \dfrac{m}{d}\right)$

or, $y = \alpha + \dfrac{m}{d}t$ for some $t \in Z$. (5)

But $t \equiv r (\text{mod } d)$ for some integer r satisfying $0 \le r < d$.

$\Rightarrow$ $t - r = ds$ for some $s \in Z$

$\Rightarrow$ $t\dfrac{m}{d} - r\dfrac{m}{d} = s \cdot m$

$\Rightarrow$ $t\dfrac{m}{d} \equiv r\dfrac{m}{d} (\text{mod } m)$.

Hence, $y \equiv \alpha + r\dfrac{m}{d} (\text{mod } m)$ [from (5)]

Thus, any solution of (1) is of the form $\alpha + r\dfrac{m}{d} (\text{mod } m)$ for $0 \le r < d$, and this is same as in (4).

This completes the proof.

Problem (3.11): Solve $25x \equiv 15(\text{mod } 120)$.

Solution: Here $(a, m) = (25, 120) = 5(= d)$

Also $d|b$, i.e., $5|15$.

So the solution exists.

Again, $25x \equiv 15(\text{mod } 120)$ can be reduced to the form $5x \equiv 3(\text{mod } 24)$ (1)

or, $5x \equiv -45(\text{mod } 24)$

or, $x \equiv -9(\text{mod } 24)$, as $(5, 24) = 1$

or, $x \equiv 15(\text{mod } 24)$

or, $x \equiv 15 + 24t,\ t = 0, 1, 2, 3, 4$ [having $d = 5$ solutions].

Therefore, $x = 15, 39, 63, 87, 111$ are 5 incongruent solutions of $25x \equiv 15(\text{mod } 120)$.

Alternative method: With the direct application of the Theorem (3.16), solutions of the given congruence are $\alpha + r\frac{m}{d}$, $r = 0, 1, 2, 3, 4$ and $\alpha = 15(< 24)$ is the only solution of $5x \equiv 3(\text{mod } 24)$.

Therefore, the solutions are

$$15, 15 + 1 \cdot \frac{120}{5},\ 15 + 2 \cdot \frac{120}{5},\ 15 + 3 \cdot \frac{120}{5} \text{ and } 15 + 4 \cdot \frac{120}{5}.$$

That is, 15, 39, 63, 87, 111.

Simultaneous linear congruences

We first go through the following examples :

Example: (i) Let us consider two simultaneous linear congruences:

$x \equiv 1(\text{mod } 3)$ (i)

$x \equiv 0(\text{mod } 6)$. (ii)

They can be solved separately [by trial method]

4, 7, 10, 13,... are the solutions of (i)

and 6, 12, 18, 24,... are the solutions of (ii).

There is no value of x which satisfies both (i) and (ii), and $(3, 6) = 3$ does not divide $1 - 0 = 1$.

(ii) Let us consider the two simultaneous linear congruences:

$x \equiv 1(\text{mod } 3)$ (1)

$x \equiv 0(\text{mod } 5).$ (2)

Solutions of (1) and (2) are respectively

7, 10, 13, 16, 19, 22, 25, 28, 31, 34,. . . ,55,. . .

and 5, 10, 15, 20, 25, 30, 35, 40, 45, 50, 55,. . .

It has been observed that (1) and (2) have common solutions 10, 25, 55,. . . and $(3, 5) = 1$.

Thus, the system of simultaneous linear congruences may or may not have common solutions according as the respective modulus m_1, m_2 (say) are prime to each other or not prime to each other.

Theorem (3.17): *If $x \equiv a(\text{mod } p)$ and $x \equiv b(\text{mod } q)$ be two simultaneous linear congruences, then to determine the values of x which satisfy both the congruences.*

Proof: Let $(p, q) = d$.

Now from $x \equiv a(\text{mod } p)$,

$x = a + py$, where y is given by $a + py \equiv b(\text{mod } q)$.

or, $py \equiv b - a(\text{mod } q)$.

If $b - a$ is not divisible by d, then the solution does not exist.

But if $d|b - a$, there is only one solution y_1 of $y < \frac{q}{d}$ which satisfies the last congruence and the general value of y is given by $y = y_1 + \frac{q}{d}t$, $t \in Z$ so that

$$x = x_1 + \frac{pq}{d}t, \quad \text{where} \quad x_1 = a + py_1.$$

Hence, $x = a + py_1 + \dfrac{pq}{d}t,\ t \in Z.$ (1)

Thus, a solution x_1 of the given congruences exists iff $b-a$ is divisible by $d = (p, q)$ and the congruences are equivalent to a single congruence $x \equiv x_1(\text{mod } l)$ where $l = [p, q]$.

Note: It should be noted that the solution always exists when $(p, q) = 1$.

Problem (3.12): Find the general values of x for $x \equiv 1(\text{mod } 6)$ and $x \equiv 4(\text{mod } 9)$.

Solution: Here $(9, 6) = 3\ (= d)$

$a = 4,\ b = 1,\ a - b = 3.$

Thus $d|a - b$. So the solution exists.

Now $x \equiv 1(\text{mod } 6)$

$\Rightarrow$ $x = 1 + 6y,$ (1)

where y is given by $1 + 6y \equiv 4(\text{mod } 9)$

or, $6y \equiv 3(\text{mod } 9),$

which has a solution $y = 2(= y_1) < \dfrac{q}{d} = \dfrac{9}{3} = 3,$

and the general values of y are given by

$$y = y_1 + \frac{q}{d}t = 2 + 3t,\ t \in Z.$$

Then $x = 1 + 6(2 + 3t)$ form (1)

or, $x = 13 + 18t,$

which gives the general values of the given congruences equivalent to a single congruence $x \equiv 13 \pmod{18}$,

where $[p, q] = [9, 6] = 18$.

Problem (3.13): Find the general solution of $19x \equiv 1 \pmod{140}$.

Solution: $19x \equiv 1 \pmod{140}$

$\Rightarrow \quad 19x \equiv 1(\text{mod } [4, 5, 7])$ is equivalent to the simultaneous equations

$\Rightarrow \quad 19x \equiv 1 \pmod 4$ (1)

$19x \equiv 1 \pmod 5$ (2)

and $\quad 19x \equiv 1 \pmod 7$. (3)

Now $\quad 19x \equiv 1 \pmod 4 \Rightarrow 19x \equiv -19 \pmod 4$

$\Rightarrow \quad x \equiv -1 \pmod 4$.

(2) is equivalent to $x \equiv -1 \pmod 5$

(3) is equivalent to $x \equiv -4 \pmod 7$

as $(19, 4) = (19, 5) = (19, 7) = 1$.

Again $x \equiv -1 \pmod 4 \Rightarrow x = -1 + 4y$, (4)

where y is given by $-1 + 4y \equiv -1 \pmod 5$

or, $\quad -1 + 4y = -1 + 5z$ or, $4y = 5z$

or, $\quad 4y \equiv 0 \pmod 5 \Rightarrow y \equiv 0 \pmod 5$

or, $\quad y = 5z$. (5)

From (4) and (5), $x = -1 + 20z$

Then $\quad -1 + 20z \equiv -4 \pmod 7 \Rightarrow 20z \equiv= -3 \pmod 7$ [from equivalent form of (3)]

$\Rightarrow \quad z \equiv 3 \pmod 7 \Rightarrow z = 3 + 7t,\ t \in Z$.

From (5), $y = 15 + 35t$

and from (4), $x = -1 + 60 + 140t = 59 + 140t$,

giving the general value of x of the given congruence.

Chinese remainder theorem

Theorem (3.15): *If* $m_1, m_2, \ldots, m_k$ *be (+)ve integers such that* $(m_i, m_j) = 1,\ i \neq j$ *and* $M = m_1 m_2 \cdots m_k$ *and* $b_i \in Z$, *then the system of congruences*

$$\left.\begin{array}{l} x_1 \equiv b_1 \pmod{m_1} \\ x_2 \equiv b_2 \pmod{m_2} \\ \ldots\ldots\ldots\ldots\ldots\ldots \\ x_k \equiv b_k \pmod{m_k} \end{array}\right\} \quad (A)$$

has exactly one solution $x_0 \pmod M$.

Proof: Let us consider the integer

$$x_0 = \frac{M}{m_1} y_1 b_1 + \frac{M}{m_2} y_2 b_2 + \cdots + \frac{M}{m_k} y_k b_k, \quad (1)$$

where the integers y_i are defined by

$$\frac{M}{m_i} y_i \equiv 1 \pmod{m_i}\ \forall\ i = 1, 2, \ldots, k.$$

We shall show that the integer x_0 is the only solution of the system.

It follows from $M = m_1 m_2 \ldots m_k$ that each of $\frac{M}{m_2}, \frac{M}{m_3}, \ldots, \frac{M}{m_k}$ is divisible by m_1.

Therefore, from (1) we have

$$x_0 = \frac{M}{m_1} y_1 b_1 + p_1 m_1 y_2 b_2 + p_2 m_1 y_3 b_3 + \cdots + p_{k-1} m_1 y_k b_k$$

for some $p_1, p_2, \ldots, p_{k-1} \in Z$

or, $\quad x_0 = \frac{M}{m_1} y_1 b_1 + m_1 t, \quad \text{where } t = p_1 y_2 b_2 + p_2 y_3 b_3 + \cdots + p_{k-1} y_k b_k \in Z$

$$\Rightarrow \quad x_0 \equiv \frac{M}{m_1} y_1 b_1 (\text{mod } m_1).$$

Again, by hypothesis $\frac{M}{m_1} y_1 \equiv 1 (\text{mod } m_1)$

and so $\quad \frac{M}{m_1} y_1 b_1 \equiv b_1 (\text{mod } m_1)$.

Therefore, $x_0 \equiv b_1 (\text{mod } m_1)$.

Similarly, we have $x_0 \equiv b_2 (\text{mod } m_2), \ldots x_0 \equiv b_k (\text{mod } m_k)$.

It follows that x_0 is a solution of the system (A) and this shows that solution of the system (A) exists.

For uniqueness of the solution of the system (A), let there be another integer r which satisfies the given system.

That is, $\quad r \equiv b_i (\text{mod } m_i) \ \forall \ i = 1, 2, \ldots, k.$

Also $\quad x_0 \equiv b_i (\text{mod } m_i), \ \forall \ i = 1, 2, \ldots, k.$

Therefore, these together imply that $m_i | r - x_0, \forall \ i = 1, 2, \ldots, k$.

Since $(m_i, m_j) = 1, \forall \ i, j, i \neq j$ and $M = m_1 m_2, \ldots, m_k$, it follows that $M | r - x_0$ and hence $r \equiv x_0 (\text{mod } M)$.

This shows that every solution of the system (A) other than x_0 is congruent to $x_0 (\text{mod } M)$.

This completes the proof of the theorem.

Problem (3.14): Find the solution of the system with the help of Chinese Method.

$$\left.\begin{aligned} x &\equiv 5(\text{mod } 4) \\ x &\equiv 3(\text{mod } 7) \\ x &\equiv 2(\text{mod } 9) \end{aligned}\right\} \quad \text{(A)}$$

Solution: Here $m_1 = 4$, $m_2 = 7$ and $m_3 = 9$. They are relatively prime to each other. So Chinese Method is applicable.

$$M = m_1 m_2 m_3 = 4 \cdot 7 \cdot 9 = 252.$$

Let the integer

$$x_0 \equiv \frac{M}{m_1} y_1 b_1 + \frac{M}{m_2} y_2 b_2 + \frac{M}{m_3} y_3 b_3 (\text{mod } M) \quad (1)$$

be the solution of the system A.

Now $\quad \frac{M}{m_1}y_1 \equiv 1 \pmod 4$

$\Rightarrow \quad 63y_1 \equiv 1 \pmod 4$

$\Rightarrow \quad y_1 \equiv -1 \pmod 4$ (i)

$\frac{M}{m_2}y_2 \equiv 1 \pmod 7 \Rightarrow 36y_2 \equiv 1 \pmod 7$

$\Rightarrow \quad y_2 \equiv 1 \pmod 7$ (ii)

$\frac{M}{m_3}y_3 \equiv 1 \pmod 9 \Rightarrow 28y_3 \equiv 1 \pmod 9$

$\Rightarrow \quad y_3 \equiv 1 \pmod 9.$ (iii)

Therefore, $x_0 \equiv 63(-1)\cdot 5 + 36(1)\cdot 3 + 28(1)(2) \pmod{252}$ [by (i), (ii) & (iii)]

or, $\quad x_0 \equiv -315 + 108 + 56 \pmod{252}$ or, $x_0 \equiv -151 \pmod{252}$

or, $\quad x_0 \equiv 101 \pmod{252}$,

which is the solution of the system (A).

Problem (3.15): Solve : $x \equiv 3 \pmod 6$, $x \equiv 5 \pmod 8$ and $x \equiv 2 \pmod{11}$.

Solution: Chinese Remainder Theorem is not applicable to the given system as $(6, 8) \neq 1$.

Now the 1st two congruences are solvable as $b - a = 5 - 3 = 2$ is divisible by $(6, 8) = 2$.

From $\quad x \equiv 3 \pmod 6$

$\Rightarrow x = 3 + 6y$, where y is given by $3 + 6y \equiv 5 \pmod 8$

or, $\quad 6y \equiv 2 \pmod 8$ or, $3y \equiv 1 \pmod 4$

or, $\quad y \equiv -1 \pmod 4$, $(3, 4) = 1$ or, $y = -1 + 4t$, $t \in Z$.

Therefore, $x = 3 + 6(-1 + 4t) = -3 + 24t$, $t \in Z$ (1)

or, $\quad x \equiv -3 \pmod{24}$

Now $\quad x \equiv -3 \pmod{24}$ and $x \equiv 2 \pmod{11}$

has a unique solution as $(24, 11) = 1$.

$x = -3 + 24t$, where t is given by $-3 + 24t \equiv 2 \pmod{11}$

or, $\quad 24t \equiv 5 \pmod{11}$ or, $24t \equiv 16 \pmod{11}$

or, $\quad 3t \equiv 2 \pmod{11}$ or, $t \equiv -3 \pmod{11}$ or, $t = -3 + 11p$, $p \in Z$. (2)

Therefore, $x = -3 + 24(-3 + 11p)$ [form (1) and (2)]

$\quad = -75 + 264p$

or, $\quad x \equiv -75 \pmod{264}$ or, $x \equiv 189 \pmod{264}$,

which is the solution of the given system.

Problem (3.16): If $x \equiv a \pmod{16}$, $x \equiv b \pmod 5$ and $x \equiv c \pmod{11}$, then show that

$$x \equiv 385a + 176b - 560c \pmod{880}.$$

Solution: $m_1 = 16$, $m_2 = 5$, $m_3 = 11$

They are relatively prime. $M = m_1m_2m_3 = 880$.

Chinese Remainder Theorem is applicable.

The integer $x_0 \equiv \frac{M}{m_1}y_1a + \frac{M}{m_2}y_2b + \frac{M}{m_3}y_3c \pmod{880}$

is the solution of the system.

Now $\quad \frac{M}{m_1}y_1 \equiv 1 \pmod{16} \Rightarrow 55y_1 \equiv 33 \pmod{16}$

$\Rightarrow \quad 5y_1 \equiv 3 \pmod{16}$

$\Rightarrow \quad y_1 \equiv 7 \pmod{16}$. (i)

$\frac{M}{m_2}y_2 \equiv 1 \pmod 5 \Rightarrow 176y_2 \equiv 1 \pmod 5$

$\Rightarrow \quad y_2 \equiv 1 \pmod 5$. (ii)

$\frac{M}{m_3}y_3 \equiv 1 \pmod{11} \Rightarrow 80y_3 \equiv 1 \pmod{11}$

$\Rightarrow \quad y_3 \equiv 4 \pmod{11}$

$\Rightarrow \quad y_3 \equiv -7 \pmod{11}$. (iii)

Therefore, $x_0 \equiv 55a \times 7 + 176b \times 1 + 80c \times (-7) \pmod{880}$ by (i), (ii) and (iii)

or, $\quad x_0 \equiv 385a + 176b - 560c \pmod{880}$.

Symbolic fraction method

If the congruence be $ax \equiv b \pmod m$, where $(a, m) = 1$, then we have $ax \equiv b + mh \pmod m$, where h is an arbitrary integer.

Now if h is so chosen that $a|(b + mh)$, then given congruence $ax \equiv b \pmod m$ is equivalent to $x \equiv \frac{b}{a} \pmod m \equiv \frac{b+mb}{a} \pmod m$, where $\frac{b+mh}{a}$ is an integer, a solution of the given congruence.

So $\frac{b}{a} \pmod m$ is a symbolic fraction representing an integer.

Example: $5x \equiv 2 \pmod{16}$

$\Rightarrow x \equiv \frac{2}{5} \pmod{16}$ or, $x \equiv \frac{2 + 3 \cdot 16}{5} \pmod{16}$

or, $x \equiv 10 \pmod{16}$.

Given congruence has a solution $10 \pmod{16}$.

Theorem (3.18): *If $(m_i, m_j) = 1 \ \forall\ i, j, i \neq j$, $b_i \in Z$ and $(a_i, m_i) = 1$ $\forall\ i = 1, 2, \ldots, r$, then the system of congruences*

$$\left.\begin{aligned} a_1x &\equiv b_1 \pmod{m_1} \\ a_2x &\equiv b_2 \pmod{m_2} \\ &\cdots\cdots\cdots\cdots \\ a_rx &\equiv b_r \pmod{m_r} \end{aligned}\right\} \quad (A)$$

has exactly one solution $x_0 \pmod M$ where $M = m_1 \cdot m_2, \ldots, m_r$.

Proof: $a_kx \equiv b_k \pmod{m_k}$

$$\Rightarrow \quad x \equiv \frac{b_k}{a_k} \pmod{m_k}$$

$$\equiv \frac{b_k + hm_k}{a_k} \pmod{m_k},$$

where h is so chosen that $a_k | b_k + hm_k$).

Then $\dfrac{b_k + hm_k}{a_k}$ = integer = c_k (say).

Therefore, $x \equiv c_k(\text{mod } m_k)$.

Putting $k = 1, 2, \ldots, r$ the given system (A) is equivalent to

$$\left.\begin{array}{l} x \equiv c_1(\text{mod } m_1) \\ x \equiv c_2(\text{mod } m_2) \\ \cdots \quad \cdots \quad \cdots \\ x \equiv c_r(\text{mod } m_r) \end{array}\right\} \qquad \text{(B)}$$

Now the system (B) can be solved by Chinese Remainder Theorem since $(m_i, m_j) = 1$, $\forall\, i, j,\ i \neq j$, and the system (B) has exactly one solution $x_0(\text{mod } M)$, where $M = m_1 m_2 \cdots m_r$.

Corollary (3.3): System of the following congruences

$$a_1 x \equiv b_1(\text{mod } m_1)$$
$$a_2 x \equiv b_2(\text{mod } m_2)$$
$$a_r x \equiv b_r(\text{mod } m_r),$$

where $(a_i, m_i) = d_i$ and $d_i | b_i\ \forall\, i = 1, 2, \ldots, r$.

Then the system has a solution $x_0(\text{mod } M)$, where $M = \dfrac{m_1}{d_1}\dfrac{m_2}{d_2}\cdots\dfrac{m_r}{d_r}$.

[Applying Theorems 3.15 and 3.16, result can be obtained as $\frac{m_k}{d_k}$ are integers $\forall\, k = 1, 2, \ldots, r$ with $\left(\frac{m_i}{d_i}, \frac{m_j}{d_j}\right) = 1$]

Fermat's Theorem (3.19): *Let a be any integer, p be prime and let p do not divide a. Then $a^{p-1} \equiv 1(\text{mod } p)$.*

Proof: We consider the first $(p - 1)$ multiples of a. That is, we consider the set of integers $a, 2a, 3a, \ldots, (p - 1)a$.

If they are divided by p, then the remainders are $1, 2, 3, 4, \ldots, (p - 1)$, taken in some order.

Then $1 \cdot a \equiv r_1(\text{mod } p)$, where r_1 is one of $1, 2, 3, \ldots, p - 1$

$2 \cdot a \equiv r_2(\text{mod } p)$, where r_2 is one of $1, 2, 3, \ldots, p - 1$ but not r_1

$\vdots$

$(p - 1)a \equiv r_{p-1}(\text{mod } p)$, where r_{p-1} is one of $1, 2, 3, \ldots, p - 1$ but $\neq r_1, r_2, \ldots, r_{p-2}$.

Then by congruence rule, $1 \cdot a \cdot 2 \cdot a \cdot 3 \cdot a \cdots (p - 1)a \equiv r_1, r_2, \ldots, r_{p-1}(\text{mod } p)$

Since $r_1, r_2, \ldots, r_{p-1}$ is a permutation of $1, 2, 3, \ldots, p - 1$ it follows that

$$a^{p-1} \cdot 1, 2, 3, \ldots, p - 1 \equiv 1 \cdot 2 \cdot 3, \ldots, p - 1(\text{mod } p)$$

$$\Rightarrow \quad a^{p-1} \cdot \lfloor p - 1 \equiv \lfloor p - 1(\text{mod } p)$$

$$\Rightarrow \quad a^{p-1} \equiv 1(\text{mod } p), \text{ since } (\lfloor p - 1, p) = 1.$$

Example: Let $a = 7,\ p = 5,\ p \nmid a$.

We consider the set of integers

$1 \cdot 7, 2 \cdot 7, 3 \cdot 7, 4 \cdot 7$ [That is, the first $(5 - 1) = 4$ multiplies of 7].

If they are divided by 5, then the remainders are

2, 4, 1, 3, i.e., 1, 2, 3, 4 in some order.

Now $1 \cdot 7 \equiv 2 (\text{mod } 5)$ [2 is one of 1, 2, 3, 4]

$2 \cdot 7 \equiv 4 (\text{mod } 5)$ [4 is one of 1, 2, 3, 4, $4 \neq 2$]

$3 \cdot 7 \equiv 1 (\text{mod } 5)$ [1 is one of 1, 2, 3, 4, $1 \neq 2, 4$]

$4 \cdot 7 \equiv 3 (\text{mod } 5)$ [3 is one of 1, 2, 3, 4, and $3 \neq 2, 4, 1$]

By congruence rule, $(1 \cdot 2 \cdot 3 \cdot 4) 7^4 \equiv 2 \cdot 4 \cdot 1 \cdot 3 (\text{mod } 5)$

or $\quad \underline{|4}\ 7^4 \equiv 1 \cdot 2 \cdot 3 \cdot 4 (\text{mod } 5) = \underline{|4}\ (\text{mod } 5)$

or $\quad 7^4 \equiv 1 (\text{mod } 5)$ $[(\underline{|4}, 5) = 1]$

or $\quad 7^{5-1} \equiv 1 (\text{mod } 5)$

Corollary (3.4): If p is prime and a is any integer, then $a^p \equiv a (\text{mod } p)$.

Proof: If p is prime and $p \nmid a$, then by Th. (3.17),

$$a^{p-1} - 1 \equiv 0 (\text{mod } p) \tag{1}$$

Now $\quad a^p - a = a[a^{p-1} - 1]$ is divisible by p,

since $a^{p-1} - 1$ is divisible by p by (1) and $a \in Z$.

Hence, $\quad a^p - a \equiv 0 (\text{mod } p)$.

That is, $\quad a^p \equiv a (\text{mod } p)$.

Now let $p | a$. Then $a = pk$ for some $k \in Z$.

Therefore, $a^p - a = a(a^{p-1} - 1) = pk(a^{p-1} - 1) = pt$, where $t = k(a^{p-1} - 1) \in Z$.

$\Rightarrow \quad a^p - a$ is divisible by p

$\Rightarrow \quad a^p - a \equiv 0 (\text{mod } p) \Rightarrow a^p \equiv a (\text{mod } p)$.

Corollary (3.5): If p is prime and $p \neq 2$, then $a^{\frac{p-1}{2}} \equiv \pm 1 (\text{mod } p)$, when $p \nmid a$.

Proof: By Fermat's Theorem (3.17), we have $a^{p-1} - 1 \equiv 0 (\text{mod } p)$

or, $\quad (a^{\frac{p-1}{2}} - 1)(a^{\frac{p-1}{2}} + 1) \equiv 0 (\text{mod } p)$ $[p \neq 2]$.

Then $\quad p | a^{\frac{p-1}{2}} - 1$ or $p | a^{\frac{p-1}{2}} + 1$, since p is prime.

Now $\quad p | a^{\frac{p-1}{2}} - 1 \Rightarrow a^{\frac{p-1}{2}} - 1 \equiv 0 (\text{mod } p)$

$$\Rightarrow \quad a^{\frac{p-1}{2}} \equiv 1 (\text{mod } p). \tag{i}$$

Again, $\quad p | a^{\frac{p-1}{2}} + \Rightarrow a^{\frac{p-1}{2}} + 1 \equiv 0 (\text{mod } p)$

$$\Rightarrow \quad a^{\frac{p-1}{2}} \equiv -1 (\text{mod } p). \tag{ii}$$

Therefore, from (i) and (ii), $a^{\frac{p-1}{2}} \equiv \pm 1 (\text{mod } p)$.

Corollary (3.6): Square of any integer is of the form $5k \pm 1$.

We have $a^{\frac{p-1}{2}} \equiv \pm 1 (\text{mod } p)$ [by Cor. [3.5]]

Then, $a^{\frac{p-1}{2}} \equiv \pm 1 + pk,\ k \in Z.$

That is, $a^{\frac{p-1}{2}} = pk \pm 1.$

When $p = 5,\ a^2 \equiv 5k \pm 1.$

Thus square of any integer is of the form: $5k \pm 1.$

Problem (3.17): $N^5 - N$ is divisible by 5 or 2 for any integer N.

Solution: Since N is any integer, $p = 5$ is prime, by Cor. (3.4) $N^5 - N$ is divisible by 5.

Now if N be an even integer, say $N = 2m,\ m \in Z,$

then $N^5 - N = (2m)^5 - 2m = 2m[(2m)^4 - 1]$ which is divisible by 2 $\forall\ m \in Z$.

Again, if N be an odd integer, say $N = 2m + 1,\ m \in Z,$

then
$$\begin{aligned} N^5 - N &= (2m+1)[(2m+1)^4 - 1] \\ &= (2m+1)[(2m)^4 + 4(2m)^3 + 6(2m)^2 + 4 \cdot 2m] \\ &= 2m(2m+1)[(2m)^3 + 4(2m)^2 + 6 \cdot 2m + 4], \end{aligned}$$

which is divisible by 2 $\forall\ m \in Z$.

$\therefore$ $N^5 - N$ is divisible by 5 or 2.

Theorem (3.20): *If p be prime, then* p_{c_r}, $0 < r < p$, *is divisible by p.*

Proof: $p(p-1)\cdots(p-r+1)$ being the product of r consecutive integers is divisible by $r!$.

That is, $r!|p(p-1)(p-2)\cdots(p-r+1)$

$\Rightarrow$ $r!|(p-1)(p-2)\cdots(p-r+1)$, since $(r!, p) = 1$

$\Rightarrow$ $(p-1(p-2)\cdots(p-r+1) = r!k$, for some $k \in Z$

$\Rightarrow$ $p(p-1)\cdots(p-r+1) = pr!k$

$\Rightarrow$ $\dfrac{p(p-1)\cdots(p-r+1)}{r!} = pk$

$\Rightarrow$ $p_{c_r} = pk, k \in Z \Rightarrow p|p_{c_r}\ \forall\ r, 0 < r < p.$

Thus p_{c_r} is divisible by p.

Theorem (3.21): *If p be prime, then* $(a+b)^p \equiv a^p + b^p \pmod p\ \forall\ a, b \in Z.$

Proof:
$$\begin{aligned} (a+b)^p &= a^p + p_{c_1}a^{p-1}b + p_{c_2}a^{p-2}b^2 + \cdots + p_{c_r}a^{p-r}b^r + \cdots + b^p \\ &= a^p + b^p + pK \text{ for some } k \in Z, \end{aligned}$$

since each p_{c_r} is divisible by $p, 1 \le r \le p-1$, by Theorem (3.18).

Therefore, $(a+b)^p \equiv a^p + b^p \pmod p.$

Theorem (3.22): [*Extension of Th.* (3.21)]:

If p be prime, then $(x_1 + x_2 + \cdots + x_a)^p \equiv x_1^p + x_2^p + \cdots + x_a^p \pmod p\ \forall\ x_i \in Z.$

Proof:
$$\begin{aligned} (x_1 + x_2 + \cdots + x_a)^p &= \{x_1 + (x_2 + \cdots + x_a)\}^p \\ &\equiv x_1^p + (x_2 + \cdots + x_a)^p \pmod p \quad \text{[by Th. (3.19)]} \\ &\equiv x_1^p + x_2^p + (x_3 + \cdots + x_a)^p \pmod p. \end{aligned}$$

Repeating the above process by $a - 1$ times, we have

$$(x_1 + x_2 + \cdots + x_a)^p \equiv x_1^p + x_2^p + \cdots + x_a^p \pmod p.$$

Corollary (3.7): $(ax)^p \equiv ax^p \pmod p$, when p is prime and x is any integer.

Proof: Let $x_1 = x_2 = \cdots = x_a = x$.

Then from above theorem,

$$(x + x + \cdots + x)^p \equiv x^p + x^p + \cdots + x^p \pmod p$$

$$\Rightarrow \quad (ax)^p \equiv ax^p \pmod p.$$

Corollary (3.8): Deduce Fermat's theorem from Theorem (3.20).

Let p be prime and a be any integer such that $p \not| a$, i.e., $(a, p) = 1$.

Now putting $x_1 = x_2 = \cdots = x_a = 1$ in Theorem (3.20),

we have $(1 + 1 + \cdots \text{ up to } a \text{ times})^p \equiv 1 + \cdots + 1 \pmod p$

or, $\quad a^p \equiv a \pmod p$

or, $\quad a^{p-1} \equiv 1 \pmod p$ [since $(a, p) = 1$, cancellation law holds]

Lemma (3.1): If p and q be the distinct primes with $a^p \equiv a \pmod q$ and $a^q \equiv a \pmod p$, then $a^{pq} \equiv a \pmod{pq}$.

Proof: $a^p \equiv a \pmod p$ by Cor. (3.4)

$\Rightarrow \quad (a^p)^q \equiv a^q \pmod p \equiv a \pmod p$, since $a^q \equiv a \pmod p$

$\Rightarrow \quad a^{pq} \equiv a \pmod p \Rightarrow p | a^{pq} - a$.

Similarly, $q | a^{pq} - a$.

These together prove that $pq | a^{pq} - a$, $\;(p, q) = 1$

Since $\quad (p, q) = 1$, $1 = px + qy$ for some $x, y, \in Z$.

$\Rightarrow \quad a^{pq} - a = (a^{pq} - a)px + (a^{pq} - a)py.$ (1)

Again, $\quad q | a^{pq} - a$ and $p | a^{pq} - a$

$\Rightarrow \quad a^{pq} - a = qr$, $a^{pq} - a = ps$, $r, s \in Z$.

From (1), $a^{pq} - a = pqrx + pqsy = pq(rx + sy) \Rightarrow pq | a^{pq} - a$.

$\Rightarrow \quad a^{pq} = a \pmod{pq}$.

Corollary (3.9): Converse of Fermat's Theorem is not true.

Solution: The result follows from the example.

We shall show that $2^{341-1} \equiv 1 \pmod{341}$.

Here $\quad$ 341 is not prime and also $341 \not| 2$.

We have, $341 = 11 \times 31$.

Now, $\quad 2^{10} \equiv 1 \pmod{31}$ $[2^{10} = 1024 = 31.33 + 1]$

$\Rightarrow \quad 2^{11} \equiv 2 \pmod{31}$. (i)

Again, $2^{10} \equiv 1 \pmod{11}$ $[2^{10} = 1024 = 11 \times 93 + 1]$

$\Rightarrow \quad 2^{30} \equiv 1 \pmod{11}$

$\Rightarrow \quad 2^{31} \equiv 2 \pmod{11}$. (ii)

From (i) & (ii) by Lemma (3.1), we have

$$2^{11 \times 31} \equiv 2 \pmod{11 \times 31} \equiv 2 \pmod{341}$$

$$\Rightarrow \quad 2^{341-1} \equiv 1 \pmod{341},$$

which shows that converse of Fermat's Theorem is not true as $p = 341$ is not prime.

Problem (3.18): Use Fermat's theorem to prove that

(i) $1^{p-1} + 2^{p-1} + 3^{p-1} + \cdots + (p-1)^{p-1} \equiv (-1)(\text{mod } p)$

(ii) $1^p + 2^p + \cdots + (p-1)^p \equiv 0(\text{mod } p)$, when p is an odd prime.

Proof: Let $p = 3$ (odd prime). Then by Using Fermat's Theorem,

$$1^{3-1} \equiv 1(\text{mod } 3)$$

$$2^{3-1} \equiv 1(\text{mod } 3) \quad [2^2 \equiv 1(\text{mod } 3).$$

Therefore, $1^{3-1} + 2^{3-1} \equiv 1 + 1(\text{mod } 3) \equiv -1(\text{mod } 3)$.

Let $p = 5$ (odd prime).

Then $1^{5-1} \equiv 1(\text{mod } 5)$

$$2^{5-1} \equiv 1(\text{mod } 5)$$

$$3^{5-1} \equiv 1(\text{mod } 5)$$

$$4^{5-1} \equiv 1(\text{mod } 5).$$

Thus $1^{5-1} + 2^{5-1} + 3^{5-1} + 4^{5-1} \equiv (1 + 1 + 1 + 1)(\text{mod } 5)$

$$\equiv 4 - 5(\text{mod } 5)$$

$$\equiv -1(\text{mod } 5).$$

So if p be an odd prime, then $1^{p-1} \equiv 1(\text{mod } p), 2^{p-1} \equiv 1(\text{mod } p)$,

$$(p-1)^{p-1} \equiv 1(\text{mod } p).$$

Therefore, $1^{p-1} + 2^{p-1} + \cdots + (p-1)^{p-1} \equiv [1 + \cdots + 1(p-1) \text{ times}] \ (\text{mod } p)$

$$\equiv (p-1)(\text{mod } p)$$

$$\equiv -1(\text{mod } p).$$

(ii) Again using the Corollary (3.4) of Fermat's Theorem [i.e., $a^p \equiv a(\text{mod } p)$, $a \in Z$, p is prime],

$$1^p \equiv 1(\text{mod } p)$$

$$2^p \equiv 2(\text{mod } p)$$

$$(p-1)^p \equiv (p-1)(\text{mod } p).$$

Therefore, $1^p + 2^p + \cdots + (p-1)^p \equiv (1 + 2 + \cdots + p - 1)(\text{mod } p)$

$$\equiv \frac{p(p-1)}{2}(\text{mod } p)$$

$$\equiv 0(\text{mod } p).$$

p being odd, $p - 1$ is a multiple of 2.

Definition (3.4): *For any (+)ve integer m, $\phi(m)$ is the number of all (+)ve integers which are less than and prime to m. $\phi(m)$ is known as Euler-Totient function or Phi-function. [See Ch. 5.]*

Example: Let $m = 21$. Then the (+)ve integers which are less than and prime to 21 are $1, 2, 4, 5, 8, 10, 11, 13, 16, 17, 19, 20$.

So $\phi(m) = \phi(21) = 12$.

Reduced system

Definition (3.5): *A set of $\phi(m)$ integers whose least residues* (mod m) *are $r_1, r_2, \ldots, r_{\phi(m)}$ in some order is called a reduced system of residues with* (mod m). *$r_1, r_2, \ldots, r_{\phi(m)}$ are prime to m and they are incongruent* (mod m) *to each other.*

Example: (i) Let $S = \{11, -1, 13, 27\}$ and $m = 10$.

Least residues of $S(\text{mod } 10)$ are $\{1, 9, 3, 7\} = S_1$ (say), that is, $\{1, 3, 7, 9\}$ in some order.

Also $S_1 = \{1, 3, 7, 9\}$ is a set of integers less than 10 and prime to it.

Therefore, S forms a reduced system with (mod 10).

(ii) Let $S = \{24, -3, 45, 47, -19, 79, 115, 432\}$ and $m = 13$.

Least residues of $S(\text{mod } 13)$ are 11, 10, 6, 8, 7, 1, 11, 3.

That is $S_1 = \{1, 3, 6, 7, 8, 10, 11\}$ in some order.

Now $S_2 = \{1, 2, 3, 4, 5, 6, 7, 8, 9, 10, 11, 12\}$, set of integers less than 13 and prime to it.

Then $S_1 \neq S_2$.

Therefore, S does not form a reduced system (mod 13).

Theorem [Euler-Fermat] (3.23): *If* $(a, m) = 1$, *then* $a^{\phi(m)} \equiv 1(\text{mod } m)$.

***Proof*:** Let $r_1, r_2, \ldots, r_{\phi(m)}$ be the reduced system modulo m. Then $ar_1, ar_2, \ldots, ar_{\phi(m)}$ is also a reduced system modulo m, since each $(ar_i, m) = 1$ and they are incongruent to each other. For if $ar_i \equiv ar_j(m)$, $i \neq j$, then $r_i \equiv r_j(\text{mod } m)$ [by cancellation law as $(a, m) = 1$], a contradiction arises.

Let the least residues of $ar_1, ar_2, \ldots, ar_{\phi(m)}$ be $p_1, p_2, \ldots, p_{\phi(m)}(\text{mod } m)$ respectively. Then $p_1, p_2, \ldots, p_{\phi(m)}$ is a permutation of $r_1, r_2, \ldots, r_{\phi(m)}$ so that $p_1 p_2 \cdots p_{\phi(m)} = r_1 r_2 \cdots r_{\phi(m)}$.

Now we have $ar_1 \equiv p_1(\text{mod } m)$

$$ar_2 \equiv p_2(\text{mod } m)$$

$$\cdots\cdots\cdots\cdots\cdots\cdots$$

$$ar_{\phi(m)} \equiv P_{\phi(m)}(\text{mod } m).$$

Then by multiplying the terms, we have

$$ar_1 ar_2 \cdots ar_{\phi(m)} \equiv p_1 p_2 \cdots p_{\phi(m)}(\text{mod } m)$$

or, $$a^{\phi(m)} \cdot r_1 r_2 \cdots r_{\phi(m)} \equiv r_1 r_2 \cdots r_{\phi(m)}(\text{mod } m)$$

or, $$a^{\phi(m)} \equiv 1(\text{mod } m),$$

since $(r_i, m) = 1$, cancellation law holds.

Therefore, $a^{\phi(m)} \equiv 1(\text{mod } m)$.

Lemma (3.2): If $a^{h_1} \equiv 1(\text{mod } m_1)$ and $a^{h_2} \equiv 1(\text{mod } m_2)$, then $a^{[h_1, h_2]} \equiv 1(\text{mod } [m_1, m_2])$.

***Proof*:** Let $[h_1, h_2] = l$ and $[m_1, m_2] = m$.

Then by definition of L.C.M., $h_1 | l$, $h_2 | l$.

Therefore, $l = h_1 k_1$, $l = h_2 k_2$ for some $k_1, k_2 \in Z$, $(k_1, k_2) = 1$.

Now $a^{h_1} \equiv 1(\text{mod } m_1) \Rightarrow (a^{h_1})^{k_1} \equiv 1(\text{mod } m_1)$

$\Rightarrow$ $a^l \equiv 1(\text{mod } m_1)$. (1)

Similarly, we can show that $a^l \equiv 1(\text{mod } m_2)$. (2)

So from (1) & (2), we have $m_1 | a^l - 1$ and $m_2 | a^l - 1$.

Then, since $m = [m_1, m_2]$, $m | a^l - 1$.

Therefore, $a^l - 1 = mk$, $k \in Z$.

$\Rightarrow$ $a^l \equiv 1(\text{mod } m) \Rightarrow a^{[h_1, h_2]} \equiv 1(\text{mod } [m_1, m_2])$.

Problem (3.19): $a^{\phi(m)} \equiv 1 \pmod{m}$ does not imply that m be a prime number.

Solution: Let $m = 21$ and $a = 2$.

Then $(a, m) = (2, 21) = 1$.

Again, $m = 21 = 3 \times 7$. Then by Fermat's Theorem we get

$$2^{3-1} \equiv 1 \pmod{3}$$

or, $2^2 \equiv 1 \pmod{3}$. (i)

Also $2^{7-1} \equiv 1 \pmod{7}$

or, $2^6 \equiv 1 \pmod{7}$. (ii)

By Lemma (3.2), from (i) & (ii) we have

$$2^{[2,6]} \equiv 1 (\text{mod } [3,7]) \quad \text{or,} \quad 2^6 \equiv 1 \pmod{21}$$

or, $2^{12} \equiv 1 \pmod{21}$

or, $2^{\phi(21)} \equiv 1 \pmod{21}$, Euler's Theorem holds.

But $m = 21$ is not a prime number.

Problem (3.20): Use Fermat's Theorem to prove that for two positive integers $a, b, a^{40} - b^{40}$ is divisible by 541 if both a and b are prime to 541.

Solution: We have, $541 = 11 \times 41 = m$.

$$(a, m) = 1 \Rightarrow (a, 11) = 1, (a, 41) = 1.$$

Taking $m = 11$, By Fermat's Theorem,

$$a^{11-1} \equiv 1 \pmod{11} \quad \text{or,} \quad a^{10} \equiv 1 \pmod{4}.$$

Also when $m = 41$, $a^{40} \equiv 1 \pmod{41}$.

Then by Lemma (3.2),

$$a^{[10,40]} \equiv 1 (\text{mod } [11, 41]) \quad \text{or,} \quad a^{40} \equiv 1 \pmod{541}.$$

Similarly, we have $b^{40} \equiv 1 \pmod{541}$

Therefore, $a^{40} - b^{40} \equiv 0 \pmod{541}$

Hence, $a^{40} - b^{40}$ is divisible by 541.

Corollary (3.10): Show that $a^{12} - b^{12} \equiv 0 \pmod{91}$ of a and b, both prime to 91.

[Proceed as in problem [3.20] when $91 = 7 \times 13$.]

Theorem (3.24): Wilson's Theorem

If p be prime, then $(p-1)! \equiv -1 \pmod{p}$.

[Go through the following matter for a prime number.]

$(2-1)! + 1 = 2$, divisible by 2

$(3-1)! + 1 = 3$, divisible by 3

$(5-1)! + 1 = 25$, divisible by 5

$(7-1)! + 1 = 721$, divisible by 7

and so on.

Now let $p = 19$, a prime number.

We shall show that $(19-1)! + 1$ is divisible by 19.

The set of (+)ve integers which are less than and prime to 19 is 1, 2, 3, 4, 5, 6, 7, 8, 9, 10, 11, 12, 13, 14, 15, 16, 17, 18.

If we leave the 1st and last number, i.e., 1 and 18, then the remaining numbers can be paired of in 8 pairs (2,10), (3,13), (4,5), (6,16), (7,11), (8,12), (9,17), (14,15) so that

$$2 \times 10 \equiv 1 (\text{mod } 19)$$
$$3 \times 13 \equiv 1 (\text{mod } 19)$$
$$4 \times 5 \equiv 1 (\text{mod } 19)$$
$$6 \times 16 \equiv 1 (\text{mod } 19)$$
$$7 \times 11 \equiv 1 (\text{mod } 19)$$
$$8 \times 12 \equiv 1 (\text{mod } 19)$$
$$9 \times 17 \equiv 1 (\text{mod } 19)$$
$$14 \times 15 \equiv 1 (\text{mod } 19).$$

Multiplying these congruences we have

$$2 \cdot 3 \cdot 4 \cdot 5 \cdot 6 \cdot 7 \cdot 8 \cdot 9 \cdot 10 \cdot 11 \cdot 12 \cdot 13 \cdot 14 \cdot 15 \cdot 16 \cdot 17 \equiv 1 (\text{mod } 19)$$

$$\Rightarrow \quad 17! \equiv 1 (\text{mod } 19) \Rightarrow 18! \equiv 18 (\text{mod } 19) \equiv -1 (\text{mod } 19).$$

That is, $(19-1)! \equiv -1 (\text{mod } 19)$.

Therefore, $(19-1)! + 1$ is divisible by 19.

Wilson's theorem holds for $p = 19$.

Proof of the theorem: For the prime number p, the set S of integers which are less than and prime to p is $\{1, 2, 3, \ldots, p-1\}$.

We denote $S = \{1, 2, 3, \ldots, p-1\}$.

Since $(a, p) = 1, a \in S$, the congruence $ax \equiv 1 (\text{mod } p)$ has a unique solution, by Th. (3.12).

That is, for $a \in s$, $\exists$ a unique element $a' \in s$ such that $a.a' \equiv 1 (\text{mod } p)$.

If $\quad a = a'$, then $a^2 \equiv 1 (\text{mod } p)$

or, $\quad (a^2 - 1) \equiv 0 (\text{mod } p)$ or, $(a+1)(a-1) \equiv 0 (\text{mod } p)$.

Then either $a - 1 \equiv 0 (\text{mod } p)$ or, $(a+1) \equiv 0 (\text{mod } p)$.

When $a - 1 \equiv 0 (\text{mod } p)$, $a = 1$ and when $a + 1 \equiv 0 (\text{mod } p)$, $a = p - 1$.

Now omitting 1 and $p-1$ from S, the remaining $p-3$ numbers of S can be paired of in $\frac{p-3}{2}$ pairs (a, a') satisfying $ax \equiv 1 (\text{mod } p)$, $a \neq a'$ and $1 < a' < p-1$. Therefore, multiplying the congruences, we get

$$2 \cdot 3 \ldots p-2 \equiv 1 (\text{mod } p) \quad \text{or,} \quad (p-2)! \equiv 1 (\text{mod } p)$$

or, $(p-1)! \equiv (p-1) (\text{mod } p)$ or, $(p-1)! \equiv -1 (\text{mod } p)$.

Converse of Wilson's Theorem (3.25): *If* $(p-1)! \equiv -1 (\text{mod } p)$, *then p is prime.*

Proof: If p is not prime, then p has a divisor, say d, i.e., $d|p, 1 < d < p$.

Now d being $\leq p-1$ is a divisor of one of the factors in $(p-1)!$, so $d|(p-1)!$.

Also $\quad p|(p-1)! + 1$ from the hypothesis.

Therefore, $d|p$ and $p|(p-1)! + 1 \Rightarrow d|(p-1)! + 1 \Rightarrow d|1$.

Thus the conclusion that $d|1$ is not true as $1 < d < p$.

Hence p must be prime.

So the converse of Wilson's Theorem holds.

Problem (3.21): For a prime p, $(p-1)! \equiv (-1)^{\frac{p-1}{2}}\left(1\cdot 2\cdots\frac{p-1}{2}\right)^2 \pmod p$.

Show that the integer $\left(\frac{p-1}{2}\right)!$ satisfies the congruence $x^2 \pm 1 \equiv 0 \pmod p$ according as $p = 4k+1$ and $p = (4k+3)$.

Solution: We consider the set of integers,

$$1, 2, 3, \frac{p-1}{2}, \frac{p+1}{2}, \ldots (p-2), (p-1) \text{ where } p \text{ is prime.}$$

Now $p-1 \equiv -1 \pmod p$

$p-2 \equiv -2 \pmod p$

$\frac{p+1}{2} \equiv -\frac{p-1}{2} \pmod p.$

$$(p-1)! = 1\cdot 2\cdot 3\cdots\frac{p-1}{2}\cdot\frac{p+1}{2}\cdots(p-2)(p-1).$$

$$\equiv \left(1\cdot 2\cdot 3\cdots\frac{p-1}{2}\right)\left(-\left(\frac{p-1}{2}\right)\cdots(-2)(-1)\right) \pmod p$$

or, $(p-1)! \equiv 1(-1)\cdot 2(-2)\cdots\left(\frac{p-1}{2}\right)\left(-\frac{p-1}{2}\right) \pmod p$

or, $(p-1)! \equiv (-1)^{\frac{p-1}{2}}\left[1.2\cdots\left(\frac{p-1}{2}\right)\right]^2 \pmod p$

or, $(p-1)! \equiv (-1)^{\frac{p-1}{2}}\left[\left(\frac{p-1}{2}\right)!\right]^2 \pmod p$, $\frac{p-1}{2} = \text{integer}$ (i)

Again, by Wilson's Theorem.

$(p-1)! \equiv (-1) \pmod p.$ (ii)

Then from (i) and (ii) $-1 \equiv (-1)^{\frac{p-1}{2}}\left(\left(\frac{p-1}{2}\right)!\right)^2 \pmod p.$ (iii)

Now if p is a prime of the form $4k+1$ for some $k \in N$ [those primes which are 5, 13, 17, 29,...],

then $-1 \equiv (-1)^{\frac{4k+1-1}{2}}\left[\left(\frac{p-1}{2}\right)!\right]^2 \pmod p$

or, $-1 \equiv 1\left[\left(\frac{p-1}{2}\right)!\right]^2 \pmod p$

or, $\left(\frac{p-1}{2}!\right)^2 \equiv -1 \pmod p$ (iv)

$[a \equiv b \pmod m \Rightarrow b \equiv a \pmod m)].$

Thus $x = \left(\frac{p-1}{2}\right)!$ satisfies the congruence $x^2 + 1 \equiv 0 \pmod p$. (v)

Again, if p is of the form $4k+3$, $k \in N$,

then $\quad -1 \equiv (-1)^{\frac{4k+3-1}{2}} \left[\left(\frac{p-1}{2}\right)!\right]^2 \pmod p$

or, $\quad -1 \equiv (-1)^{2k+1} \left[\left(\frac{p-1}{2}\right)!\right]^2 \pmod p$

or, $\quad 1 \equiv 1 \left[\left(\frac{p-1}{2}\right)!\right]^2 \pmod p$

or, $\quad \left[\left(\frac{p-1}{2}\right)!\right]^2 \equiv 1 \pmod p.$

So $x = \left(\frac{p-1}{2}\right)!$ satisfies the congruence $x^2 - 1 \equiv 0 \pmod p$. (vi)

The result follows from (v) and (vi).

Problem (3.22): If p be an odd prime, then show that

(1) $1^2 \cdot 3^2 \cdot 5^2 \cdots (p-2)^2 \equiv (-1)^{\frac{p+1}{2}} \pmod p$

(2) $2^2 \cdot 4^2 \cdot 6^2 \cdots (p-1)^2 \equiv (-1)^{\frac{p-1}{2}} \pmod p$.

Solution: (1) We use the result, $p - k \equiv -k \pmod p$ and $k \equiv -(p-k) \pmod p$. (i)

$$(p-1)! = 1 \cdot 2 \cdot 3 \cdot 4 \cdot 5 \cdots \frac{p-1}{2}\,\frac{p+1}{2} \cdots (p-4)\cdot(p-3)\cdot(p-2)\cdot(p-1)$$

$$= \{1\cdot(p-1)\}\{(3\cdot(p-3)\}\{(5\cdot(p-5)\}\cdots\{(4\cdot(p-4)\}\{(2\cdot(p-2)\}$$

$$\equiv 1(-1)\cdot 3(-3)\cdot 5(-5)\cdots\{-(p-4)\cdot(p-4)\}\{-(p-2)\cdot(p-2)\} \pmod p$$

or, $\quad (p-1)! \equiv (-1)^{\frac{p-1}{2}} 1^2 \cdot 3^2 \cdot 5^2 \cdots (p-2)^2 \pmod p.$ (ii)

Again, by Wilson's Theorem,

$(p-1)! \equiv -1 \pmod p.$ (iii)

From (ii) & (iii), we have

$$-1 \equiv (-1)^{\frac{p-1}{2}} 1^2 \cdot 3^2 \cdot 5^2 \cdots (p-2)^2 \pmod p$$

$\Rightarrow \quad 1^2 \cdot 3^2 \cdot 5^2 \cdots (p-2)^2 \equiv (-1)(-1)^{\frac{p-1}{2}} \pmod p$

$\Rightarrow \quad 1^2 \cdot 3^2 \cdot 5^2 \cdots (p-2)^2 \equiv (-1)^{\frac{p+1}{2}} \pmod p.$

(2) It can be obtained from

$$(p-1)! = 2(p-2)4(p-4)6(p-6)\cdots 3\cdot(p-3)1\cdot(p-1),$$

by using the results of (i).

Problem (3.23): If p be a prime number, then show that

$$(p-1)! \equiv p - 1 \pmod{(1 + 2 + 3 + \cdots + p - 1)}.$$

Solution: By Wilson's Theorem,

$$(p-1)! \equiv (p-1) \pmod p$$

$\Rightarrow \quad p \mid (p-1)! - (p-1)$

Also, $\quad \frac{p-1}{2} \mid (p-1)! - (p-1)$, since $p-1$ is even.

hence, $\quad p\frac{(p-1)}{2} \mid (p-1)! - (p-1)$

$$\Rightarrow \quad (p-1)! - (p-1) \equiv 0\left[\text{mod } \frac{p(p-1)}{2}\right]$$

$$\Rightarrow \quad (p-1)! \equiv p-1[\text{mod } (1+2+\cdots+p-1)]$$

$$\left[\text{since } 1+2+\cdots+p-1 = \frac{p(p-1)}{2}\right].$$

Problem (3.24): Show that $\underline{|28} + 233 \equiv 0(\text{mod } 899)$.

Solution: $899 = 29 \times 31$

By Wilson's Theorem ($p = 29$, a prime number),

$\underline{|28} + 1 \equiv 0(\text{mod } 29)$

or, $\underline{|28} \equiv -1(\text{mod } 29)$. (1)

Again, $233 \equiv 1(\text{mod } 29)$. (2)

Adding (1) & (2), $\underline{|28} + 233 \equiv 0(\text{mod } 29)$ (3)

$p = 31$, a prime number.

So by Wilson's Theorem, $\underline{|30} + 1 \equiv 0(\text{mod } 31)$. (4)

Also $233 \equiv 16(\text{mod } 31)$

or, $233 - 16 \equiv 0(\text{mod } 31)$. (5)

From (4), $30 \cdot 29\, \underline{|28} + 1 \equiv 0(\text{mod } 31)$

or, $30 \cdot 29\, \underline{|28} \equiv -1(\text{mod } 31) \equiv 30(\text{mod } 31)$

$\Rightarrow \quad 29\, \underline{|28} \equiv 1(\text{mod } 31)\ [(30, 31) = 1]$

$\Rightarrow \quad (-2)29\, \underline{|28} \equiv -2(\text{mod } 31) \equiv +29(\text{mod } 31)$

$\Rightarrow \quad (-2)\underline{|28} \equiv 1(\text{mod } 31)\quad [(29, 31) = 1]$

$\equiv 32(\text{mod } 31)$

$\Rightarrow \quad \underline{|28} \equiv -16(\text{mod } 31)$

$\Rightarrow \quad \underline{|28} + 16 = 0(\text{mod } 31)$. (6)

Adding (5) and (6), we get

$\underline{|28} + 233 \equiv 0(\text{mod } 31)$. (7)

Therefore, from (3) & (7), $\underline{|28} + 233 \equiv 0(\text{mod } [29, 31])$

or, $\underline{|28} + 233 \equiv 0(\text{mod } 899)$.

Corollary (3.11): For a prime p of the form $4k + 3$, $\left(\frac{p-1}{2}\right)! \equiv 1(\text{mod } p)$ or $\left(\frac{p-1}{2}\right)! \equiv -1(\text{mod } p)$.

[The result follows from Problem 3.21].

Lagrange's Theorem (3.26): *Let $f(x) = a_n x^n + a_{n-1}x^{n-1} + a_{n-2}x^{n-2} + \cdots + a_1 x + a_0$ be a polynomial of degree $n \geq 1$ with integral coefficient a_n and let m be a prime number such that $a_n \not\equiv 0(\text{mod } m)$, then $f(x) \equiv 0(\text{mod } m)$ has at most n incongruent solutions modulo m.*

Proof: If $f(x)$ be a polynomial of degree 1, then $f(x) = a_1 x + a_0$.

Then $f(x) \equiv 0 (\text{mod } m)$

$\Rightarrow \quad a_1 x \equiv -a_0 (\text{mod } m)$

has a unique solution since $(a_1, m) = 1$ as $a_1 \not\equiv 0(\text{mod } m)$.

So the theorem holds good for a polynomial of degree 1.

For applying induction method, let the theorem hold for a polynomial of degree $n-1$.

That is, a polynomial of degree $n-1$ has at most $(n-1)$ incongruent solutions modulo m. We further assume that $f(x) \equiv 0(\text{mod } m)$ has at most $(n+1)$ incongruent solutions modulo m, say, $x_0, x_1, x_2, \ldots, x_n$.

That is $f(x_k) \equiv 0(\text{mod } m)$ for $k = 0, 1, 2, \ldots, n$.

Now from the identity,

$$\begin{aligned} f(x) - f(x_0) &= (a_n x^n + a_{n-1}x^{n-1} + \cdots + a_1 x + a_0) \\ &\qquad - (a_n x_0^n + a_{n-1}x_0^{n-1} + \cdots + a_1 x_0 + a_0) \\ &= a_n(x^n - x_0^n) + a_{n-1}(x^{n-1} - x_0^{n-1}) + \cdots + a_1(x - x_0) \\ &= (x - x_0)[a_n(x^{n-1} + x^{n-2}x_0 + \cdots + x_0^{n-1}) \\ &\qquad + a_{n-1}(x^{n-2} + x^{n-3}x_0 + \cdots + x_0^{n-2}) + \cdots + a_1] \\ &= (x - x_0)[a_n x^{n-1} + (a_n x_0 + a_{n-1})x^{n-2} + \cdots + a_1] \\ &= (x - x_0)h(x), \end{aligned}$$

where $h(x)$ is a polynomial of degree $n-1$, with leading coefficient a_n.

Now for $x = x_k$, $f(x_k) - f(x_0) = (x_k - x_0)h(x_k)$ (1)

Also $\quad f(x_k) \equiv 0(\text{mod } m)$ and $f(x_0) \equiv 0(\text{mod } m)$

$\Rightarrow \quad f(x_k) - f(x_0) \equiv 0(\text{mod } m)$

$\Rightarrow \quad (x_k - x_0)h(x_k) \equiv 0(\text{mod } m)$, by (1)

$\Rightarrow \quad h(x_k) \equiv 0(\text{mod } m)$, since $x_k \not\equiv x_0(\text{mod } m)$, $k \neq 0$.

$\Rightarrow \quad h(x_k) \equiv 0(\text{mod } m)$, for $k = 1, 2, \ldots, n$.

Therefore, $h(x) \equiv 0(\text{mod } m)$, where $h(x)$ is a polynomial of degree $n-1$, has n incongruent solutions modulo m, contradicting our inductive hypothesis.

Hence the theorem.

Application of Lagrange's Theorem

Theorem (3.27): *If $f(x) = a_n x^n + a_{n-1}x^{n-1} + \cdots + a_1 x + a_0$ be a polynomial of degree n, $a_n \in Z$ and if $f(x) \equiv 0(\text{mod } m)$, where m is a prime number, has more than n solutions, then $m | a_k \ \forall\ k = 0, 1, 2, \ldots, n$.*

Proof: If possible, let $\exists$ a coefficient a_k, the largest index $k (\leq n)$ such that $m \nmid a_k$.

Then by Lagrange's Theorem,

$$a_k x^k + a_{k-1}x^{k-1} + \cdots + a_1 x + a_0 \equiv 0(\text{mod } m)$$

has at most k solutions, $\alpha_1, \alpha_2, \ldots, \alpha_k$ (say).

That is, $a_k(x - \alpha_1)(x - \alpha_2)\cdots(x - \alpha_k) \equiv 0(\text{mod } m)$. (1)

Now if there exists more than k solutions, then let α_{k+1} be such solution distinct from $\alpha_1, \alpha_2, \ldots, \alpha_k$. That is, $\alpha_{k+1} \not\equiv \alpha_p(\text{mod } m)$, for $p = 1, 2, \ldots, k$.

Hence from (1),

$$a_k(\alpha_{k+1} - \alpha_1)(\alpha_{k+1} - \alpha_2)\cdots(\alpha_{k+1} - \alpha_k) \equiv 0 (\text{mod } m)$$

$\Rightarrow \quad a_k \equiv 0(\text{mod } m)$

$\Rightarrow \quad m|a_k$, a contradiction.

Hence the theorem.

Theorem (3.28): *For any prime m, all the coefficients of the polynomial $f(x) = (x-1)(x-2)\cdots(x-m+1) - x^{m-1} + 1$ are divisible by m.*

[Go through the following Example.]

Let $\quad m \equiv 5$ (prime) and $g(x) = (x-1)(x-2)(x-3)(x-4)$.

The roots of $g(x) = 0$ are 1, 2, 3, 4.

Hence they satisfies $g(x) \equiv 0(\text{mod } 5)$.

Again, let $h(x) = x^{5-1} - 1 = x^4 - 1$.

Then $\quad x^4 - 1 \equiv 0(\text{mod } 5)$ for $x = 1, 2, 3, 4$.

$$f(x) = g(x) - h(x) = (x-1)(x-2)(x-3)(x-4) - x^4 + 1 \equiv 0(\text{mod } 5). \quad (1)$$

$f(x)$ is a polynomial of degree 3 having 4 solutions 1, 2, 3, 4. That is $f(x) \equiv 0(\text{mod } 5)$, where $f(x)$ is a polynomial of degree 3, has more than 3 solutions and so

$$\begin{aligned} f(x) &= (x^2 - 3x + 2)(x^2 - 7x + 12) - x^4 + 1 \text{ [from (1)]} \\ &= -10x^3 + 35x^2 - 50x + 25, \end{aligned}$$

shows that each coefficient of $f(x)$ is divisible by 5.

Also $25 = 24 + 1 = \underline{|4} + 1 = \underline{|5-1} + 1$ is divisible by 5 $\Rightarrow$ Wilson's theorem holds for $m = 5$.

Proof: Let $g(x) = (x-1)(x-2)\cdots(x-(m-1))$

The roots of $g(x)$ are $1, 2, \ldots, m-1$.

Hence they satisfy $g(x) \equiv 0(\text{mod } m)$.

Again, let $h(x) = x^{m-1} - 1$.

Then $\quad x^{m-1} - 1 \equiv 0(\text{mod } m)$ for $x = 1, 2, \ldots, m-1$.

$$\begin{aligned} \text{Therefore, } f(x) &= g(x) - h(x) \\ &= (x-1)(x-2)\cdots(x-(m-1)) - x^{m-1} + 1 \quad (1) \\ &\equiv 0(\text{mod } m). \end{aligned}$$

$f(x)$ is a polynomial of degree $m-2$, but the congruence $f(x) \equiv 0(\text{mod } m)$ has $m-1$ solutions, i.e., more than $(m-2)$ solutions. Therefore, by Theorem (3.25), each coefficient of $f(x)$ is divisible by m.

Hence the theorem.

Corollary (3.12): Deduction of Wilson's Theorem from Th. (3.28).

Proof: The constant term in (1) from above theorem is $1 \cdot 2 \cdot \ldots (m-1) + 1$ and it is divisible by m.

That is, $\underline{|m-1} + 1 \equiv 0(\text{mod } m)$, which is Wilson's Theorem for any prime m.

Polynomial congruence modulo m

Definition (3.5): *An expression of the form $a_n x^n + a_{n-1}x^{n-1} + \cdots + a_0$, where x is a variable and $a_n \in R$ is independent of x and n is a (+)ve integer, is called a polynomial of degree n, provided $a_n \neq 0$. We denote it by $f(x)$.*

Example: (i) $2x^4 + \sqrt{2}x^3 + \dfrac{3}{2}$ (ii) $-3x^5 + 2x^4 + \dfrac{5}{8}x^2 - 8x + 9$ are polynomials in x.

Definition (3.6): Integral polynomial: *A polynomial $f(x) = a_n x^n + a_{n-1}x^{n-1} + \cdots + a_1 x + a_0$ is called an integral polynomial of degree n, if $a_n \in Z\ \forall\ n,\ a_n \neq 0$.*

Example: $f(x) = 2x^3 + 3x^2 - 5x - 1$ is a polynomial. Here we shall take $f(x)$ **as an integral polynomial, as each co-efficient is an integer.**

Henceforth we shall consider only integral polynomials.

Definition (3.7): Polynomial congruence modulo m: *If $f(x) = a_n x^n + a_{n-1}x^{n-1} + \cdots + a_1 x + a_0$ such that $a_n \equiv 0 \pmod{m}\ \forall\ n$, then $f(x) \equiv 0 \pmod{m}$. We say that $f(x)$ is identically congruence to 0 (mod m).*

Further, if $\phi(x)$ be an integral polynomial of degree $k(\leq n$ or $> n)$ such that $f(x) - \phi(x) \equiv 0 \pmod{m}$, i.e., $f(x) \equiv \phi(x) \pmod{m}$, then $f(x)$ is said to be identically congruent to $\phi(x)$ modulo m.

Example: (i) Let $f(x) = 5x^3 + 8x^2 + 6x + 19$

and $\quad \phi(x) = 3x^2 + x + 4 \ (k = 2 < 3)$

$$= 0 \cdot x^3 + 3x^2 + x + 4.$$

$$f(x) - \phi(x) = (5-0)x^3 + (8-3)x^2 + (6-1)x + (19-4).$$

Here $\quad a_n - b_n \equiv 0 \pmod{5}\ \forall\ n = 3, 2, 1, 0.$

So $\quad f(x) - \phi(x) \equiv 0 \pmod{5}$

i.e. $\quad f(x) \equiv \phi(x) \pmod{5}$, identically.

(ii) Let $f(x) = 5x^3 + 8x^2 + 6x + 19$

$$\phi(x) = 20x^5 + 15x^4 - 10x^3 - 2x^2 - 4x + 4 \quad (k = 5 > 3).$$

Then $\quad f(x) - \phi(x) = (0-20)^5 + (0-15)x^4 + (5+10)x^3 + (8+2)x^2 + (6+4)x + (19-4).$

Here also $a_n - b_n \equiv 0 \pmod{5}\ \forall\ n = 5, 4, 3, 2, 1, 0.$

Then $\quad f(x) \equiv \phi(x) \pmod{5}$, identically.

(iii) $f(x) = 45x^5 + 30x^4 + 5x^3 + 8x^2 + 6x + 19$

$$\phi(x) = 20x^5 + 15x^4 - 10x^3 - 2x^2 - 4x + 4 \quad (k = 5).$$

Here also $f(x) - \phi(x) \equiv 0 \pmod{5}$

i.e., $\quad f(x) \equiv \phi(x) \pmod{5}$, identically.

Thus we can define as under:

Definition (3.8): *If $f(x)$ and $\phi(x)$ be two polynomials of the form $a_n x^n + a_{n-1}x^{n-1} + \cdots + a_1 x + a_0$ and $b_n x^n + b_{n-1}x^{n-1} + \cdots + b_1 x + b_0$ such that $a_i \equiv b_i \pmod{m}\ \forall\ i = 0, 1, 2, \ldots, n$, then $f(x)$ is said to be identically congruent to $\phi(x)$ modulo m, i.e., $f(x) \equiv \phi(x) \pmod{m}$, identically, which holds $\forall\ x$.*

But the converse of it does not hold.

Example: By Fermat's Theorem, $x^5 \equiv x (\text{mod } 5)\ \forall\ x$. Let $f(x) = x^5$ and $\phi(x) = x$.

Then $\quad f(x) \equiv \phi(x)(\text{mod } 5)$.

For $\quad f(x),\ a_5 = 1,\ a_i = 0\ \forall\ i = 4, 3, 2, 1, 0$

and $\quad \phi(x),\ b_1 = 1,\ b_i = 0\ \forall\ i = 5, 4, 3, 2, 0$

$a_5 - b_5 = 1 \not\equiv 0(\text{mod } 5)$

$a_1 - b_1 = -1 \not\equiv 0(\text{mod } 5)$. So $a_i \not\equiv b_i(\text{mod } 5)\ \forall i$,

which shows that the converse is not true.

Definition (3.9): *$f(x)$ is said to be completely divisible by $\phi(x)(\text{mod } m)$ if degree of $\phi(x) \leq$ degree $f(x)$ and if there exists a polynomial $g(x)$ such that $f(x) \equiv \phi(x)g(x)(\text{mod } m)$, identically.*

Example: (i) Let $f(x) = 5x^2 + 3x + 16$

and $\quad \phi(x) = 3x - 2$.

We divide $f(x)$ by $\phi(x)(\text{mod } 7)$.

$$\begin{array}{cc|ccc} 3 & -2 & 5 & 3 & 16 \\ \hline 4 & -1 & 12 & 3 & 16 \\ & & 12 & -8 & \\ \hline & & & 11 & 16 \\ & & & -3 & 2 \\ & & & -3 & 2 \\ \hline & & & 0 & 0 \end{array}$$

where $5 \equiv 12(\text{mod } 7)$

$11 \equiv -3(\text{mod } 7)$

$16 \equiv 2(\text{mod } 7)$

$g(x) = 4x - 1$.

Then $\quad f(x) = (3x - 2)(4x - 1)(\text{mod } 7)$.

(ii) Let $f(x) = 5x^3 + 43x^2 + 5x + 3$

$\phi(x) = 7x^2 - 1$.

We divide $f(x)$ by $\phi(x)(\text{mod } 8)$.

$$\begin{array}{ccc|cccc} & & & 5 & 43 & 5 & 3 \\ 7 & 0 & -1 & 21 & 3 & -3 & 3 \\ \hline 3 & & -3 & 21 & 0 & -3 & \\ \hline & & & & 3 & 0 & 3 \\ & & & & -21 & 0 & 3 \\ & & & & -21 & 0 & 3 \\ \hline & & & & 0 & 0 & 0 \end{array}$$

where $5 \equiv 21(\text{mod } 8)$

$43 \equiv 3(\text{mod } 8)$

$5 \equiv -3(\text{mod } 8)$

$g(x) = 3x - 3 = 3(x - 1)$.

Then $\quad f(x) \equiv 3(7x^2 - 1)(x - 1)(\text{mod } 8)$.

Definition (3.10): *If $f(x)$ is not completely divisible by $\phi(x)$ modulo m, then there exists a remainder $r(x)$ where degree of $r(x)$ < degree of $\phi(x)$ such that $f(x) \equiv \phi(x)g(x) + r(x)(\text{mod } m)$.*

Theorem (3.29): *If $f(x) \equiv \phi(x)g(x) + r(x)(\text{mod } m)$, then $g(x)$ and $r(x)$ are unique with* mod (m).

Proof: If possible, let $f(x) \equiv \phi(x)g_1(x) + r_1(x)(\text{mod } m)$

and $\quad f(x) \equiv \phi(x)g_2(x) + r_2(x)(\text{mod } m)$.

Then we have $\phi(x)[g_1(x) - g_2(x)] \equiv (r_1(x) - r_2(x)](\text{mod } m)$. $\quad$ (1)

Now if $g_1(x) \not\equiv g_2(x) (\text{mod } m)$, then $\phi(x)[g_1(x) - g_2(x)]$ possesses a degree not less than that of $\phi(x)$ while the degree of $[r_1(x) - r_2(x)]$ is less than the degree of $\phi(x)$ and consequently the identically congruence (1) does not hold.

Therefore, $g_1(x) \equiv g_2(x) (\text{mod } m)$ and subsequently $r_1(x) \equiv r_2(x) (\text{mod } m)$.

Hence, $g_1(x), g_2(x)$ are unique and so also $r_1(x), r_2(x)$.

Therefore, $f(x)$ can be expressed uniquely in the form $f(x) \equiv \phi(x)g(x) + r(x) (\text{mod } m)$.

Solution of $f(x) \equiv 0(\text{mod } m)$

m is a prime number and degree of $f(x) \geq m$.

Case(1): Let $f(x)$ not be completely divisible by $x^m - x(\text{mod } m)$ and degree of $f(x) \geq m$.

Then $f(x) \equiv (x^m - x)g(x) + r(x)(\text{mod } m)$, (1)

where degree of $r(x) < m$.

By Fermat's Theorem, $x^m - x \equiv 0(\text{mod } m) \ \forall\ x$.

Hence from (1), $f(x) \equiv r(x)(\text{mod } m)$.

So the solutions of $f(x) \equiv 0(\text{mod } m)$ are given by $r(x) \equiv 0(\text{mod } m)$.

Case(2): Let $f(x)$ be completely divisible by $x^m - x(\text{mod } m)$, then

$f(x) \equiv (x^m - x)g(x)(\text{mod } m), g(x)$ is a poly of degree $n - m$.

$\equiv 0(\text{mod } m) \ \forall\ x$.

Hence the solutions of $f(x) \equiv 0(\text{mod } m)$ are given by $x \equiv 0, 1, 2, \cdots m - 1(\text{mod } m)$.

Problem (3.25): Find the solutions of $f(x) \equiv 0(\text{mod } 5)$, where $f(x) = 6x^7 + 3x^6 + 3x^5 + x^4 + x^2 + 2x + 1$.

Solution: Let us consider $\phi(x) = x^5 - x$

such that the degree of $\phi(x) <$ degree of $f(x)$. So $f(x)$ is divisible by $\phi(x)$.

1 0 0 0 −1 0	6	3	3	1	0	1	2	1	$6 \equiv 1(\text{mod } 5)$
	1	−2	−2	1	0	1	2	1	$3 \equiv -2(\text{mod } 5)$
1 −2 −2	1	0	0	0	−1	0			
$= g(x)$		−2	−2	1	1	1	2		
		−2	0	0	0	2	0		
			−2	1	1	−1	2	1	
			−2	0	0	0	2	0	
				1	1	−1	0	1	$= r(x)$

Then $f(x) \equiv (x^5 - x)(x^2 - 2x - 2) + x^4 + x^3 - x^2 + 1(\text{mod } 5)$.

Solutions of $f(x) \equiv 0(\text{mod } 5)$ are given by $r(x) \equiv 0(\text{mod } 5)$;

i.e., $x^4 + x^3 - x^2 + 1 \equiv 0(\text{mod } 5)$

or, $(x + 1)(x^3 - x + 1) \equiv 0(\text{mod } 5)$

or, $x + 1 \equiv 0(\text{mod } 5)$ and $x^3 - x + 1 \equiv 0(\text{mod } 5)$

$\Rightarrow$ $x \equiv -1(\text{mod } 5)$ and $x \equiv 3(\text{mod } 5)$

$\equiv 4(\text{mod } 5)$

$\Rightarrow$ $x \equiv 3, 4(\text{mod } 5)$ are only two solutions of $f(x) \equiv 0(\text{mod } 5)$.

The solution of $f(x) \equiv 0 (\text{mod } m)$, when the degree of $f(x) < m$, m is prime.

Lemma (3.3): Let $f(x) = a_n x^n + a_{n-1}x^{n-1} + \cdots + a_1 x + a_0, a_i \in Z$ and let α be a solution of $f(x) \equiv 0(\text{mod } m)$, then $f(x) \equiv (x - \alpha)g(x)(\text{mod } m)$ identically, where $g(x)$ is a polynomial of degree $n - 1$.

Proof: Since α is a solution of $f(x) \equiv 0(\text{mod } m)$, we have $f(\alpha) \equiv 0(\text{mod } m)$.

Then $f(x) \equiv f(x) - f(\alpha)(\text{mod } m) \equiv [(a_n x^n + a_{n-1}x^{n-1} + \cdots + a_1 x + a_0)$
$- (a_n \alpha^n + a_{n-1}\alpha^{n-1} + \cdots + a_n \alpha + a_0)(\text{mod } m)]$
$\equiv [a_n(x^n - \alpha^n) + a_{n-1}(x^{n-1} - \alpha^{n-1}) + \cdots + a_1(x - \alpha)(\text{mod } m)]$

i.e., $f(x) \equiv (x - \alpha)[a_n(x^{n-1} + x^{n-2}\alpha + \cdots + \alpha^{n-1}) + a_{n-1}(x^{n-2} + x^{n-3}\alpha + \cdots$
$+ \alpha^{n-2}) + a_1(\text{mod } (m)]$

i.e., $f(x) \equiv (x - \alpha)g(x)(\text{mod } m)$, where $g(x)$ is a polynomial of degree $n - 1$.

Lemma (3.4): If α_1, α_2 be two incongruent solutions of $f(x) \equiv 0(\text{mod } m)$, m is prime, then $f(x) \equiv (x - \alpha_1)(x - \alpha_2)h(x)(\text{mod } m)$, where $h(x)$ is a polynomial of degree $(n - 2)$.

Proof: Since α_1 is a solution of $f(x) \equiv 0(\text{mod } m)$, by Lemma [3.9], we have

$$f(x) \equiv (x - \alpha_1)g(x)(\text{mod } m), \tag{1}$$

where $g(x)$ is polynomial of degree $(n - 1)$.

Again $f(\alpha_2) \equiv 0(\text{mod } m)$.

Then $f(\alpha_2) \equiv (\alpha_2 - \alpha_1)g(\alpha_2)(\text{mod } m)$ [form (1)]
$\equiv 0(\text{mod } m)$

$\Rightarrow$ either $\alpha_2 - \alpha_1 \equiv 0(\text{mod } m)$

or, $g(\alpha_2) \equiv 0(\text{mod } m)$.

But $\alpha_2 \equiv \alpha_1(\text{mod } m)$ is not possible, as they are incongruent (mod m).

Therefore, $g(\alpha_2) \equiv 0(\text{mod } m)$.

Again, by Lemma [3.3], $g(x) \equiv (x - \alpha_2)h(x)(\text{mod } m)$, (2)

where $h(x)$ is a polynomial of degree $(n - 2)$.

So $f(x) \equiv (x - \alpha_1)(x - \alpha_2)h(x)(\text{mod } m)$ [by (1) & (2)].

Lemma (3.5): If $\alpha_1, \alpha_2, \ldots, \alpha_n$ be n incongruent solutions of $f(x) \equiv 0(\text{mod } m)$, m is prime and

$$f(x) = a_n x^n + a_{n-1}x^{n-1} + \cdots + a_1 x + a_0,$$

then $f(x) \equiv a_n(x - \alpha_1)(x - \alpha_2)\cdots(x - \alpha_n)(\text{mod } m)$.

Proof: Follows from Lemmas 3.4 and 3.5.

Theorem (3.30): *Every congruence $f(x) \equiv 0(\text{mod } m)$, where $f(x) = a_n x^n + a_{n-1}x^{n-1} + \cdots + a_1 x + a_0$, m is prime such that $n < m$, cannot have more than n incongruent solutions unless $f(x) \equiv 0(\text{mod } m)$ identically.*

Proof: Let $\alpha_1, \alpha_2, \ldots, \alpha_n$ be n incongruent solutions of

$$f(x) \equiv 0(\text{mod } m), \tag{1}$$

then by Lemma [3.5],

$$f(x) \equiv a_n(x-\alpha_1)(x-\alpha_2)\cdots(x-\alpha_n)(\text{mod } m). \quad (2)$$

If possible, let α_{n+1} be another solution of (1) such that $\alpha_{n+1} \not\equiv \alpha_r(\text{mod } m)$ for $r = 1, 2, \ldots, n$.

Then $\quad f(\alpha_{n+1}) \equiv a_n(\alpha_{n+1}-\alpha_1)(\alpha_{n+1}-\alpha_2)\cdots(\alpha_{n+1}-\alpha_n)(\text{mod } m)$.

Also $\quad f(\alpha_{n+1}) \equiv 0(\text{mod } m)$.

Therefore, $a_n(\alpha_{n+1}-\alpha_1)\cdots(\alpha_{n+1}-\alpha_n) \equiv 0(\text{mod } m)$

and hence $a_n \equiv 0(\text{mod } m)$, since $\alpha_{n+1} \not\equiv \alpha_r(\text{mod } m)$ for $r = 1, 2, \ldots, n$.

Then $f(x) \equiv 0(\text{mod } m)$ reduced to the form

$$a_{n-1}x^{n-1} + a_{n-2}x^{n-2} + \cdots + a_1^x + a_0 \equiv 0(\text{mod } m),$$

which has more than $(n-1)$ incongruent solutions (mod m). Then proceeding as above we can show that $a_{n-1} \equiv 0(\text{mod } m), \cdot a_{n-2} \equiv 0(\text{mod } m), \ldots, a_0 \equiv 0(\text{mod } m)$ and hence $f(x) \equiv 0(\text{mod } m)$ identically.

This completes the proof.

Corollary (3.13): $f(x) \equiv 0(\text{mod } m)$ may have more than n incongruent solutions when $f(x) = a_nx^n + a_{n-1}x^{n-1} + \cdots + a_1x + a_0$ and m is not a prime number.

Problem (3.26): Solve the congruence $x^2 - 9 \equiv 0(\text{mod } 16)$.

Solution: Solutions of $x^2 - 9 \equiv 0(\text{mod } 16)$ belong to S.

$$S = \{0, 1, 2, \ldots, 15\}$$

Now $\quad x = 3 \Rightarrow 3^2 - 9 \equiv 0(\text{mod } 16)$

$x = 13 \Rightarrow 13^2 - 9 \equiv 0(\text{mod } 16)$

$x = 5,\ 5^2 - 9 \equiv 0(\text{mod } 16)$

$x = 11,\ 121 - 9 \equiv 0(\text{mod } 16)$

$\therefore \quad x \equiv 3, 5, 11, 13(\text{mod } 16) \equiv 3, 5, 11, 13(\text{mod } 16)$ are four solutions.

The congruence $x^2 - 9 \equiv 0(\text{mod } 16)$ has more than 2 solutions.

Problem (3.27): Show that the congruence $x^2 \equiv 1(\text{mod } 8)$ has exactly four solutions.

Solution: $x^2 - 1 \equiv 0(\text{mod } 8)$

8 is a composite number.

Solutions $x \in S = \{0, 1, 2, 3, 4, 5, 6, 7\}$

Here $\quad x = 1, 3, 5, 7$

$\Rightarrow \quad 1^2 - 1 \equiv 0(\text{mod } 8)$

$3^2 - 1 \equiv 0(\text{mod } 8)$

$5^2 - 1 \equiv 0(\text{mod } 8)$

$7^2 - 1 \equiv 0(\text{mod } 8)$.

$\therefore\ x \equiv 1, 3, 5, 7(\text{mod } 8)$ are four solutions.

Theorem (3.31): *If $f(x) = a_nx^n + a_{n-1}x^{n-1} + \cdots + a_1x + a_0$, then $f(x) \equiv 0(\text{mod } m)$, where m is prime and degree of $f(x) < m$ has exactly n solutions, if and only if $f(x)|x^m - x$ mod (m).*

Proof: Let $x^m - x$ be completely divisible by $f(x)$ with modulo m. Then there exists a polynomial $g(x)$ of degree $(m-n)$ such that $x^m - x \equiv f(x)g(x)(\text{mod } m)$.

Now $x^m - x \equiv 0 (\text{mod } m)$ has m solutions $0, 1, 2, \ldots, m-1$.

$g(x) \equiv 0(\text{mod } m)$ has at most $(m-n)$ incongruent solutions by Theorem [3.28].

As a consequence of these two, $f(x) \equiv 0(\text{mod } m)$ has exactly $m-(m-n) = n$ solutions.

Conversely, let $f(x) \equiv 0(\text{mod } m)$ have exactly n incongruent solutions.

If possible, let $x^m - x$ not be completely divisible by $f(x)$ with (mod m). Then by Remainder Theorem,

$$x^m - x \equiv f(x)g(x) + r(x)(\text{mod } m), \tag{1}$$

where $r(x)$ is a polynomial of degree $< n$.

Let a be a solution of $f(x) \equiv 0(\text{mod } m)$. Then $f(a) \equiv 0(\text{mod } m)$. Also $a^m - a \equiv 0$ (mod m).

Therefore, from (1) it follows that $r(a) \equiv 0(\text{mod } m)$, which shows that every solution of $f(x) \equiv 0(\text{mod } m)$ is a solution of $r(x) \equiv 0(\text{mod } m)$ and it also follows that $r(x) \equiv 0(\text{mod } m)$ has n solutions by our assumption. But the degree of $r(x) < n \Rightarrow r(x) \equiv 0(\text{mod } m)$ identically and hence from (1) it follows that $x^m - x$ is completely divisible by $f(x)$ with (mod m).

Problem (3.28): Show that $2x^3 + x^2 + 3x - 1 \equiv 0(\text{mod } 5)$ has exactly 3 solutions.

Solution: By Fermat's Theorem $x^5 - x \equiv 0(\text{mod } 5)$ has 5 incongruent solutions 0, 1, 2, 3, 4. Now dividing $x^5 - x$ by $f(x) = 2x^3 + x^2 + 3x - 1(\text{mod } 5)$, we have

2 1 3 −1	1	0	0	0	−1	0			$1 \equiv 6(\text{mod } 5)$
3 1	6	0	0	0	−1	0			
	6	3	9	−3					$-3 \equiv 2(\text{mod } 5)$
		−3	−9	3	−1				$-9 \equiv 1(\text{mod } 5)$
		2	1	3	−1				
$\phi(x) = 3x^2 + x$		2	1	3	−1				
		0	0	0	0				

Then $(2x^3 + x^2 + 3x - 1)(3x^2 + x) \equiv (x^5 - x)(\text{mod } 5)$.

Therefore, $x^5 - x$ is completely divisible by $f(x)(\text{mod } 5)$. So $f(x) \equiv 0(\text{mod } 5)$ has exactly 3 solutions. These 3 solutions $\in S = \{0, 1, 2, 3, 4\}$.

Here $x = 1, 2, 4$, satisfies $f(x) \equiv 0(\text{mod } 5)$.

Hence, $x \equiv 1, 2, 4(\text{mod } 5)$ are its only solutions.

Exercise 3

1. Show that $1! + 2! + 3! + \cdots + 100! \equiv 9(\text{mod } 12)$.
2. Show that $5n^3 + 7n^5 \equiv 0(\text{mod } 12)$ for all integers n.
3. If $N = a_0 + a_1 t + a_2 t^2 + \cdots + a_n t^n$,

   $S = a_0 - a_1 + a_2 - \cdots + (-1)^n a_n$,

   where $t = 10\,000$ and $0 \le a_r < t$.

   Prove that $N \equiv S(\text{mod } 73)$ or $N \equiv S(\text{mod } 137)$.
4. If p and q are different primes, then $p^{q-1} + q^{p-1} \equiv 1(\text{mod } pq)$.
5. Show that $18! + 1 \equiv 0(\text{mod } 437)$.
6. The nineth power of any number n is of the form $19m$, $19m \pm 1$.

7. Show that $5^{2n} + 3 \cdot 2^{5n-2} \equiv 0 \pmod 7$, $n \geq 1$.

8. If $a \equiv b \pmod n$ and the integers a, b, n are all divisible by $d (> 0)$, then $\frac{a}{d} \equiv \frac{b}{d} \left(\text{mod } \frac{m}{d}\right)$.

9. Give an example that $a^2 \equiv b^2 \pmod n$ need not imply $a \equiv b \pmod n$.

10. If $a \equiv b \pmod n$, prove that $(a, n) = (b, n)$.

11. If $(a, n) = 1$, then prove that the integers $c, c + a, c + 2a, \ldots, c + (n-1)a$ form a complete set of residues modulo n for any c.

12. If $a \equiv b \pmod{n_1}$, $a \equiv c \pmod{n_2}$, then prove that $b \equiv c \pmod n$ where $n = n_1 \cdot n_2$.

13. Show that

(a) $a^{21} \equiv a \pmod{15}\ \forall\, a$

(b) $a^{13} \equiv a \pmod{3.7.13}\ \forall a$.

14. If $7 \nmid a$, then prove that either $a^3 + 1$ or $a^3 - 1$ is divisible by 7.

15. Prove that the odd prime divisors of the integer $n^2 + 1$ are of the form $4k + 1$.

16. Prove that an integer n is divisible by 3, if and only if the sum of the digits is divisible by 3.

17. Prove that the 8th power of any integer is of the form $17k$ or $17k \pm 1$.

18. If p is prime, show that $2\,\underline{|p-3} + 1$ is divisible by p.

19. Solve :

(a) $29x \equiv 1 \pmod{13}$.

(b) $2x \equiv 3 \pmod 7$ and $3x \equiv 5 \pmod{11}$.

20. Arrange the numbers 2, 3, 4, 5,. . . ,15 in pairs (a, a') such that for each pair (a, a'), $a \cdot a' \equiv 1 \pmod{17}$.

21. Show that the congruence $x^2 \equiv 2 \pmod 5$ has no solution.

Chapter 4

Diophantine Equation

Linear Diophantine Equation and its Solution

Definition (4.1): *A linear equation of the form $ax + my = b$ where $a, m, b \in Z$ and a and m are not both zero, having only the integral solutions of x and y is known as 'Linear Diophantine equation'.*

Example: $8x + 4y = 12$ is a linear Diophantine equation.

It can have number of solutions, followed by

$$\begin{aligned}
8 \cdot 1 + 4 \cdot 1 = 12 &\Rightarrow (x, y) = (1, 1) \\
8 \cdot 4 + 4(-5) = 12 &\Rightarrow (x, y) = (4, -5) \\
8 \cdot 3 + 4(-3) = 12 &\Rightarrow (x, y) = (3, -3) \\
8(-2) + 4 \cdot 7 = 12 &\Rightarrow (x, y) = (-2, 7)
\end{aligned}$$

and many others.

Again, the linear Diophantine equation of the form: $2x + 4y = 11$ has no solution, since the L.H.S. is always an even number for any values of $x, y \in Z$, while the R.H.S. is an odd number and this is not possible.

Thus we observe that a linear Diophantine equation may or may not have solutions. So we shall discuss the situation of the existence and non-existence of the solution of a linear Diophantine equation and the ways of finding the solutions, if it exists.

Theorem (4.1): *If $(a, m) = d$, then a Diophantine equation $ax + my = b$ is solvable if and only if $d|b$.*

Proof: Since $(a, m) = d$, $\exists\ x_1, y_1 \in Z$ such that

$$ax_1 + my_1 = d. \tag{1}$$

Now multiplying both sides of (1) by $\frac{b}{d}$ we get

$$a \cdot \frac{b}{d}x_1 + m \cdot \frac{b}{d}y_1 = b. \tag{2}$$

Let $\quad d|b.$

Then $b = dk$ for some $k \in Z$.

From (2), $a \cdot \frac{dk}{d}x_1 + m \cdot \frac{dk}{d}y_1 = b \Rightarrow akx_1 + mky_1 = b.$

Therefore, $x = kx_1$, $y = ky_1$ are the integral solution of the given equation.

Next, let (x_0, y_0) be the integral solution of the given equation $ax + my = b$.

Then $\quad ax_0 + my_0 = b. \quad (3)$

Since $\quad (a, m) = d,\ d|a \quad \text{and} \quad d|m.$

Therefore, $a = da'$ and $m = dm'$ for some

$$a', m' \in Z \quad \text{and} \quad (a', m') = 1.$$

The equation (3) is reduced to the form $da'x_0 + dm'y_0 = b$

$$da'x_0 + dm'y_0 = b$$
$$\Rightarrow \quad d(a'x_0 + m'y_0) = b$$
$$\Rightarrow \quad dp = b, \quad \text{where } p = a'x_0 + m'y_0 \in Z.$$
$$\Rightarrow \quad d|b.$$

So the solution exists, when $d|b$.

Note: We, therefore, conclude that the linear equation $ax+my = b$ is always solvable, if $(a, m) = 1$.

So the general form a linear Diophantine equation can be put in the form:

$$ax + my = b, \quad (a, m) = 1.$$

Theorem (4.2): *If (x_0, y_0) be a solution of $ax+my = b$, $(a, m) = d$, then the general solution of it is given by $\left(x_0 + \frac{m}{d}t,\ y_0 - \frac{a}{d}t\right)$, where $t \in Z$ and conversely.*

Proof: Let (x_0, y_0) be a solution of $ax + my = b$, $(a, m) = d$.

$$\Rightarrow \quad d|a \quad \text{and} \quad d|m$$
$$\Rightarrow \quad a = da',\ m = dm',\ a', m' \in Z \quad \text{and} \quad (a', m') = 1.$$

Then $\quad ax_0 + my_0 = b. \quad$ (1)

Now let (u_0, v_0) be another solution of $ax + my = b$ differing from the given solution (x_0, y_0).

Then $\quad au_0 + mv_0 = b = ax_0 + my_0 \quad$ [by (1)]

$$\Rightarrow \quad da'(u_0 - x_0) = dm'(y_0 - v_0)$$
$$\Rightarrow \quad a'(u_0 - x_0) = m'(y_0 - v_0)$$
$$\Rightarrow \quad m'|a'(u_0 - x_0)$$
$$\Rightarrow \quad m'|(u_0 - x_0), \quad \text{since } (a', m') = 1$$
$$\Rightarrow \quad u_0 - x_0 = m't \text{ for some } t \in Z$$
$$\Rightarrow \quad u_0 = x_0 + m't, t \in Z. \quad (2)$$

Again, $a'(u_0 - x_0) = m'(y_0 - v_0) \ \Rightarrow\ a'm't = m'(y_0 - v_0)$

$$\Rightarrow \quad a't = y_0 - v_0 \ \Rightarrow\ v_0 = y_0 - a't, t \in Z. \quad (3)$$

Therefore, $(u_0, v_0) = (x_0 + m't,\ y_0 - a't)$ which is equal to $\left(x_0 + \dfrac{m}{d}t,\ y_0 - \dfrac{a}{d}t\right), t \in Z$.

So the general solution is given by $x_0 + \dfrac{m}{d}t,\ y_0 - \dfrac{a}{d}t$.

Converse: Putting $x = x_0 + \dfrac{m}{d}t,\ y = y_0 - \dfrac{a}{d}t$ in $ax + my = b$, we get

$$a\left(x_0 + \frac{m}{d}t\right) + m\left(y_0 - \frac{a}{d}t\right) = b \quad \text{or,} \quad ax_0 + my_0 = b,$$

which shows that (x_0, y_0) is a solution of $ax + my = b, (a, m) = d$.

Note: The general solution of $ax + my = b$, when $(a, m) = 1$ may be expressed as $x_0 + mt,\ y_0 - at$ by taking $d = 1$ in the above result.

Theorem (4.3): *If $x = x_0$ be the solution of the Diophantine equation $ax + my_0 = b, (a, m) = 1$, for some integer y_0, then it is also a solution of the congruence $ax \equiv b (\text{mod } m)$ and conversely.*

Proof: Since $x = x_0$ is a solution of $ax + my_0 = b$ for some integer y_0, we have

$$ax_0 + my_0 = b \quad \text{or.} \quad ax_0 = b - my_0$$

or, $\quad ax_0 \equiv b (\text{mod } m), \quad - y_0 \in Z,$

which shows that x_0 is a solution of $ax \equiv b (\text{mod } m)$, $(a, m) = 1$.

Conversely, let x_0 be a solution of $ax \equiv b (\text{mod } m)$.

Then $\quad ax_0 \equiv b (\text{mod } m)$

$\Rightarrow \quad ax_0 - b$ is divisible by m.

$\Rightarrow \quad ax_0 - b = m(-y_0)$ for some integer y_0.

$\Rightarrow \quad ax_0 + my_0 = b,$

which shows that $x = x_0$ is a solution of $ax + my_0 = b$ for some integer y_0.

Note: In finding the solution of a Diophantine equation $ax + my = b$, $(a, m) = 1$, we follow the following methods:

1. Substitution method.
2. Simple continued fraction method.
3. Euclidean algorithm method.

Problem (4.1): Solve : Find the integral solution of $68x - 157y = 1$.
[Method of substitution]

Solution: Here $(68, 157) = 1$

$$x = \frac{157y + 1}{68} = \frac{(68 \times 2 + 21)y + 68 - 67}{68} = \frac{68(2y + 1) + 21y - 67}{68}$$

$$= 2y + 1 + \frac{21y - 67}{68}.$$

$2y + 1 \in Z.$

We now put an integer $u = \dfrac{21y - 67}{68}$

$$\Rightarrow \quad y = \frac{68u + 67}{21} = \frac{(21 \times 3 + 5)u + 63 + 4}{21}$$

$$= \frac{21(3u + 3) + 5u + 4}{21} = 3u + 3 + \frac{5u + 4}{21}.$$

$3u + 3 \in Z.$

We put an integer $v = \dfrac{5u + 4}{21}$.

$$\Rightarrow \quad u = \frac{21v - 4}{5} = \frac{(5 \times 4 + 1)v - 5 + 1}{5}$$

$$= \frac{5(4v - 1) + v + 1}{5} = 4v - 1 + \frac{v + 1}{5}.$$

$4v - 1 \in Z.$

We put an integer $w = \dfrac{v + 1}{5}$.

$\Rightarrow \quad v = 5w - 1.$

Now for $w = 0$, $v = -1$ (we get the integral value of v)

So when $v = -1$, $u = \dfrac{21v - 4}{5} = -5$

$$u = -5,\ y = \frac{68u + 67}{21} = \frac{-340 + 67}{21} = -13$$

$$y = -13,\ x = \frac{157y + 1}{68} = \frac{-157 \times 13 + 1}{68} = -30$$

$(x, y) = (-30, -13)$.

General solution: $\left.\begin{aligned} x &= -30 - 157t \\ y &= -13 - 68t \end{aligned}\right\} t \in Z$

For $t = -1$, we get the (+)ve integral solution,

$$x = -30 + 157 = 127$$

$$y = -13 + 68 = 55.$$

[**Note:** Substitution is continued until we get the integral values of the variables]

Next, **Solution of $68x - 157y = 1$** (1)

by the method of simple continued Fraction.

We express $\dfrac{68}{157}$ as a S.C.F. into even number of quotients.

0	68	157	2
	0	136	
3	68	21	4
	63	20	
5	5	1	
	5		
	0		

$$\frac{68}{157} = 0 + \frac{1}{2+}\ \frac{1}{3+}\ \frac{1}{4+}\ \frac{1}{5}$$

$$= 0 + \frac{1}{2+}\ \frac{1}{3+}\ \frac{1}{4+}\ \frac{1}{4+}\ \frac{1}{1}$$

$$\frac{p_5}{q_5} = 0 + \frac{1}{2+}\ \frac{1}{3+}\ \frac{1}{4+}\ \frac{1}{4} = \frac{55}{127} = \frac{y_0}{x_0}.$$

Then $68 \times 127 - 157 \times 55 = 1$.

Thus $(x_0, y_0) = (127, 55)$ is a (+)ve integral solution of (1)

General solution, is $x = x_0 - 157t = 127 - 157t$

$$y = y_0 - 68t = 55 - 68t$$

$$t = -1,\ x = -30,\ y = -13,\ (x, y) = (-30, -13)$$

So x, y are equivalent to $x = -30 - 157t$, $y = -13 - 68t$.

Again, **solution of (1) by the Method of Euclidean Algorithm:**

By divisional algorithm, we have

$$157 = 68 \times 2 + 21$$

$$68 = 21 \times 3 + 5$$

$$21 = 5 \times 4 + 1$$

$$5 = 5 \times 1 + 0.$$

Now $21 = 5 \times 4 + 1$

$\Rightarrow \quad 1 = 21 - 5 \times 4 = 21 - 4(68 - 21 \times 3) = 21 \times 13 - 4 \times 68$

$= 13 \times (157 - 68 \times 2) - 68 \times 4 = 13 \times 157 - 30 \times 68$

$= -30 \times 68 - 157(-13),$

which shows that $x = -30,\ y = -13$ satisfies (1).

$\therefore\ (x, y) = (-30, -13)$ is an integral solution of (1).

General solution: $x = -30 - 157t$

$y = -13 - 68t.$

for $t = -1,\ x = 127,\ y = 55.\ (x, y) = (127, 55)$ is a (+)ve integral solution of (1).

Problem (4.2): Solve $13x - 17y = 5$ by S.C.F. method; $(a, m) = (13, 17) = 1$.

Solution: $\frac{13}{17} = 0 + \frac{1}{1+}\ \frac{1}{3+}\ \frac{1}{4}$ containing even number of quotients.

$\frac{p_3}{q_3} = 0 + \frac{1}{1+}\ \frac{1}{3} = \frac{3}{4} = \frac{y_0}{x_0}.$

Then $13 \cdot 4 - 17 \cdot 3 = 1$.

Now multiplying by 5,

$13 \cdot 20 - 17 \cdot 15 = 5,$

which shows that $x = 20,\ y = 15$ satisfies the given equation.

$(x, y) = (20, 15)$ is a (+)ve integral solution of it.

General solution is $\left.\begin{aligned} x &= 20 - 17t \\ y &= 15 - 13t \end{aligned}\right\}\ t \in Z$

$t = 1,\ x = 3,\ y = 2$

$(x, y) = (3, 2)$ is the least (+)ve integral solution.

Problem (4.3): Solve : $56x + 72y = 40$ by E. Alg. method.

Solution: $(a, m) = (56, 72) = 8(= d)$

Also $d|40$. Solution exists.

$72 = 56 \times 1 + 16$

$56 = 16 \times 3 + 8$

$16 = 8 \times 2 + 0$

$8 = 56 - 16 \times 3 = 56 - 3 \times (72 - 56 \times 1) = 56 - 3 \times 72 + 56 \times 3$

$= 56 \times 4 - 3 \times 72 = 56 \times 4 - 72 \times 3.$

Multiplying by 5,

$40 = 56 \times 20 - 72 \times 15.$

Then $x = 20,\ y = -15$ satisfies the given equation.

$(x, y) = (20, -15)$ is an integral solution of it.

General solution is $x = 20 + \frac{72}{8}t = 20 + 9t$

$y = -15 - -\frac{56}{8}t = -15 - 7t.$

Problem (4.4): Solve : $8x - 27y = 125$.

Solution: Here $(a, m) = (8, 27) = 1$.

$$x = \frac{27y + 125}{8} = \frac{(8 \times 3 + 3)y + 8 + 117}{8}$$

$$= \frac{8(3y + 1) + 3y + 117}{8} = 3y + 1 + \frac{3y + 117}{8}.$$

$3y + 1 \in Z$. We put an integer $u = \dfrac{3y + 117}{8}$

$$\therefore \quad y = \frac{8u - 117}{3} = \frac{(3 \times 2 + 2)u + 3 - 120}{3}$$

$$= \frac{3(2u + 1) + 2u - 120}{3} = 2u + 1 + \frac{2u - 120}{3}.$$

$2u + 1 \in Z$. We put an integer $v = \dfrac{2u - 120}{3}$.

So $\quad u = \dfrac{3v + 120}{2} = \dfrac{(2 \times 1 + 1)v + 2 + 118}{2} = (v + 1) + \dfrac{v + 118}{2}$.

$v + 1 \in Z$, we put an integer $w = \dfrac{v + 118}{2}$

That is $v = 2w - 118$.

$$w = 0 \Rightarrow v = -118$$

$$v = -118, \ u = \frac{3v + 120}{2} = \frac{-354 + 120}{2} = -117$$

$$u = -117, \ y = \frac{8u - 117}{3} = \frac{8(-117) - 117}{3} = \frac{9 \times (-117)}{3}$$

$$= 3 \times (-117) = -351.$$

$$x = \frac{27y + 125}{8} = \frac{27(-351) + 125}{8} = \frac{-9352}{8} = -1169.$$

$(x, y) = (-1169, -351)$ is an integral solution of the given equation.

General solution is $\left.\begin{aligned} x &= -1169 - 27t \\ y &= -351 - 8t \end{aligned}\right\} t \in Z$

Now for $t = -44$, $x = -1169 + 27 \times 44 = -1169 + 1188 = 19$

$$y = -351 + 8 \times 44 = -351 + 352 = 1$$

Least (+)ve integral solution is $(x, y) = (19, 1)$.

Problem (4.5): The sum of two positive integers is 100. If one is divided by 7 the remainder is 1, and if the other is divided by 9 the remainder is 7. Find the numbers.

Solution: Let p and q be two positive integers such that $p + q = 100$. (1)

When 'p is divided by 7 the remainder is 1'.

We have $p = 7x + 1$, when x is a (+)ve integer. (2)

When 'q is divided by 9 the remainder is 7'.

Then $\quad q = 9y + 7$, where y is a (+)ve integer. (3)

Therefore, from (1) by (2) & (3), we have $7x + 1 + 9y + 7 = 100$

or, $7x + 9y = 92,\ (7,9) = 1,$ (4)

which is a Linear Diophantine equation. Applying Euclidean Algorithm method,

$$9 = 1 \cdot 7 + 2$$
$$7 = 3 \cdot 2 + 1$$
$$2 = 1 \cdot 2 + 0$$
$$1 = 7 - 3 \cdot 2 = 7 - 3[9 - 1 \cdot 7] = 7 \cdot 4 - 9 \cdot 3$$

Multiplying by 92,

$$7 \cdot 368 + 9(-276) = 92,$$

which shows that $x = 368 (= x_0)$

$$y = -276 (= y_0)$$

is a solution of (4).

General solution is $x = x_0 + mt = 368 + 9t$

$$y = y_0 - at = -276 - 7t$$

$$t = -40,\quad x = 368 - 360 = 8$$
$$y = -276 + 280 = 4.$$

$(x, y) = (8, 4)$ are the least (+)ve integral solution of (4).

Hence $p = 7x + 1 = 7 \cdot 8 + 1 = 57$

$$q = 9y + 7 = 9 \cdot 4 + 7 = 43.$$

That is, numbers are 57 and 43.

Exercise 4

1. Show that $(3, -5)$ is a solution of $7x - 5y = 46$. Find the general solution of it.
2. Are the following Diophantine equations solvable? If not, state the reason.

   (a) $6x + 51y = 22$

   (b) $221x + 35y = 11$.
3. A certain number of sixes and nines is added to give a sum of 126. If the number of sixes and nines is interchanged, the new sum is 114. How many sixes and nines were there originally?
4. Applying Euclidian algorithm find the general solution of $112x + 70y = 168$.
5. Apply simple continued fraction method to determine the solution of $24x + 138y = 18$.
6. Apply Chinese remainder theorem to solve the system,

$$x \equiv 3 \pmod 6),\ x \equiv 5 \pmod 7,\ x \equiv 2 \pmod{11}.$$

7. Show that the solution of the system $x \equiv a \pmod{21}$, $x \equiv b \pmod{16}$ is $x \equiv 64a - 63b \pmod{336}$.
8. Solve the system of congruences $x \equiv 1 \pmod 3$, $x \equiv 2 \pmod 4$, $x \equiv 3 \pmod 5$.

Chapter 5

Number Theoretic Functions

Definition (5.1): *Let $N = \{x,\ x$ is any $(+)$ve integer$\}$ and $R = \{x;\ x$ is any real number$\}$, $C = \{x,\ x$ is any complex number$\}$. Then $f : N \to R$ (or C) is defined to be a Number theoretic or Arithmetic function.*

It is to be noted that the value of a number theoretic function is not necessarily a (+)ve integer or an integer only, it may assume any real or complex values.

Most of the number theoretic functions, which we shall discuss here, assume only integral values.

Here we shall discuss two important number theoretic (Arithmetic) functions.

(i) Euler-Totient function or Phi function.

(ii) Mobius function.

Definition (5.2): *A number theoretic function f is said to be multiplicative, if $f(m_1, m_2) = f(m_1) \cdot f(m_2)$, whenever $(m_1, m_2) = 1$.*

Definition (5.3): *A number theoretic function f is said to be completely multiplicative, if $f(m_1, m_2) = f(m_1) \cdot f(m_2)\ \forall\ m_1, m_2 \in N$.*

Theorem (5.1): *If f be a multiplicative function which is not identically zero, then $f(1) = 1$.*

Proof: Since f is not identically zero, $\exists\ n \in Z^+$ such that $f(n) \neq 0$.

But $f(n) = f(n \cdot 1) = f(n) \cdot f(1)$.

Cancelling $f(n)$, we have $f(1) = 1$.

Hence the theorem.

The functions $\tau(n)$ and $\sigma(n)$

Definition (5.4): *Let $n \in N$. Then $\tau(n)$ and $\sigma(n)$ are defined by*

(i) $\tau(n)$ denotes the number of (+)ve divisors of n

and (ii) $\sigma(n)$ denotes the sum of the (+)ve divisors of n.

Example: If $n = 20$,

then values of d, $d|20$ are 1, 2, 4, 5, 10, 20 ($d \in Z^+$).

Therefore, $\tau(20) = 6$,

while $\sigma(20) = 1 + 2 + 4 + 5 + 10 + 20 = 42$.

Also $\tau(1) = 1,\ \Sigma(1) = 1$.

From the definitions of τ and σ, they may be represented by

$$\tau(n) = \sum_{d|n} 1,\ \sigma(n) = \sum_{d|n} d$$

The notation $\sum_{d|n} 1$, in particular, expresses that we are to add together as many 1's as there are (+)ve divisors of n. That is, we sum up by writing 1 to each divisor of n.

$$\tau(20) = \sum_{d|20} 1 = 1+1+1+1+1+1 = 6.$$

[We write 1 for each d, $d = 1, 2, 4, 5, 10, 20$]

Theorem (5.2): *If $n = p_1^{a_1} p_2^{a_2} \dots P_k^{a_k}$ $(n > 1)$, where p_i are prime numbers and $a_i \in Z^+$. Then*

(i) $\tau(n) = (a_1+1)(a_2+1)\dots(a_k+1)$

(ii) $\sigma(n) = \dfrac{p_1^{a_1+1}-1}{p_i-1} \cdot \dfrac{p_2^{a_2+1}-1}{p_2-1} \cdots \dfrac{p_k^{a_k+1}-1}{p_k-1}$

***Proof*:** See Th. (2.8).

Theorem (5.3): *Both the functions τ and σ are multiplicative.*

***Proof*:** Let $(m, n) = 1$.

Case(I): If $m = 1$, $n = 1$,

$$\tau(mn) = \tau(1) = 1 = 1 \cdot 1 = \tau(1) \cdot \tau(1) = \tau(m) \cdot \tau(n)$$

$$\sigma(mn) = \sigma(1) = 1 = 1 \cdot 1 = \sigma(m) \cdot \sigma(n).$$

Case(II): Let $m > 1$ and $n > 1$.

Then we write $m = p_1^{a_1} p_2^{a_2} \dots p_k^{a_k}$

and $n = q_1^{b_1} q_2^{b_2} \dots q_l^{b_l}$

where p_i, q_i are primes and no p_i may be equal to q_i for any i. Also $a_i \geq 0$, $b_i \geq 0$.

$$mn = p_1^{a_1} p_2^{q_2} \dots p_k^{a_k} \cdot q_1^{b_1} q_2^{b_2} \dots q_l^{b_l}$$

$$\tau(mn) = [(a_1+1)(a_2+1)\dots(a_k+1)][(b_1+1)(b_2+1)\dots(b_l+1)]$$

[by Th. (5.2)]

$$= \tau(m) \cdot \tau(n).$$

Again, $\sigma(mn) = \left(\dfrac{p_1^{a_1+1}-1}{p_1-1} \cdot \dfrac{p^{a_2+1}-1}{p_2-1} \cdots \dfrac{p_k^{a_{k+1}}-1}{p_k-1}\right) \cdot \left(\dfrac{q_1^{b_1+1}-1}{q_1-1} \cdots \dfrac{q^{b_l+1}-1}{q_l-1}\right)$

[by Th. (5.2)]

$$= \sigma(m) \cdot \sigma(n).$$

Euler-Totient function (or Phi-function)

Definition (5.5): *If $n \geq 1$, then the Euler-Totient function, denoted by $\phi(n)$, is defined by the numbers of (+)ve integers which are less than and prime to n.*

That is, $\phi(n) = \sum_{k=1}^{n} 1$, where 1 is taken for each k satisfying $(k, n) = 1$ and $k < n$.

Example: $n = 12$

$k = 1, 5, 7, 11$, where $k < n$ and $(k, n) = 1$.

Then $\phi(12) = 1+1+1+1+ = 4$ $\left[\sum_{k=1}^{12} 1 = \phi(12),\ k < n,\ (k, n) = 1\right]$

A short table for $\phi(n)$, given as follows :

n	1	2	3	4	5	6	7	8	9	10	11	12	13	14	15
$\phi(n)$	1	1	2	2	4	2	6	4	6	4	10	4	12	6	8

It is to be noted that when n is prime, $\phi(n) = n - 1$.

Theorem (5.4): *Euler-Totient function has the following properties:*

1. $\phi(p^a) = p^a\left(1 - \frac{1}{p}\right)$ *for any prime p and* $a \geq 1$.
2. $\phi(mn) = \phi(m)\phi(n)$, $(m, n) = 1$.
3. *If the integer* $n(> 1)$ *is of the form* $n = p_1^{a_1} P_2^{a_2} \ldots p_k^{a_k}$, *where each* $p_i, i = 1, 2, \ldots, k$, *is distinct prime number,*

   then $\phi(n) = n\left(1 - \dfrac{1}{p_1}\right)\left(1 - \dfrac{1}{p_2}\right)\cdots\left(1 - \dfrac{1}{p_k}\right) = n\displaystyle\prod_{r=1}^{k}\left(1 - \frac{1}{p_r}\right).$
4. $\phi(mn) = \phi(m)\phi(n)\dfrac{d}{\phi(d)}$ *when* $(m, n) = d$.
5. $a|b \Rightarrow \phi(a)|\phi(b)\ \forall\ a, b \in Z$.

(Go through the following Example.)

Let $a = 2$ and $p = 7$ (prime number)

Then the integers which are not prime to $7^2 = 49$ are 7, 14, 21, 28, 35, 42, 49 which we write in the forms $1 \cdot 7, 2 \cdot 7, 3 \cdot 7, 4 \cdot 7, 5 \cdot 7, 6 \cdot 7, 7 \cdot 7$.

Therefore, the numbers which are less than and prime to 7^2 are $7^2 - 7^{2-1} = 7^2\left(1 - \frac{1}{7}\right)$. Thus $\phi(7^2) = 7^2\left(1 - \frac{1}{7}\right)$.

***Proof*: (1)** The integers which are not prime to p^a are p, $2p$, $3p$, $\ldots$, $p^{a-1} \cdot p$.

These numbers are p^{a-1}.

Therefore, the numbers which are less than and prime to $p^a = p^a - p^{a-1}$.

That is $\phi(p^a) = p^a\left(1 - \dfrac{1}{p}\right)$, $a \geq 1$.

***Proof*: (2) Case(1):** $m = 1$, $n = 1$

$$\phi(mn) = \phi(1) = 1 = \phi(1)\phi(1) = \phi(m)\phi(n).$$

Case(2): Let $m > 1$, $n > 1$, and $(m, n) = 1$.

In this case we arrange the integers from 1 to mn in m-columns and n-rows such that the integers in each column and in each row are in A.P. series. Elements in the rth column is of the form $mq + r, q = 0, 1, 2, \ldots, (n - 1)$.

We have $(mq + r, m) = (r, m)$, iff $(r, m) = 1$ [by Th.(1.16), Ch-1]

Therefore, the numbers in the rth column are relatively prime to m, iff $(r, m) = 1$.

Let $(r, m) = 1$. Then $\phi(m)$ column contains integers which are relatively prime to m. Again, each of the $\phi(m)$ columns satisfying $(r, m) = 1$, contains exactly $\phi(n)$ integers relatively prime to n.

Thus, there would be $\phi(m) \cdot \phi(n)$ numbers in the table given that are relatively prime to both m and n and consequently $\phi(m) \cdot \phi(n)$ numbers are relatively prime to mn [by Lemma 1.17A-Ch1]. Therefore, $\phi(m)\phi(n)$ numbers are less than and prime to mn.

Hence, $\phi(m)\phi(n) = \phi(mn)$.

Table (Arrangement of integers)

1	2	r	m
$m+1$	$m+2$	$m+r$	$2m$
$2m+1$	$2m+2$	$2m+r$	$3m$
$\cdots$	$\cdots$	$\cdots$	$\cdots$
$(n-1)m+1$	$(n-1)m+2$	$(n-1)m+r$	nm

Example: Let $m = 4$ and $n = 5$.

We arrange the set of integers from 1 to 20 in 4 columns and in 5 rows. That is,

1	2	3	4
$4+1$	$4+2$	$4+3$	$4+4$
$2\cdot4+1$	$2\cdot4+2$	$2\cdot4+3$	$2\cdot4+4$
$3\cdot4+1$	$3\cdot4+2$	$3\cdot4+3$	$3\cdot4+4$
$4\cdot4+1$	$4\cdot4+2$	$4\cdot4+3$	$4\cdot4+4$

That is,

1	2	3	4
5	6	7	8
9	10	11	12
13	14	15	16
17	18	19	20

$r = 1, 3$ where $(r, m) = 1$

$\phi(m) = \phi(4) = 2$, each member of Ist and 3rd columns is relatively prime to 4.

Again, each column contains

$\phi(n) = \phi(5) = 4$ integers relatively prime to 5.

In the Ist column, 1, 9, 13, 17 are relatively prime to 5 and 3rd column, 3, 7, 11, 19 are relatively prime to 5.

Therefore, $\phi(4)\phi(5) = 2 \cdot 4 = 8$.

Also $\phi(20) = 8$, numbers less than and prime to 20.

That is, the numbers in the 1st and the 3rd columns are 1, 9, 13, 17, 3, 7, 11, 19, which are relatively prime to both 4 and 5 and so prime to 20.

Proof: (3) Since each p_i is distinct prime numbers, $(p_i, p_j) = 1$, which follows $(p_i^{a_i}, p_j^{a_j}) = 1 \;\forall\; i, j$.

Thus $(p_1^{a_1} p_2^{a_2} \cdots p_{k-1}^{a_{k-1}}, p_k^{a_k}) = 1$

$\Rightarrow \quad \phi(n) = \phi(p_1^{a_1} p_2^{a_2} \cdots p_{k-1}^{a_{k-1}})\phi(p_k^{a_k})$ [by (2)]

Again $(p_1^{a_1} p_2^{a_2} \cdots p_{k-2}^{a_{k-2}}, p_{k-1}^{a_{k-1}}) = 1$.

So $\quad \phi(n) = \phi(p_1^{a_1} p_2^{a_2} \cdots p_{k-2}^{a_{k-2}})\phi(p_{k-1}^{a_{k-1}})\phi(p_k^{a_k})$.

Continuing the rule $(p_i^{a_i}, p_j^{a_j}) = 1$, we have

$$\phi(n) = \phi(p_1^{a_1})\phi(p_2^{a_2}) \cdots \phi(p_{k-1}^{a_{k-1}})\phi(p_k^{a_k})$$

$$= p_1^{a_1}\left(1 - \frac{1}{p_1}\right) p_2^{a_2}\left(1 - \frac{1}{p_2}\right) \cdots p_{k-1}^{a_{k-1}}\left(1 - \frac{1}{p_{k-1}}\right) p_k^{a_k}\left(1 - \frac{1}{p_k}\right)$$

[by (1)]

$$= p_1^{a_1} p_2^{a_2} \ldots p_k^{a_k}\left(1 - \frac{1}{p_1}\right)\left(1 - \frac{1}{p_2}\right)\cdots\left(1 - \frac{1}{p_k}\right)$$

$$= n\left(1 - \frac{1}{p_1}\right)\cdots\left(1 - \frac{1}{p_k}\right) = n\prod_{r=1}^{k}\left(1 - \frac{1}{p_r}\right).$$

Example: Let $n = 2520$. Then Find $\phi(2520)$.

Here $\quad n = 2520 = 2^3 \cdot 3^2 \cdot 5 \cdot 7$

$$\begin{aligned}\phi(2520) &= \phi(2^3\cdot 3^2\cdot 5\cdot 7) = \phi(2^3)\phi(3^2)\phi(5)\phi(7)\\ &= 2^3\cdot\left(1-\frac{1}{2}\right)3^2\left(1-\frac{1}{3}\right)5\left(1-\frac{1}{5}\right)7\left(1-\frac{1}{7}\right)\\ &= 2^3\cdot 3^2\cdot 5\cdot 7\cdot\frac{1}{2}\cdot\frac{2}{3}\cdot\frac{4}{5}\cdot\frac{6}{7}\\ &= 2^6\cdot 3^2 = 64\times 9 = 576.\end{aligned}$$

Proof: **(4)** Let $m = \prod_{i=1}^{k} p_i^{a_i}$ and $n = \prod_{j=1}^{l} q_j^{b_j}$, where p_i, q_j are distinct primes and a_i, b_j are non-negative integral powers of p_i and q_i respectively, and whatever may be k and l (that is, $k \geq l$ or, $k < l$).

Also let $(m, n) = d$.

If $p_r, p_s, p_t, \ldots$ be the divisors of d, then there exist $q_u, q_v, q_w, \ldots$ such that $p_r = q_u,\ p_s = q_v,\ p_t = q_w$, since d is the common divisor of m and n. $1 \leq r, s, t \leq k,\ 1 \leq u, v, w \leq l$. In the product $\phi(m)\cdot\phi(n)$ each of the factors $1-\frac{1}{p_r}, 1-\frac{1}{p_s}, 1-\frac{1}{p_t}\cdots$ will occur twice.

We have

$$\phi(m) = m\left(1-\frac{1}{p_1}\right)\cdots\left(1-\frac{1}{p_r}\right)\cdots\left(1-\frac{1}{p_s}\right)\cdots\left(1-\frac{1}{p_t}\right)\cdots\left(1-\frac{1}{p_k}\right)$$

$$\phi(n) = n\left(1-\frac{1}{q_1}\right)\cdots\left(1-\frac{1}{q_k}\right)\cdots\left(1-\frac{1}{q_v}\right)\cdots\left(1-\frac{1}{q_w}\right)\cdots\left(1-\frac{1}{q_l}\right)$$

$$1 \leq r, s, t \leq k,\ 1 \leq u, v, w \leq l.$$

$$\begin{aligned}\phi(m)\cdot\phi(n) = mn&\left(1-\frac{1}{p_1}\right)\cdots\left(1-\frac{1}{p_r}\right)^2\cdots\left(1-\frac{1}{p_s}\right)^2\cdots\left(1-\frac{1}{p_t}\right)^2\cdots\left(1-\frac{1}{p_k}\right)\\ &\times\left(1-\frac{1}{q_1}\right)\cdots\left(1-\frac{1}{q_{u-1}}\right)\left(1-\frac{1}{q_{u+1}}\right)\cdots\left(1-\frac{1}{q_{v-1}}\right)\left(1-\frac{1}{q_{v+1}}\right)\\ &\times\left(1-\frac{1}{q_{w-1}}\right)\left(1-\frac{1}{q_{w+1}}\right)\cdots\left(1-\frac{1}{q_l}\right) \quad (1)\end{aligned}$$

[since $q_u = p_r,\ q_v = p_s,\ q_w = p_t$]

Also, $d = p_r p_s p_t \Rightarrow \phi(d) = d\left(1-\frac{1}{p_r}\right)\left(1-\frac{1}{p_s}\right)\left(1-\frac{1}{p_t}\right)$

$$\Rightarrow \quad \frac{\phi(d)}{d} = \left(1-\frac{1}{p_r}\right)\left(1-\frac{1}{p_s}\right)\left(1-\frac{1}{p_t}\right). \quad (2)$$

Dividing (1) by (2), we get

$$\begin{aligned}\frac{\phi(m)\cdot\phi(n)}{\frac{\phi(d)}{d}} &= mn\left(1-\frac{1}{p_1}\right)\cdots\left(1-\frac{1}{p_r}\right)\cdots\left(1-\frac{1}{p_s}\right)\cdots\left(1-\frac{1}{p_t}\right)\cdots\\ &\times\left(1-\frac{1}{p_k}\right)\cdots\left(1-\frac{1}{q_1}\right)\cdots\left(1-\frac{1}{q_{u-1}}\right)\left(1-\frac{1}{q_{u+1}}\right)\cdots\left(1-\frac{1}{q_{v-1}}\right)\\ &\times\left(1-\frac{1}{q_{v+1}}\right)\cdots\left(1-\frac{1}{q_{w-1}}\right)\left(1-\frac{1}{q_{w+1}}\right)\cdots\left(1-\frac{1}{q_l}\right). \quad (3)\end{aligned}$$

$$\begin{aligned}\text{Now}\quad mn = p_1^{a_1}p_2^{a_2}&\ldots p_r^{a_r+b_u}\ldots p_s^{a_s+b_v}\ldots p_t^{a_t+b_w}\ldots p_k^{a_k}\\ &\times q_1^{a_1}q_2^{a_2}\ldots q_{u-1}^{b_{u-1}}q_{u+1}^{b_{u+1}}\ldots q_{v-1}^{b_{v-1}}q_{v+1}^{b_{v+1}}\ldots q_{w-1}^{b_{w-1}}q_{w+1}^{b_{w+1}}\ldots q_l^{b_l}\end{aligned}$$

$$\phi(mn) = mn\left(1-\frac{1}{p_1}\right)\cdots\left(1-\frac{1}{p_r}\right)\cdots\left(1-\frac{1}{p_s}\right)\cdots\left(1-\frac{1}{p_t}\right)\cdots\left(1-\frac{1}{p_k}\right)$$
$$\times\left(1-\frac{1}{q_1}\right)\cdots\left(1-\frac{1}{q_{u-1}}\right)\left(1-\frac{1}{q_{u+1}}\right)\cdots\left(1-\frac{1}{q_{v-1}}\right)\left(1-\frac{1}{q_{v+1}}\right)\cdots$$
$$\times\left(1-\frac{1}{q_{w-1}}\right)\left(1-\frac{1}{q_{w+1}}\right)\cdots\left(1-\frac{1}{q_l}\right). \quad (4)$$

From (3) and (4), we have $\dfrac{\phi(m)\cdot\phi(n)}{\frac{\phi(d)}{d}} = \phi(mn)$.

That is, $\phi(mn) = \phi(m)\phi(n)\cdot\dfrac{d}{\phi(d)}$.

Example: Let $m = 2\cdot 3^2\cdot 5^3\cdot 7\cdot 9\cdot 13^3\cdot 19$

$n = 5\cdot 7^2\cdot 17\cdot 19^3\cdot 21$

Then $d = (m,n) = 5\cdot 7\cdot 19$

$$\frac{\phi(d)}{d} = \left(1-\frac{1}{5}\right)\left(1-\frac{1}{7}\right)\left(1-\frac{1}{19}\right) \quad (1)$$

$$\phi(m) = m\cdot\left(1-\frac{1}{2}\right)\left(1-\frac{1}{3}\right)\left(1-\frac{1}{5}\right)\left(1-\frac{1}{7}\right)\left(1-\frac{1}{9}\right)\left(1-\frac{1}{13}\right)\left(1-\frac{1}{19}\right)$$

$$\phi(n) = n\left(1-\frac{1}{5}\right)\left(1-\frac{1}{7}\right)\left(1-\frac{1}{17}\right)\left(1-\frac{1}{19}\right)\left(1-\frac{1}{21}\right)$$

$$\phi(m)\cdot\phi(n) = mn\left(1-\frac{1}{2}\right)\left(1-\frac{1}{3}\right)\left(1-\frac{1}{5}\right)^2\left(1-\frac{1}{7}\right)^2\left(1-\frac{1}{9}\right)\left(1-\frac{1}{13}\right)$$
$$\times\left(1-\frac{1}{19}\right)^2\left(1-\frac{1}{17}\right)\left(1-\frac{1}{21}\right). \quad (2)$$

$$(2)\div(1): \frac{\phi(m)\cdot\phi(n)}{\frac{\phi(d)}{d}} = mn\left(1-\frac{1}{2}\right)\left(1-\frac{1}{3}\right)\left(1-\frac{1}{5}\right)\left(1-\frac{1}{7}\right)\left(1-\frac{1}{9}\right)\left(1-\frac{1}{13}\right)$$
$$\times\left(1-\frac{1}{19}\right)\cdot\left(1-\frac{1}{17}\right)\cdot\left(1-\frac{1}{21}\right) \quad (3)$$

Again, $mn = 2\cdot 3^2\cdot 5^4\cdot 7^3\cdot 9 : 13^3\cdot 19^4\cdot 17.21$

$$\phi(mn) = mn\left(1-\frac{1}{2}\right)\left(1-\frac{1}{3}\right)\left(1-\frac{1}{5}\right)\left(1-\frac{1}{7}\right)\left(1-\frac{1}{9}\right)\left(1-\frac{1}{13}\right)$$
$$\times\left(1-\frac{1}{19}\right)\left(1-\frac{1}{17}\right)\left(1-\frac{1}{21}\right). \quad (4)$$

From (3) and (4) we have $\dfrac{\phi(m)\cdot\phi(n)}{\frac{\phi(d)}{d}} = \phi(mn)$.

That is, $\phi(mn) = \dfrac{\phi(m)\phi(n)d}{\phi(d)}$.

Example: Let $m = 6;\ n = 120$

$6 = 2\cdot 3,\ 120 = 2^3\cdot 3\cdot 5$

$(m,n) = (6,120) = 6 = d = 2\cdot 3$

$$\phi(6) = 6\left(1 - \frac{1}{2}\right) \cdot \left(1 - \frac{1}{3}\right)$$

$$\phi(120) = 120\left(1 - \frac{1}{2}\right)\left(1 - \frac{1}{3}\right)\left(1 - \frac{1}{5}\right)$$

$$\phi(6) \cdot \phi(120) = 720\left(1 - \frac{1}{2}\right)^2\left(1 - \frac{1}{3}\right)^2\left(1 - \frac{1}{5}\right) \quad (1)$$

$$\phi(d) = \phi(6) = 6\left(1 - \frac{1}{2}\right)\left(1 - \frac{1}{3}\right)$$

$$\frac{\phi(d)}{d} = \frac{\phi(6)}{6} = \left(1 - \frac{1}{2}\right)\left(1 - \frac{1}{3}\right). \quad (2)$$

(1) $\div$ (2) gives $\dfrac{\phi(6) \cdot \phi(120)}{\frac{\phi(6)}{6}} = 720\left(1 - \frac{1}{2}\right)\left(1 - \frac{1}{3}\right)\left(1 - \frac{1}{5}\right).$ (3)

Again, $mn = 2^4 \cdot 3^2 \cdot 5$

$$\phi(mn) = \phi(720) = 720\left(1 - \frac{1}{2}\right)\left(1 - \frac{1}{3}\right)\left(1 - \frac{1}{5}\right). \quad (4)$$

From (3) and (4), $\dfrac{\phi(6) \cdot \phi(120) \times 6}{\phi(6)} = \phi(720).$

***Proof* (5):** $a|b \Rightarrow b = a \cdot c$ **for some** $c \in Z$, $1 \le c \le b$.

Case(1): (i) $c = 1 \Rightarrow b = a \Rightarrow \phi(b) = \phi(a)$

$\therefore \quad a|b \Rightarrow \phi(a)|\phi(b)$

(ii) $c = b \Rightarrow a = 1 \Rightarrow \phi(a) = \phi(1) = 1$

$\therefore \quad 1|b \Rightarrow 1|\phi(b)$

So $a|b \Rightarrow \phi(a)|\phi(b)$.

Case(2): Let $(a, c) = 1,\ 1 < c < b$.

Then $b = ac \Rightarrow \phi(b) = \phi(ac) = \phi(a) \cdot \phi(c)$ **[by Prop.(2)]**

$\Rightarrow \quad \phi(a)|\phi(b)$, **since** $\phi(c) \in Z$.

Case(3): Let $(a, c) = d(> 1)$ **and let the result be true for any** $c < b$.

Then $d|c \Rightarrow \phi(d)|\phi(c)$

$\Rightarrow \quad \phi(c) = k\phi(d)$ for some $k \in Z$. (1)

Again, $\phi(b) = \phi(a)\phi(c)\dfrac{d}{\phi(d)}$ [Property 4]

$$= \phi(a)k\phi(d) \cdot \frac{d}{\phi(d)} \text{ [by (1)]}$$

$$= kd\phi(a) = \phi(a)k',\ k' = k \cdot d \in Z.$$

$\Rightarrow \quad \phi(a)|\phi(b)$.

The result is true $\forall\ c,\ 1 \le c \le b$.

Example: $a = 7,\ b = 21$. Then $a|b$.

Also $\phi(7) = 6, \phi(21) = 12 \Rightarrow \phi(7)|\phi(21)$.

So $a|b \Rightarrow \phi(a)|\phi(b)$.

Theorem (5.5): *If $f(n)$ be a function of n, then $\sum_{d|n} f(d) = \sum_{d|n} f\left(\frac{n}{d}\right)$, where n is a (+)ve integer ≥ 1.*

***Proof*:** Let $d_1(=1), d_2, d_3, \ldots, d_t$ be the divisors of n, taken in ascending order of magnitude.

Now $d_1|n \Rightarrow n = d_1 d_t \Rightarrow \frac{n}{d_1} = d_t$

$d_2|n \Rightarrow n = d_2 d_{t-1} \Rightarrow \frac{n}{d_2} = d_{t-1}$

$d_3|n \Rightarrow n = d_3 d_{t-2} \Rightarrow \frac{n}{d_3} = d_{t-2}$

....

$d_t|n \Rightarrow n = d_t d_1 \Rightarrow \frac{n}{d_t} = d_1$.

Now $$\sum_{d|n} f(d) = f(d_1) + f(d_2) + \cdots + f(d_t) = \sum_{i=1}^{d_t} f(d_i). \quad (1)$$

Also $$\sum_{d|n} f\left(\frac{n}{d}\right) = f\left(\frac{n}{d_1}\right) + f\left(\frac{n}{d_2}\right) + \cdots + f\left(\frac{n}{d_t}\right)$$

$$= f(d_t) + f(d_{t-1}) + \cdots + f(d_1)$$

$$= f(d_1) + f(d_2) + \cdots + f(d_t)$$

$$= \sum_{i=1}^{d_t} f(d_i). \quad (2)$$

So from (1) and (2), $\sum_{d|n} f(d) = \sum_{d|n} f\left(\frac{n}{d}\right)$.

Example: Let $n = 15$.

Then $d = 1, \quad 3, \quad 5, \quad 15,$ where $d|n$

$\downarrow d_1 \ \downarrow d_2 \ \downarrow d_3 \ \downarrow d_4$.

Now $1|15 \Rightarrow 15 = 1 \cdot 15 = 1 \cdot d_4 \Rightarrow \frac{15}{1} = d_4$

$3|15 \Rightarrow 15 = 3 \cdot 5 = 3 \cdot d_3 \Rightarrow \frac{15}{3} = d_3$

$5|15 \Rightarrow 15 = 5 \cdot 3 = 5 \cdot d_2 \Rightarrow \frac{15}{5} = d_2$

$15|15 \Rightarrow 15 = 15 \cdot 1 = 15 \cdot d_1 \Rightarrow \frac{15}{15} = d_1$

$$\sum_{d|15} f(d) = f(1) + f(3) + f(5) + f(15) = \sum_{d=1,3,5,15} f(d). \quad (1)$$

$$\sum f\left(\frac{n}{d}\right) = \sum_{d|15} f\left(\frac{15}{d}\right) = f\left(\frac{15}{1}\right) + f\left(\frac{15}{3}\right) + f\left(\frac{15}{5}\right) + f\left(\frac{15}{15}\right)$$

$$= f(15) + f(5) + f(3) + f(1)$$

$$= \sum_{d=1,3,5,15}^{1} f(d). \tag{2}$$

Hence, $\sum_{d|15} f(d) = \sum_{d|15} f\left(\frac{15}{d}\right)$.

Theorem (5.6): *For an integer* $n \geq 1$, $n = \sum_{d|n} \phi(d)$.

***Proof*:** Let $S = \{1, 2, \ldots, n-1, n\}$

We define $S_d = \{x;\ (x, n) = d,\ 1 \leq d \leq n\}$.

Then S_d are non-empty pairwise disjoint subsets of S. So S_d forms a partition of S.

Hence, $S = \bigcup S_d$.

Here the suffix d represents those (+)ve integers which divide n. For other values of d, $S_d = \phi$ i.e., $S_d = \phi$, when $d \not| n$.

$n(s)$ denotes the number of elements in S.

We have $S = \bigcup_{d|n} S_d$. Then $n(s) = n\left[\bigcup_{d|n} S_d\right]$

$$\Rightarrow \quad n = \sum_{d|n} n(S_d) = n(S_{d_1}) + n(S_{d_2}) + \cdots + (S_{d_t}), \tag{1}$$

where $\quad d_i | n,\ i = 1, 2, \ldots, t$.

Now $(x, n) = d$ if and only if $\left(\frac{x}{d}, \frac{n}{d}\right) = 1$.

Thus, S_d contains those (+)ve integers which are relatively prime to $\frac{n}{d}$ and not exceeding $\frac{n}{d}$. In other words,

$$n(S_{d_1}) = \phi\left(\frac{n}{d_1}\right), n(S_{d_2}) = \phi\left(\frac{n}{d_2}\right), \cdots, n(S_{d_t}) = \phi\left(\frac{n}{d_t}\right).$$

From (1), $n = \phi\left(\frac{n}{d_1}\right) + \phi\left(\frac{n}{d_2}\right) + \cdots + \phi\left(\frac{n}{d_t}\right) = \sum_{d|n} \phi\left(\frac{n}{d}\right)$

$$= \sum_{d|n} \phi(d) \quad \text{[by Th. 5.5]}.$$

Example: Let $n = 12$. We consider the set $S = \{1, 2, \ldots, 12\}$

$\therefore \quad n(S) = 12,\ d = 1, 2, 3, 4, 6, 12, \text{ where } d|12$

$S_1 = \{x; (x, n) = 1\} = \{1, 5, 7, 11\}$

$n(S_1) = 4.\quad$ Also $n(S_1) = \phi\left(\frac{n}{1}\right) = \phi(12)$

$S_2 = \{x, (x, n) = 2\} = \{2, 10\}$

$n(S_2) = 2.\quad$ Also $n(S_2) = \phi\left(\frac{n}{2}\right) = \phi(6)$

$S_3 = \{x, (x, n) = 3\} = \{3, 9\}$

$n(S_3) = 2.$ Also $n(S_3) = \phi\left(\frac{n}{3}\right) = \phi(4)$

$S_4 = \{x, (x, n) = 4\} = \{4, 8\}$

$n(S_4) = 2.$ Also $n(S_4) = \phi\left(\frac{n}{4}\right) = \phi(3)$

$S_6 = \{x, (x, n) = 6\} = \{6\}$

$n(S_6) = 1.$ Also $n(S_6) = \phi\left(\frac{n}{6}\right) = \phi(2) = 1$

$S_{12} = \{x, (x, n) = 12\} = \{12\}$

$n(S_{12}) = 1.$ Also $n(S_{12}) = \phi\left(\frac{n}{12}\right) = \phi(1).$

[Note that $S_5, S_7, S_8, S_9, S_{10}, S_{11} = \phi$]

$$S = \bigcup_{d|n} S_d = S_1 \cup S_2 \cup S_3 \cup S_4 \cup S_6 \cup S_{12}$$

$$n(S) = n(S_1) + n(S_2) + n(S_3) + n(S_4) + n(S_6) + n(S_{12})$$
$$= 4 + 2 + 2 + 2 + 1 + 1 = 12.$$

Therefore, $n(S) = \phi(12) + \phi(6) + \phi(4) + \phi(3) + \phi(2) + \phi(1)$

i.e., $12 = \sum_{d|12} \phi(d).$

Problem (5.1): If $n(> 2)$ has r distinct odd prime factors, then show that $2^r | \phi(n)$.

Solution: Let $n = p_1^{a_1} p_2^{a_2} \dots p_r^{a_r}$, where p_is are distinct odd primes.

Then $\phi(n) = p_1^{a_1} p_2^{a_2} \cdots p_r^{a_r} \left(1 - \frac{1}{p_1}\right)\left(1 - \frac{1}{p_2}\right) \cdots \left(1 - \frac{1}{p_r}\right)$

$$= p_1^{a_1-1} p_2^{a_2-1} \cdots p_r^{a_r-1} (p_1 - 1) \cdots (p_r - 1)$$

$$= \prod_{i=1}^{r} p_i^{a_i-1}(p_i - 1). \qquad (1)$$

Since p_i is odd, $p_i - 1$ is even. So each $(p_i - 1)$ contributes a factor 2 in the product (1).

Therefore, $\phi(n) = 2^r \cdot p_1^{a_1-1} p_2^{a_2-1} \cdots p_r^{a_r-1} k,$

where k is the product of (+)ve integers and may contain a factor 2, and r is the minimum integral power of 2.

$\Rightarrow \quad \phi(n) = 2^r \cdot t, t \in Z \;\Rightarrow\; 2^r | \phi(n).$

Example: (i) Let $n = 3^2 \cdot 5^3 \cdot 7 \cdot 11^3 \cdot 13$, containing 5 prime factors.

Then $\phi(n) = 3^2 \cdot 5^2 \cdot 7 \cdot 11^3 \cdot 13 \left(1 - \frac{1}{3}\right)\left(1 - \frac{1}{5}\right)\left(1 - \frac{1}{7}\right)\left(1 - \frac{1}{11}\right)\left(1 - \frac{1}{13}\right)$

$$= 3 \cdot 5^2 \cdot 11^2 \cdot (2 \cdot 4 \cdot 6 \cdot 10 \cdot 12) = 3 \cdot 5^2 \cdot 11^2 \cdot 2^5 (1 \cdot 2 \cdot 3 \cdot 5 \cdot 6)$$

$$= 2^5 \cdot k,\ k = 2^2 \cdot 3^4 \cdot 5^3 \cdot 11^2 \in Z^+, \text{ containing a factor 2.}$$

$\Rightarrow \quad 2^5 | \phi(n)$

(ii) Let $n = 3^2 \cdot 7 \cdot 11$, containing 3 prime factors.

Then $\phi(n) = 3 \cdot (2 \cdot 6 \cdot 10) = 3 \cdot 2^3 \cdot (1 \cdot 3 \cdot 5) = 2^3(1 \cdot 3^2 \cdot 5)$

$= 2^3 \cdot k,\ k = 1 \cdot 3^2 \cdot 5 \cdot \in Z^+$ does not contain a factor 2.

$\Rightarrow \quad 2^3 | \phi(n)$.

Note: r is the minimum integral power of 2 in the product expression of $\phi(n)$.

Theorem (5.7): *Phi-function is multiplicative but not completely multiplicative.*

***Proof*:** If $(m_1, m_2) = 1$, then from Th. 5.4(2), it follows that $\phi(m_1, m_2) = \phi(m_1) \cdot \phi(m_2)$.

So ϕ is multiplicative.

But $\forall\, m_1, m_2$, i.e., when $(m_1, m_2) \neq 1$,

$\phi(m_1, m_2) \neq \phi(m_1) \cdot \phi(m_2)$, follows from the example.

Let $m_1 = 2,\ m_2 = 4\ (m_1, m_2) = 2$

$m_1 \cdot m_2 = 8$

$$\phi(m_1 m_2) = \phi(8) = 4 \tag{1}$$

$\phi(m_1) = \phi(2) = 1$

$\phi(m_2) = \phi(4) = 2.$

So $$\phi(m_1) \cdot \phi(m_2) = 2. \tag{2}$$

From (1) and (2) it follows that $\phi(m_1 m_2) \neq \phi(m_1)\phi(m_2)$. So ϕ is not completely multiplicative.

Problem (5.2): Show that $\phi(3n) = 2\phi(n)$, iff $3 \not| n$.

Solution: $3 \not| n$, iff $n = 2^{a_1} 5^{a_2} 7^{a_3} \cdots$

$$\phi(3n) = \phi(2^{a_1} 3 \cdot 5^{a_2} \cdot 7^{a_3} \cdots)$$

$$= \phi(3) \cdot 2^{a_1} \cdot 5^{a_2} \cdot 7^{a_3} \cdots \left(1 - \frac{1}{2}\right)\left(1 - \frac{1}{5}\right)\left(1 - \frac{1}{7}\right)$$

$$= 2 \cdot \phi(n).$$

Since $\phi(n) = 2^{a_1} \cdot 5^{a_2} \cdot 7^{a_3} \left(1 - \frac{1}{2}\right)\left(1 - \frac{1}{5}\right)\left(1 - \frac{1}{7}\right) \cdots$

Problem (5.3): If p and $2p + 1$ be both odd primes, and $n = 4p$, then show that $\phi(n+2) = \phi(n) + 2$.

Solution: $n = 4p$

Then $$\phi(n+2) = \phi(4p+2) = \phi(2) \cdot \phi(2p+1) = 1 \cdot 2p. \tag{1}$$

Since $\phi(2) = 1$ and $2p + 1$ is odd primes,

$\phi(2p+1) = (2p+1) - 1 = 2p.$

Again, $\phi(n) = \phi(4p) = \phi(4) \cdot \phi(p)\ [(4, p) = 1]$

$= 2 \cdot (p - 1) = 2p - 2$

$$\phi(n) + 2 = 2p - 2 + 2 = 2p. \tag{2}$$

From (1) and (2), we have $\phi(n+2) = \phi(n) + 2$.

Problem (5.4): Show that $\phi(3n) = 3\phi(n)$, iff $3|n$.

Solution: $3|n$, iff $n = 2^{a_1} \cdot 3^{a_2} \cdot 5^{a_3} \cdots$

where $a_2 \geq 1,\ a_i \in Z^+,\ i = 1, 3, 4, \ldots$

Then $\phi(n) = n \cdot \left(1 - \frac{1}{2}\right)\left(1 - \frac{1}{3}\right)\left(1 - \frac{1}{5}\right)$ (1)

$$\begin{aligned}\phi(3n) &= \phi(2^{a_1}.3^{a_2+1}.5^{a_3}\ldots) \\ &= 2^{a_1}.3^{a_2+1}.5^{a_3}\cdots\left(1-\frac{1}{2}\right)\left(1-\frac{1}{3}\right)\left(1-\frac{1}{5}\right)\cdots \\ &= 3.2^{a_1}.3^{a_2}.5^{a_3}\cdots\left(1-\frac{1}{2}\right)\left(1-\frac{1}{3}\right)\left(1-\frac{1}{5}\right)\cdots \\ &= 3 \cdot n\left(1-\frac{1}{2}\right)\left(1-\frac{1}{3}\right)\left(1-\frac{1}{5}\right)\cdots \\ &= 3 \cdot \phi(n), \text{ by (1).}\end{aligned}$$

Theorem (5.8): *If $n > 2$, then $\phi(n)$ is even.*

***Proof*:** For $n > 2$, n is either some integral power of 2 or n is of the form $p^c q$, where p is an odd prime, $c, q \in Z^+$.

Case(1): Let $n = 2^a,\ a \in Z^+,\ a > 0$.

Then $\phi(n) = \phi(2^a) = 2^a\left(1 - \frac{1}{2}\right) = 2^{a-1} = \text{even } \forall\ a \in Z^+$.

Case(2): Let $n = p^c q,\ p$ is an odd prime, $c, q \in Z^+$.

Then $\phi(n) = \phi(p^c q) = \phi(p^c)\phi(q) \quad [(p^c, q) = 1]$

$$= p^{c-1}(p-1)\phi(q).$$

Since p is odd, $p - 1$ is even and therefore, $\phi(n)$ is even.

Example: $\phi(375),\ n > 2$

$$\phi(375) = \phi(3 \times 5^3) = \phi(3)\phi(5^3) = 2 \cdot 5^2 \cdot 4 = \text{even.}$$

Theorem (5.9): *If $n \geq 2$, the sum of the integers less than and prime to n is $\frac{1}{2}n\phi(n)$.*

***Proof*:** For $n = 2$, $\phi(2) = 1 = \frac{1}{2} \cdot 2 \cdot \phi(2)$.

$\Rightarrow$ sum of the integers less than and prime to 2 is $\frac{1}{2} \cdot 2 \cdot \phi(2)$.

For $n > 2$, $\phi(n)$ is even by Th. (5.8).

Now if x is any integer less than and prime to n, then $n - x$ is so also.

Since $\phi(n)$ is even, the integers less than and prime to n can be arranged into pairs as $(1, n-1), (p_1, n-p_1), \ldots, (p_n, n-p_n)$, where each $p_i < n$ and $(p_i, n) = 1$, such that each pair has a sum n.

No. of pairs is $\frac{1}{2}\phi(n)$.

Sum of the integers in each pair is n.

Therefore, sum of the integers less than and prime to $n = \frac{1}{2}n\phi(n)$.

Problem (5.5): If n be a composite number, then show that $\phi(n) \leq n - \sqrt{n}$.

Solution: Let $n = p_1^{a_1} p_2^{a_2} \ldots p_k^{a_k}$, where each p_i is prime and p_1 is the smallest prime.

Then $\phi(n) = n\left(1 - \frac{1}{p_1}\right)\left(1 - \frac{1}{p_2}\right)\cdots\left(1 - \frac{1}{p_k}\right) \leq n\left(1 - \frac{1}{p_1}\right)$ as $1 - \frac{1}{p_i} \leq 1 \; \forall\, i$.

Again, p_1 is a divisor of n, $p_1 < \sqrt{n} \Rightarrow 1 - \frac{1}{p_1} < 1 - \frac{1}{\sqrt{n}}$.

$$\therefore \quad \phi(n) \leq n\left(1 - \frac{1}{p_1}\right) < n\left(1 - \frac{1}{\sqrt{n}}\right) = n - \sqrt{n}.$$

Definition (5.6): *A fraction $\frac{a}{n} (\leq 1)$ is called proper fraction and is called irreducible fraction, if $(a, n) = 1$.*

Theorem (5.10): *The number of irreducible proper fractions with denominator n is $\phi(n)$.*

Proof: Let $r_1, r_2, \ldots, r_n \in Z^+$ such that $r_i \angle n$ and $(r_i, n) = 1 \; \forall i$.

Then $\frac{r_1}{n}, \frac{r_2}{n}, \ldots, \frac{r_n}{n}$ are the only irreducible proper fractions.

If possible let $\frac{a}{n} (a < n)$ be any other proper fraction such that $\frac{a}{n} \neq \frac{r_i}{n}$ for any i.

Then $(a, n) \neq 1$ and consequently $\frac{a}{n}$ is not irreducible.

Hence the no. of irreducible proper fraction $\frac{r_i}{n}$, $i = 1, 2, \ldots, n$, with denominator n is equal to $\phi(n)$.

Problem (5.6): Verify that $\phi(n) = \phi(n+1) = \phi(n+2)$ if $n = 5186$.

Solution: $\phi(n) = \phi(5186) = \phi(2 \times 2593) = \phi(2)\phi(2593)$

$= 1 \cdot 2592 = 2592$ [2593 is prime]. (i)

$\phi(n+1) = \phi(5187) = \phi(3 \times 7 \times 13 \times 19) = \phi(3) \cdot \phi(7) \cdot \phi(13) \cdot \phi(19)$

$= 2 \cdot 6 \cdot 12 \cdot 18 = 2592.$ (ii)

$\phi(n+2) = \phi(5188) = \phi(4 \times 1297) = \phi(4) \times \phi(1297)$

$= 2 \times 1296 = 2592.$ (iii)

From (i), (ii) and (iii), it follows that $\phi(n) = \phi(n+1) = \phi(n+2)$.

Problem (5.7): If every prime that divides n also divides m, then show that $\phi(mn) = n\phi(m)$. In particular, $\phi(n^2) = n\phi(n)$ for every (+)ve integer n.

Solution: Let $n = p_1^{a_1} p_2^{a_2} \ldots p_k^{a_k}$, where p_i are primes and let $m = p_1^{b_1} p_2^{b_2} \cdots p_k^{b_k}$ such that $p_i | n$ and $p_i | m$.

$$mn = p_1^{b_1 + a_1} p_2^{b_2 + a_2} \ldots p_k^{b_k + a_k}$$

$$\phi(mn) = p_1^{b_1 + a_1} p_2^{b_2 + a_2} \ldots p_k^{b_k + a_k} \left(1 - \frac{1}{p_1}\right)\left(1 - \frac{1}{p_2}\right)\cdots\left(1 - \frac{1}{p_k}\right)$$

$$= mn\left(1 - \frac{1}{p_1}\right)\cdots\left(1 - \frac{1}{p_k}\right)$$

$$= n\phi(m), \text{ since } \phi(m) = m\left(1 - \frac{1}{p_1}\right)\cdots\left(1 - \frac{1}{p_k}\right).$$

In particular, $m = n$, $\phi(n^2) = n\phi(n)$.

Möbius function $\mu(n)$

We now introduce another arithmetical function, the Möbius μ function on the set of (+)ve integers.

Definition (5.7): *The Möbius μ function is defined as under:*

for $n = 1$, $\mu(1) = 1$

for $n > 1$, let $n = p_1^{a_1} p_2^{a_2} \dots p_k^{a_k}$, p_i are distinct primes.

Then, $\mu(n) = (-1)^k$, when $a_i = 1 \ \forall \ i$

$= 0$, when $a_i \geq 2$ for at least one i.

[i.e., when $p^2 | n$ for any prime p]

Example: $\mu(2) = (-1)^1 = 1$, $\mu(4) = \mu(2^2) = 0$

$\mu(3) = (-1)^1 = -1$, $\mu(6) = \mu(2 \cdot 3) = (-1)^2 = 1$

$\mu(54) = \mu(2 \cdot 3^3) = \mu(2 \cdot 3 \cdot 3^2) = 0$ [$\because 3^2 | 54$]

A short table is given:

$n =$	1	2	3	4	5	6	7	8	9	10	11	12	13	14	15	16
$\mu(n) =$	1	−1	−1	0	−1	1	−1	0	0	1	−1	0	−1	1	1	0

Theorem (5.11): *$\mu(n)$ is multiplicative.*

That is, $\mu(m_1, m_2) = \mu(m_1)\mu(m_2)$, if $(m_1, m_2) = 1$.

Proof: **Case(1):** $m_1 = m_2 = 1$.

Then $\mu(1) = 1 = 1.1 = \mu(1)\mu(1)$.

Case(2): Let $m_1 = p_1 p_2 \cdots p_k$, $m_2 = q_1 q_2 \cdots q_l$,

where $(p_i, q_j) = 1$, $p_i \neq q_j$, since $(m_1, m_2) = 1$.

Then $\mu(m_1) = (-1)^k$, $\mu(m_2) = (-1)^l$.

Also $m_1 m_2 = p_1 p_2 \dots p_k \, q_1 q_2 \cdots q_l$

$\mu(m_1 m_2) = (-1)^{k+l} = (-1)^k \cdot (-1)^l = \mu(m_1) \cdot \mu(m_2)$.

Case(3): If $p^2 | m_1$, then $\mu(m_1) = 0$.

$p^2 | m_1 \Rightarrow p^2 | m_1 m_2 \Rightarrow \mu(m_1 m_2) = 0$.

$\therefore \mu(m_1 m_2) = 0 = 0.\mu(m_2) = \mu(m_1)\mu(m_2)$.

Corollary (5.1): $\mu(n)$ is not completely multiplicative.

That is, $\mu(m_1 m_2) \neq \mu(m_1)\mu(m_2)$, when $(m_1, m_2) \neq 1$.

Proof: Let $m_1 = m_2 = 2$, $(m_1, m_2) = 2 \neq 1$.

$\mu(m_1) = -1$, $\mu(m_2) = -1$

$\mu(m_1) \cdot \mu(m_2) = 1$, but $\mu(m_1 m_2) = \mu(2^2) = 0$.

$\therefore \mu(m_1 m_2) \neq \mu(m_1) \cdot \mu(m_2)$.

Lemma (5.1): If $f(n)$ be multiplicative not identically zero, then

1. $f(1) = 1$.
2. If $n = p_1^{a_1} p_2^{a_2} \dots p_k^{a_k}$, p_i are distinct primes, $a_i \geq 0 \ \forall \ i$. Then

(a) $\sum_{d|n} f(d) = \{1+f(p_1)+f(p_1^2)+\cdots+f(p_1^{a_1})\} \times \{1+f(p_2)+f(p_2^2)+\cdots+f(p_2^{a_2})\} \times$
$\cdots \times \{1+f(p_k)+f(p_k^2)+\cdots+f(p_k^{a_k})\}$.

(b) $\sum_{d|n} \mu(d)f(d) = (1-f(p_1))(1-f(p_2))\cdots(1-f(p_k))$.

Proof: **1.** $f(n)$ is not identically zero, so $\exists$ a (+)ve integer n, such that $f(n) \neq 0$.

Now $f(n) = f(n.1) = f(n)\cdot f(1)$

$\Rightarrow$ $f(1) = 1$, $[f(n) \neq 0]$, by cancellation.

2(a). Let $d_1, d_2, \ldots, d_t$ be the divisors of n,

then $\sum_{d|n} f(d) = f(d_1) + f(d_2) + \cdots + f(d_t)$.

Also every divisor d of n is of the form $\Pi p_i^{x_i}$, $0 \le x_i \le a_i$, $\forall\, i$.

Let $A = \{f(d), d|n\}$ and $B = \{f(p_1^{x_1})f(p_2^{x_2})\cdots f(p_k^{x_k}),\ 0 \le x_i \le a_i\ \forall\, i\}$.

Now $x \in A \Rightarrow x = f(d) = f(p_1^{x_1}p_2^{x_2}\cdots p_k^{x_k})$

$= f(p_1^{x_1})f(p_2^{x_2})\cdots f(p_k^{x_k})$, since f is multiplicative.

$\Rightarrow$ $x \in B$, $\therefore$ $A \subseteq B$.

Also $y \in B \Rightarrow y = f(p_1^{x_1})\cdot f(p_2^{x_2})\cdots f(p_k^{x_k}) = f(p_1^{x_1}p_2^{x_2}\ldots p_k^{x_k}) = f(d) \in A$.

$\therefore$ $B \subseteq A$.

Again, each term of A and B is distinct.

Therefore, $A = B$.

$$\text{So}\quad \sum_{d|n} f(d) = \sum_{0\le x_i \le a_i} \{f(p_1^{x_1})f(p_2^{x_2})\cdots f(p_k^{x_k})\}$$
$$= \{f(1)+f(p_1)+f(p_1^2)+\cdots+f(p_1^{a_1})\}\{f(1)+f(p_2)+f(p_2^2)+\cdots+f(p_2^{a_2})\} \times \{f(1)+f(p_k)+(p_k^2)+\cdots+f(p_k^{a_k})\}$$
$$= \{1+f(p_1)+f(p_1^2)+\cdots+f(p_1^{a_1})\} \times \cdots \times \{1+f(p_k)+\cdots+f(p_k^{a_k})\}$$

2(b). Let $F(n) = \mu(n)f(n)$.

Then for any prime p, $F(p) = \mu(p)f(p) = -f(p)$ $[\because\ \mu(p) = (-1)^1 = -1]$. (i)

$F(p^a) = \mu(p^a)\cdot f(p^a)$, $a > 1$

$= 0\cdot f(p^a) = 0$. (ii)

$$\text{Now}\ \sum_{d|n}\mu(d)f(d) = \sum_{d|n} F(d) = \{1+F(p_1)+F(p_1^2)+\cdots+F(p_1^{a_1})\}$$
$$\times\{1+F(p_2)+F(p_2^2)+\cdots+F(p_2^{a_2})\} \times \{1+F(p_k) + F(p_k^2)+\cdots+F(p_k^{a_k})\}\quad [\text{by } 2(\text{a})]$$
$$= (1-f(p_1))(1-f(p_2))\cdots(1-f(p_k))\quad [\text{by (i) and (ii)}]$$

Theorem (5.12): *Relation between Phi-function $\phi(n)$ and Möbius function $\mu(n)$ is*

$$\phi(n) = n\sum_{d|n}\frac{\mu(d)}{d},\ n = \prod_{i=1}^{k} p_i^{a_i},\ a_i \ge 0.$$

Proof: We have by 2(b) of Lemma [5.1].

$$\sum_{d|n} \mu(d)f(d) = (1 - f(p_1))(1 - f(p_2)) \cdots (1 - f(p_k)). \quad (1)$$

Let $f(n) = \dfrac{1}{n}$.

Then $f(n)$ is multiplicative, since

$$f(mn) = \frac{1}{mn} = \frac{1}{m} \cdot \frac{1}{n} = f(m) \cdot f(n) \quad \text{and} \quad f(1) = 1.$$

$$\therefore \quad \sum_{d|n} \mu(d)\frac{1}{d} = \left(1 - \frac{1}{p_1}\right)\left(1 - \frac{1}{p_2}\right)\cdots\left(1 - \frac{1}{p_k}\right). \quad (2)$$

Also $\phi(n) = n\left(1 - \dfrac{1}{p_1}\right)\left(1 - \dfrac{1}{p_2}\right)\cdots\left(1 - \dfrac{1}{p_k}\right)$.

$$\therefore \quad \phi(n) = n\sum_{d|n} \frac{\mu(d)}{d} \quad \text{[by (2).]}$$

Theorem (5.13): *If* $n \geq 1$, *then* $\sum_{d|n} \mu(d) = \left[\dfrac{1}{n}\right] = 1,$ *when* $n = 1$

$= 0,$ *when* $n > 1$.

Proof: The result holds when $n = 1$.

For $\quad n = 1,\ d = 1.$

Therefore, $\sum_{d|n} \mu(d) = \sum_{d|1} \mu(d) = \mu(1) = 1 = \left[\dfrac{1}{n}\right]$.

Now when $n > 1$, let $n = \prod_{i=1}^{k} p_i^{a_i}$, p_i are distinct primes, $a_i \geq 0$.

Then

$$\sum_{d|n} \mu(d) = \sum_{1|n} \mu(d) + \sum_{p_i|n} \mu(d) + \sum_{p_ip_j|n} \mu(d) + \sum_{p_ip_jp_k|n} \mu(d) + \cdots + \sum_{p_1p_2\ldots p_k|n} \mu(d)$$

Since $\mu(d) = 0$ for other divisors of n. That is, $\mu(d) = 0$, when $d = p_1^{a_1} p_2^{a_2} \cdots p_k^{a_k}$, $a_i > 1$ for at least one i, $0 \leq i \leq k$.

So

$$\begin{aligned}\sum_{d|n} \mu(d) &= \mu(1) + \{\mu(p_1) + \mu(p_2) + \cdots + \mu(p_k)\} + \{\mu(p_1p_2) + \mu(p_1p_3) \\ &\quad + \cdots + \mu(p_1p_k)\} + \{\mu(p_1p_2p_3) + \mu(p_1p_3p_4) + \cdots \\ &\quad + \mu(p_1p_2p_k)\} + \cdots + \mu(p_1p_2p_3\ldots p_k) \\ &= 1 + k_{c_1}(-1)^1 + k_{c_2}(-1)^2 + k_{c_3}(-1)^3 + \cdots + (-1)^k,\end{aligned}$$

since $\mu(p_i)$ occurs k_{c_1} times, $\mu(p_ip_j)$ occurs k_{c_2} times and so on and $\mu(p_i) = (-1)^1$, $\mu(p_ip_j) = (-1)^2 \cdots$ and $\mu(p_1p_2\ldots p_k) = (-1)^k$

$$\begin{aligned}&= 1 - k_{c_1} + k_{c_2} - k_{c_3} + \cdots + (-1)^k \\ &= (1 - 1)^k = 0.\end{aligned}$$

Therefore, $\sum_{d|n} \mu(d) = 0 = \left[\frac{1}{n}\right]$, when $n > 1$.

Hence the theorem.

Corollary (5.2): If $n = p_1^{a_1} p_2^{a_2} \dots p_k^{a_k}$, then $\sum_{d|n} |\mu(d)| = 2^k$.

Proof: From above theorem,

$$\sum_{d|n} |\mu(d)| = |\mu(1)| + k_{c_1}|(-1)^1| + k_{c_2}|(-1)^2| + k_{c_3}|(-1)^3| + \dots + |(-1)^k|$$

$$= 1 + k_{c_1}1 + k_{c_2}1 + \dots + 1 = (1+1)^k = 2^k.$$

Example: Expression of Phi-function and Möbius functions in other forms:

First of all we go through the examples.

Example: (i) Let $n = 12$.

The integers not exceeding 12 and prime to 12 are 1, 5, 7, 11.

$\phi(12) =$ no. of integers not exceeding 12 and prime to 12

$$= 1 + 1 + 1 + 1 = \sum_{k=1}^{12} 1,$$

at the time of counting the integers we write 1 in place of all k, $k \le 12$, $(k, 12) = 1$.

(ii) Let $n = 25$.

The integers not exceeding 25 and prime to 25 are 1, 2, 3, 4, 6, 7, 8, 9, 11, 12, 13, 14, 16, 17, 18, 19, 21, 22, 23, 24.

$$\phi(25) = 1+1+1+1+1+1+1+1+1+1+1+1+1+1+1+1 \\ +1+1+1+1 = \sum_{k=1}^{25} 1 = 20,$$

where $k \le 25$, $(k, 25) = 1$.

Thus, Phi-function $\phi(n)$ can be written as

$$\phi(n) = \sum_{k=1}^{n} 1,\ k \le n,\ (k, n) = 1, \tag{1}$$

which can be further written as

$$\phi(n) = \sum_{k=1}^{n} \left[\frac{1}{(n,k)}\right],\ k \le n,\ (k, n) = 1. \tag{2}$$

Since $\left[\frac{1}{(n,k)}\right]$, greatest integer not exceeding $\frac{1}{(n,k)}$ is $1\ \forall\ k$,

Möbius function μ is defined by Th. [5.13]

$$\sum_{d|n} \mu(d) = \left[\frac{1}{n}\right] = 1,\quad n = 1$$

$$= 0,\quad n > 1,$$

which can be written as

$$\sum_{d|(n,k)} \mu(d) = \left[\frac{1}{(n,k)}\right]. \tag{3}$$

Theorem (5.14): [*Independent proof of the relationship between $\phi(n)$ and $\mu(n)$*]

If $n \geq 1$, then $\phi(n) = n\sum_{d|n} \frac{\mu(d)}{d}$.

***Proof*:** From the results (2) and (3) of the above example, we have

$$\phi(n) = \sum_{k=1}^{n}\left[\frac{1}{(n,k)}\right] = \sum_{k=1}^{n}\sum_{d|(n,k)} \mu(d) = \sum_{k=1}^{n}\sum_{\substack{d|n \\ d|k}} \mu(d).$$

For a fixed divisor d of n we take the sum only for those values of k such that $1 \leq k \leq n$ and $d|n$.

Now $\quad d|n \Rightarrow n = dx,\ x \in Z^+$.

So $\quad 1 \leq k \leq dx,\ x \in Z^+$.

Thus the values of k are multiples of d.

Let $\quad k = d \cdot u,\ u \in Z^+$.

Then $\quad 1 \leq k \leq n \iff 1 \leq u \leq \frac{n}{d}$

as $\quad du \leq n \Rightarrow u \leq \frac{n}{d}$.

$$\text{Therefore, } \phi(n) = \sum_{d|n}\sum_{u=1}^{\frac{n}{d}} \mu(d) = \sum_{d|n}\mu(d)\sum_{u=1}^{\frac{n}{d}} 1$$

$$= \sum_{d|n}\mu(d)\cdot\frac{n}{d} = n\sum_{d|n}\frac{\mu(d)}{d}.$$

Example: Let $n = 15$.

$\phi(n) = \phi(15) = 8,\ d = 1, 3, 5, 15$ [divisors of 15]

$$\text{Now } \sum_{d|n}\frac{\mu(d)}{d} = \frac{\mu(1)}{1} + \frac{\mu(3)}{3} + \frac{\mu(5)}{5} + \frac{\mu(15)}{15}$$

$$= 1 - \frac{1}{3} - \frac{1}{5} + \frac{1}{15} \quad [\because\ \mu(15) = \mu(3\cdot5) = (-1)^2 = 1]$$

$$= 1 - \frac{8}{15} + \frac{1}{15} = \frac{7}{15} + \frac{1}{15} = \frac{8}{15}. \tag{i}$$

$$\text{Also } \frac{\phi(n)}{n} = \frac{8}{15}. \tag{ii}$$

From (i) and (ii), $\frac{\phi(n)}{n} = \sum_{d|n}\frac{\mu(d)}{d}$, i.e., $\phi(n) = n\sum_{d|n}\frac{\mu(d)}{d}$.

Theorem (5.15): $\sum_{a=1}^{n}\mu(a)\left[\frac{n}{a}\right] = 1$, **where $\left[\frac{n}{a}\right]$ is the greatest integer not exceeding $\frac{n}{a}$.**

Go through the example.

Let $n = 9$, a varies from 1 to 9.

We have proved, $\sum_{d|n} \mu(d) = 1$ [Th. 5.13]

$$\text{Now } \sum_{d|1} \mu(d) = \mu(1)$$

$$\sum_{d|2} \mu(d) = \mu(1) + \mu(2)$$

$$\sum_{d|3} \mu(d) = \mu(1) \qquad + \mu(3)$$

$$\sum_{d|4} \mu(d) = \mu(1) + \mu(2) \qquad + \mu(4)$$

$$\sum_{d|5} \mu(d) = \mu(1) \qquad + \mu(5)$$

$$\sum_{d|6} \mu(d) = \mu(1) + \mu(2) + \mu(3) \qquad + \mu(6)$$

$$\sum_{d|7} \mu(d) = \mu(1) \qquad + \mu(7)$$

$$\sum_{d|8} \mu(d) = \mu(1) + \mu(2) \qquad + \mu(4) \qquad + \mu(8)$$

$$\sum_{d|9} \mu(d) = \mu(1) \qquad + \mu(3) \qquad + \mu(9)$$

$$\text{Then } \sum_{d|1} \mu(d) + \sum_{d|2} \mu(d) + \cdots + \sum_{d|9} \mu(d)$$

$$= 9\mu(1) + 4\mu(2) + 3\mu(3) + 2\mu(4) + \mu(5) + \mu(6) + \mu(7) + \mu(8) + \mu(9)$$

$$= \mu(1)\left[\frac{9}{1}\right] + \mu(2)\left[\frac{9}{2}\right] + \mu(3)\left[\frac{9}{3}\right] + \mu(4)\left[\frac{9}{4}\right] + \mu(5)\left[\frac{9}{5}\right] + \mu(6)\left[\frac{9}{6}\right]$$

$$+ \mu(7)\left[\frac{9}{7}\right] + \mu(8)\left[\frac{9}{8}\right] + \mu(9)\left[\frac{9}{9}\right]$$

$$= \sum_{a=1}^{9} \mu(a)\left[\frac{9}{a}\right].$$

$$\text{Also } \sum_{a=1}^{9} \mu(a)\left[\frac{9}{a}\right] = 9\cdot 1 + 4(-1) + 3(-1) + 2\cdot 0 + (-1) + 1 + (-1) + 0 + 0$$

$$= 9 - 4 - 3 - 1 + 1 - 1 = 1.$$

Proof: We have $\sum_{d|n} \mu(d) = 1,\ n \geq 1$.

$$\text{That is, } \sum_{d|1} \mu(1) + \sum_{d|2} \mu(d) + \cdots + \sum_{d|n} \mu(d) = 1. \tag{1}$$

It can be found that in the expressions of the L.H.S. of (1),

(i) $\mu(1)$ occurs in every term of (1) once only. That is, $\left[\frac{n}{1}\right]$ times in all.

(ii) $\mu(2)$ occurs once in every term of $\sum_{d|2}\mu(d), \sum_{d|4}\mu(d), \sum_{d|6}\mu(6), \ldots, \sum_{d|[\frac{n}{2}]2}\mu(d)$.

That is, $\left[\frac{n}{2}\right]$ times in all, since $\forall n = 2m = \left[\frac{2m}{2}\right]\cdot 2,\ m = 1, 2, 3, \ldots$.

(iii) $\mu(3)$ occurs once in every term of $\sum_{d|3}\mu(d), \sum_{d|6}\mu(d), \ldots, \sum_{d|[\frac{n}{3}]3}\mu(d)$.

That is, $\left[\frac{n}{3}\right]$ times in all, since $\forall n = 3m = \left[\frac{3m}{3}\right]\cdot 3$.

(iv) Finally, $\mu(n)$ occurs only in $\sum_{d|n}\mu(d)$ and it occurs $\left[\frac{n}{n}\right]$ times only.

Summing up the above results, we get

$$\mu(1)\left[\frac{n}{1}\right] + \mu(2)\left[\frac{n}{2}\right] + \mu(3)\left[\frac{n}{3}\right] + \cdots + \mu(n)\left[\frac{n}{n}\right] = 1 \text{ or, } \sum_{a=1}^{n}\mu(a)\left[\frac{n}{a}\right] = 1.$$

Problem (5.8): For $n \geq 1$, show that $\mu(n)\mu(n+1)\mu(n+2)\mu(n+3) = 0$.

Solution: For every integer $n \geq 1$, at least one term of the product contains a prime number p having power $m \geq 2$ and hence $\mu(p^m) = 0$.

For,
$$n = 1,\ \mu(n+3) = \mu(4) = \mu(2^2) = 0.$$
$$n = 13,\ \mu(n+3) = \mu(16) = \mu(2^4) = 0.$$
$$n = 20,\ \mu(n) = \mu(2^2.5) = \mu(2^2)\cdot\mu(5) = 0.$$

Therefore, $\mu(n)\mu(n+1)\mu(n+2)\mu(n+3) = 0\ \forall\ n \geq 1$.

Problem (5.9): Show that, $\sum_{k=1}^{n}\mu(k!) = 1$.

Solution: $\sum_{k=1}^{3}\mu(k!) = \mu(1!) + \mu(2!) + \mu(3!) = \mu(1) + \mu(2) + \mu(3\cdot 2) = 1 - 1 + 1 = 1.$

$$\begin{aligned}\sum_{k=1}^{4}\mu(k!) &= \mu(1!) + \mu(2!) + \mu(3!) + \mu(4!)\\ &= \mu(1) + \mu(2) + \mu(3\cdot 2) + \mu(4\cdot 3\cdot 2\cdot 1)\\ &= 1 - 1 + 1 + 0 \quad [\because\ \mu(2^3\cdot 3) = 0]\\ &= 1\end{aligned}$$

Each term of $\mu(n!)$ after the 3rd term contains a power of 2 and hence $\mu(n!) = 0\ \forall\ n \geq 4$.

Therefore, $\sum_{k=1}^{n}\mu(k!) = \mu(1) + \mu(2!) + \mu(3!) + 0 = 1 - 1 + 1 = 1.$

Problem (5.10): If $n(> 1)$ is of the form $p_1^{a_1}p_2^{a_2}\cdots p_k^{a_k}$, where p_i are distinct primes and $a_i \geq 0$, then show that

(i) $\sum_{d|n}\mu(d)\tau(d) = (-1)^k$

(ii) $\sum_{d|n}\mu(d)\sigma(d) = (-1)^k p_1p_2\cdots p_k$

Solution: (i) We have $\sum_{d|n} \mu(d)f(d) = (1-f(p_1))(1-f(p_2))\cdots(1-f(p_k))$

[by Lemma 5.1-2(b)]

Taking $f(d) = \tau(d)$, we get

$$\begin{aligned}\sum_{d|n} \mu(d)\tau(d) &= (1-\tau(p_1))(1-\tau(p_2))\cdots(1-\tau(p_k)) = (1-2)(1-2)\cdots(1-2)\\ &= (-1)\cdot(-1)\cdots(-1),\ k \text{ times}\\ &= (-1)^k.\end{aligned}$$

[$\tau(p_k)$ = no. of divisors of $p_k = 2 \cdot 1$ and p_k itself $\forall k$.]

(ii) Now taking $f(d) = \sigma(d)$, we get

$$\begin{aligned}\sum_{d|n} \mu(d)\sigma(d) &= (1-\sigma(p_1))(1-\sigma(p_2))\cdots(1-\sigma(p_k))\\ &= (1-(p_1+1))(1-(p_2+1))\cdots(1-(p_k+1))\\ &= -(p_1)\cdot(-p_2)\cdots(-p_k)\\ &= (-1)^k \cdot p_1 \cdot p_2 \ldots p_k\end{aligned}$$

[Divisors of p_k are 1 and $p_k \Rightarrow \sigma(p_k)$ = sum of the divisors $= 1 + p_k\ \forall\ k$]

Problem (5.11): Verify that $\sum_{d|n} \mu(d)\tau(d) = (-1)^k$

and $\sum_{d|n} \mu(d)\sigma(d) = (-1)^k \cdot p_1 \cdot p_2 \cdots p_k$,

when (i) $n = 10$ (ii) $n = 2^2 \cdot 3^2 \cdot 5^2$.

Solution: (i) $n = 10 = 2 \cdot 5$

$$\begin{aligned}\sum_{d|n} \mu(d)\tau(d) = \sum_{d|n} \mu(d)\tau(d) &= \mu(1)\tau(1) + \mu(2)\tau(2) + \mu(5)\tau(5) + \mu(10)\tau(10)\\ &= 1\cdot 1 + (-1)\cdot 2 + (-1)\cdot 2 + 1\cdot 4\\ &= 1 - 4 + 4 = 1 = (-1)^2 \quad [k = 2, \text{ no. of distinct primes}]\end{aligned}$$

$$\begin{aligned}\sum_{d|n} \mu(d)\sigma(d) = \sum_{d|10} \mu(d)\sigma(d) &= \mu(1)\sigma(1) + \mu(2)\sigma(2) + \mu(5)\sigma(5) + \mu(10)\sigma(10)\\ &= 1\cdot 1 + (-1)\cdot 3 + (-1)6 + 1\cdot 18\\ &= 1 - 3 - 6 + 18 = 10 = (-1)^2 \cdot 2 \cdot 5\end{aligned}$$

(ii) $$\begin{aligned}\sum_{d|n} \mu(d)\tau(d) = \sum_{d|2^2\cdot 3^3\cdot 5^2} \mu(d)\tau(d) &= \mu(1)\tau(1) + \mu(2)\tau(2) + \mu(3)\tau(3) + \mu(5)\tau(5)\\ &\quad + \mu(6)\tau(6) + \mu(10)\tau(10) + \mu(15)\tau(15) + \mu(30)\tau(30) + 0\end{aligned}$$

[for other divisors of n, $\mu(d) = 0$]

$$\begin{aligned}&= 1\cdot 1 + (-1)\cdot 2 + (-1)\cdot 2 + (-1)\cdot 2 + 1\cdot 4 + 1\cdot 4 + 1\cdot 4 + (-1)\cdot 8\\ &= 1 - 2 - 2 - 2 + 12 - 8 = -1 = (-1)^3.\end{aligned}$$

[$k = 3$, no. of distinct primes]

$$\sum_{d|n}\mu(d)\sigma(d) = \sum_{d|2^2 3^2 5^2}\mu(d)\sigma(d) = \mu(1)\sigma(1)+\mu(2)\sigma(2)+\mu(3)\sigma(3)+\mu(5)\sigma(5)$$
$$+\mu(6)\sigma(6)+\mu(10)\sigma(10)+\mu(15)\sigma(15)+\mu(30)\sigma(10)+0$$
$$[\text{for other divisors } \mu(d)=0]$$
$$= 1\cdot 1+(-1)\cdot 3+(-1)\cdot 4+(-1)\cdot 6+1\cdot 12+1\cdot 18+1\cdot 24+(-1)\cdot 72$$
$$= 1-3-4-6+12+18+24-72$$
$$= -85+55 = -30 = (-1)^3\cdot 2\cdot 3.5.$$

Theorem (5.16): *If n be composite, then* $(n, \phi(n)) > 1$.

***Proof*:** Let $n = p_1^{a_1}p_2^{a_2}\dots p_r^{a_r}\dots p_k^{a_k}$ whose each p_i is distinct prime and $a_i \geq 0\ \forall\ i = 1, 2, \dots, k$.

$$a_i \neq 0\ \forall\ i,\ \text{ since } n > 1.$$

$$\phi(n) = n\left(1-\frac{1}{p_1}\right)\left(1-\frac{1}{p_2}\right)\cdots\left(1-\frac{1}{p_r}\right)\cdots\left(1-\frac{1}{p_k}\right)$$
$$= \frac{n(p_1-1)(p_2-1)\cdots(p_r-1)\cdots(p_k-1)}{p_1\cdot p_2\dots p_k}$$
$$= p_1^{a_1-1}p_2^{a_2-1}\cdots p_r^{a_r-1}\cdots p_k^{a_k-1}\prod_{i=1}^{k}(p_i-1).$$

p_i is prime, so (p_i-1) is even and $\prod_{i=1}^{k}(p_i-1)$ becomes multiple of 2, i.e., of the form $2^x\prod_{i=1}^{k}p_i^{b_i}, b_i \geq 0, x > 0$.

Let $\prod_{i=1}^{k}(p_i-1) = 2^x p_r^{b_r}p_s^{b_s}p_t^{b_t}\quad 1 \leq r, s, t \leq k.$

Then $\phi(n) = 2^x p_1^{a_1-1}p_2^{a_2-1}\cdots p_r^{a_r-1+b_r}\cdots p_s^{a_s-1+b_s}\cdots p_t^{a_t-1+b_t}\cdots p_k^{a_k-1}.$

Hence $(n, \phi(n))$

$$= (p_1^{a_1}p_2^{a_2}\cdots p_r^{a_r}\cdots p_s^{a_s}\cdots p_t^{a_t}\cdots p_k^{a_k},\ 2^x p_1^{a_1-1}p_2^{a_2-1}\cdots p_r^{a_r-1+b_r}\cdots p_s^{a_s-1+b_s}\cdots p_t^{a_t-1+b_t}\cdots p_k^{a_k-1})$$
$$= p_1^{a_1-1}p_2^{a_2-1}\cdots p_r^{a_r}\cdots p_s^{a_s}\cdots p_t^{a_t}\cdots p_k^{a_k-1} > 1.$$

Example: Let $n = 2^4\cdot 3^2\cdot 7^2\cdot 11^3$.

Then $\phi(n) = n\left(1-\frac{1}{2}\right)\left(1-\frac{1}{3}\right)\left(1-\frac{1}{7}\right)\left(1-\frac{1}{11}\right)$

$$= \frac{n\cdot 1\cdot 2\cdot 6\cdot 10}{2\cdot 3\cdot 7\cdot 11}$$
$$= 2^3\cdot 3\cdot 7\cdot 11^2\ (2\cdot 60)$$
$$= 2^3\cdot 3\cdot 7\cdot 11^2\ (2^3\cdot 3\cdot 5)\quad [\because\ 2\cdot 60 = 120 = 2^3\cdot 3\cdot 5]$$
$$= 2^6\cdot 3^2\cdot 5\cdot 7\cdot 11^2$$
$$(n, \phi(n)) = (2^4\cdot 3^2\cdot 7^2\cdot 11^3,\ 2^6\cdot 3^2\cdot 5\cdot 7\cdot 11^2)$$
$$= 2^4\cdot 3^2\cdot 7\cdot 11^2 > 1.$$

Example: Let $n = 3^3 \cdot 5^2 \cdot 11^3$.

$$\phi(n) = n\left(1-\frac{1}{3}\right)\left(1-\frac{1}{5}\right)\left(1-\frac{1}{11}\right) = \frac{n \cdot 2 \cdot 4 \cdot 10}{3 \cdot 5 \cdot 11}$$
$$= 3^2 \cdot 5 \cdot 11^2(2 \cdot 40) = 3^2 \cdot 5 \cdot 11^2(2^4 \cdot 5) \quad [\because\ 2 \cdot 40 = 80 = 2^4 \cdot 5]$$
$$= 2^4 \cdot 3^2 \cdot 5^2 \cdot 11^2.$$
$$(n, \phi(n)) = (3^3 \cdot 5^2 \cdot 11^3,\ 2^4 \cdot 3^2 \cdot 5^2 \cdot 11^2) = 3^2 \cdot 5^2 \cdot 11^2 > 1.$$

Problem (5.12): If the same primes divide m and n, then show that $n\phi(m) = m\phi(n)$.

Solution: Let $m = p_1^{a_1} p_2^{a_2} \ldots p_k^{a_k}$

and $\quad n = p_1^{b_1} p_2^{b_2} \ldots p_k^{b_k}$,

m, n having the same prime (p_i) in factorisation.

Then $\quad \phi(m) = m\left(1-\dfrac{1}{p_1}\right)\cdots\left(1-\dfrac{1}{p_k}\right) \qquad (1)$

$\quad \phi(n) = n\left(1-\dfrac{1}{p_1}\right)\cdots\left(1-\dfrac{1}{p_k}\right). \qquad (2)$

(1) ÷ (2) gives $\dfrac{\phi(m)}{\phi(n)} = \dfrac{m}{n}, \quad \therefore\ n\phi(m) = m\phi(n).$

Möbius inversion formula

Theorem (5.17): *Let f and F be two number theoretic functions such that* $F(n) = \sum_{d|n} f(d)$,

then $\quad f(n) = \sum_{d|n} \mu(d) F\left(\frac{n}{d}\right) = \sum_{d|n} \mu\left(\frac{n}{d}\right) F(n).$

[Go through the examples.]

Let $n = 15$.

$d =$ (+)ve divisors of $15 = 1, 3, 5, 15$.

Then $\quad \sum_{d|n} \mu(d) F\left(\dfrac{n}{d}\right) = \sum_{d|15} \mu(d) F\left(\dfrac{15}{d}\right)$

$\quad = \mu(1)F(15) + \mu(3)F(5) + \mu(5)F(3) + \mu(15)F(1) \qquad (1)$

Again, $\quad F(n) = \sum_{d|n} f(d)$

$$\left.\begin{aligned} F(15) &= \sum_{d|15} f(d) = f(1) + f(3) + f(5) + f(15) \\ F(5) &= \sum_{d|5} f(d) = f(1) + f(5) \\ F(3) &= \sum_{d|3} f(d) = f(1) + f(3) \\ f(1) &= \sum_{d|1} f(1) = f(1) \end{aligned}\right\} \qquad (2)$$

From (1),

$$\sum_{d|n} \mu(d) F\left(\frac{n}{d}\right) = 1 \cdot F(15) - F(5) - F(3) + F(1)$$

$$= f(1) + f(3) + f(5) + f(15) - f(1) - f(5) - f(1)$$

$$- f(3) + f(1) \quad \text{[by (2)]}$$

$$= f(15) = f(n). \tag{3}$$

Also we write, $\sum_{d|n} \mu(d) F\left(\frac{n}{d}\right) = \sum_{d|n} \mu(d) \sum_{c|\frac{n}{d}} f(c)$ as $F(n) = \sum_{d|n} f(d)$

$$\Rightarrow \quad F\left(\frac{n}{d}\right) = \sum_{c|\frac{n}{d}} f(c) \quad \left[\text{replacing } n \text{ by } \frac{n}{d}\right]$$

Then $\sum_{d|n} \mu(d) F\left(\frac{n}{d}\right) = \sum_{d|n} \mu(d) \sum_{c|\frac{n}{d}} f(c) = \sum_{cd|n} \mu(d) f(c)$

$d|n \Rightarrow d = 1, 3, 5, 15,$ when $n = 15$

$c|\frac{n}{d} \Rightarrow c = 1, 3, 5, 15,$ when $d = 1$

$\Rightarrow c = 1, 5,$ when $d = 3$

$\Rightarrow c = 1, 3,$ when $d = 5$

$\Rightarrow c = 1,$ when $d = 15.$

$$\sum_{d|n} \mu(d) \sum_{c|\frac{n}{d}} f(c) = \mu(1)[f(1) + f(3) + f(5) + f(15)] + \mu(3)[f(1) + f(5)]$$

$$+ \mu(5)[f(1) + f(3) + \mu 15 f(1)$$

$$= f(15) = \sum_{d|n} \mu(d) F\left(\frac{n}{d}\right) \text{ [by (1)]}$$

Also $\sum_{cd|n} \mu(d) f(c).$

Here c, d assume those values only when $cd|n$.

List of values of c, d for which $cd|n$ are as under:

$d = 1,\ c = 1, 3, 5, 15 \Rightarrow cd = 1, 3, 5, 15 \Rightarrow cd|n$

$d = 3,\ c = 1, 5 \Rightarrow cd = 3, 15 \Rightarrow cd|n$

$d = 5,\ c = 1, 3 \Rightarrow cd = 5, 15 \Rightarrow cd|n$

$d = 15,\ c = 1 \Rightarrow cd = 15 \Rightarrow cd|n.$

Thus, cd has the values 1, 3, 5, 15.

$$\sum_{cd|n} \mu(d) f(c) = \mu(1)[f(1) + f(3) + f(5) + f(15) + \mu(3)[f(1) + f(5)]$$

$$+ \mu(5)[f(1) + f(3)] + \mu(15)[f(1)]$$

$$= f(15).$$

Therefore, we have $\sum_{d|n} \mu(d) \sum_{c|\frac{n}{d}} f(c) = \sum_{cd|n} \mu(d) f(c).$

Proof: The equivalent form of $F(n) = \sum_{d|n} f(d)$ is $F\left(\frac{n}{d}\right) = \sum_{c|\frac{n}{d}} f(c)$ obtained by replacing n by $\frac{n}{d}$.

Then $\sum_{d|n} \mu(d) F\left(\frac{n}{d}\right) = \sum_{d|n} \mu(d) \sum_{c|\frac{n}{d}} f(c).$

Since each term of the sum is of the form $\mu(d)f(c)$, when $d|n$ and $c\left|\frac{n}{d}\right.$.

Also $c\left|\frac{n}{d}\right. \Rightarrow \frac{n}{d} = ck$ for some $k \in Z$.

$\Rightarrow$ $n = dck \Rightarrow cd|n.$

We take those values of c and d so that $cd|n$ and subsequently

$$\sum_{d|n} \mu(d) \sum_{c|\frac{n}{d}} f(c) = \sum_{cd|n} \mu(d) f(c). \tag{1}$$

Again, $\sum_{d|n} \mu(d) F\left(\frac{n}{d}\right) = \sum_{c|n} f(c) \sum_{d|\frac{n}{c}} \mu(d)$

$$= \sum_{dc|n} f(c)\mu(d) = \sum_{cd|n} f(c)\mu(d). \tag{2}$$

Therefore, from (1) and (2) it follows that

$$\sum_{d|n} \mu(d) F\left(\frac{n}{d}\right) = \sum_{c|n} f(c) \sum_{d|\frac{n}{c}} \mu(d).$$

Now $\sum_{d|\frac{n}{c}} \mu(d) = 1, \quad \text{when } \frac{n}{c} = 1, \text{ i.e., } c = n$

$= 0, \quad \text{when } \frac{n}{c} > 1, \text{ i.e., } c < n.$

Hence, $\sum_{d|n} \mu(d) F\left(\frac{n}{d}\right) = f(n)\mu(1) + \sum_{\substack{d|r \\ r>1}} \mu(d) f(r)$

$= f(n)\mu(1) + 0$ [by Th. (5.13)]

$= f(n).1 = f(n).$

Exercise 5

1. Find all the integers such that

   (i) $\phi(n) = \frac{n}{2}$ **(ii)** $\phi(n) = \phi(2n)$ **(iii)** $\phi(n) = 12$.

2. **(a)** If $(m, n) = 1$, then does $(\phi(m), \phi(n)) = 1$.
   Justify with reason.

   (b) Show that if $n = 2^k$ or $2^k \cdot 3^l$ with k and l are positive integers, then $\phi(n)|n$.

3. If $n = \prod_{i=1}^{k} p_i^{a_i}$, where each p_i is distinct prime and $a_i \geq 0$, then show that

   (i) $\sum_{d|n} \mu(d)/d = \prod_{i=1}^{k} \left(1 - \frac{1}{p_i}\right)$ **(ii)** $\sum_{d|n} d\mu(d) = \prod_{i=1}^{k} (p_i - 1)$.

4. If $n > 1$ has k distinct prime factors, then $\phi(n) \geq n/2^k$.

5. If n and $n+2$ a pair of twin primes, then $\phi(n+2) = \phi(n) + 2$. Also verify it for $n = 12$, 14 and 20.

6. If $n > 1$ and $n = \prod_{i=1}^{k} p_i^{a_i}$, then show that

(a) $\sum_{d|n} \mu(d)\phi(d) = \prod_{i=1}^{k}(2 - p_i)$

(b) $\sum_{d|n} \frac{\phi(d)}{d} = \left(1 + \frac{a_1(p_1 - 1)}{p_1}\right)\left(1 + \frac{a_2(p_2 - 1)}{p_2}\right) \times \cdots \times \left(1 - \frac{a_k(p_k - 1)}{p_k}\right)$.

Chapter 6

Primitive Roots

It has been observed in Euler-Fermat's Theorem that $a^{\phi(m)} \equiv 1 \pmod{m}$, when $(a, m) = 1$. Now if we go through the following examples we shall look that for some (+)ve integer $r < \phi(m) a^r \equiv 1 \pmod{m}$.

Example: (1) Let us consider $m = 11,\ a = 2,\ \phi(m) = 10$

$2^1 \equiv 2 \pmod{11}$ $\quad 2^5 \equiv 10 \pmod{11}$ $\quad 2^9 \equiv 6 \pmod{11}$

$2^2 \equiv 4 \pmod{11}$ $\quad 2^6 \equiv 9 \pmod{11}$ $\quad 2^{10} \equiv 1 \pmod{11} \Rightarrow 2^{\phi(n)} \equiv 1 \pmod{11}$

$2^3 \equiv 8 \pmod{11}$ $\quad 2^7 \equiv 11 \pmod{11}$

$2^4 \equiv 5 \pmod{11}$ $\quad 2^8 \equiv 3 \pmod{11}$

(2) $m = 12, \phi(m) = \phi(12) = \phi((3 \cdot 2^2) = \phi(3) \cdot \phi(4) = 4.$

$a = 5,\ (a, m) = (5, 12) = 1$

$5^1 \equiv 5 \pmod{12}$ $\quad 5^3 \equiv 5 \pmod{12}$

$5^2 \equiv 1 \pmod{12}$ $\quad 5^4 \equiv 1 \pmod{12} \Rightarrow 5^{\phi(m)} \equiv 1 \pmod{12}$.

(3) $m = 7,\ a = 2,\ \phi(m) = 6$

$2^1 \equiv 2 \pmod 7$ $\quad 2^4 \equiv 2 \pmod 7$

$2^2 \equiv 4 \pmod 7$ $\quad 2^5 \equiv 4 \pmod 7$

$2^3 \equiv 1 \pmod 7$ $\quad 2^6 \equiv 1 \pmod 7 \Rightarrow 2^{\phi(m)} \equiv 1 \pmod 7$.

In Ex. (1), $\phi(m)$ is the power of 2, where $2^{\phi(m)} \equiv 1 \pmod{11}$.

In Ex. (2), $\exists$ a (+)ve integer $r = 2 < 4 = \phi(12)$ such that $5^2 \equiv 1 \pmod{12}$, though $5^{\phi(12)} \equiv 1 \pmod{12}$.

In Ex. (3), $\exists$ a (+)ve integer $r = 3 < 6 = \phi(7)$ such that $2^3 \equiv 1 \pmod 7$, though $2^{\phi(7)} \equiv 1 \pmod 7$.

Order (or Exponent)

Definition (6.1): *If there exists a smallest (+)ve integer k $(1 \le k \le \phi(m))$, $m > 1$ such that $a^k \equiv 1$ (mod m), then k is called the order of a modulo m (or exponent of a, denoted by $k = \exp_m(a)$), denoted by $k = O$ (a (mod m)) we shall use O (a (mod m)), as order of a (mod m)*

In Ex. (1) $O\ (2 \pmod{11}) = 10$

Ex. (2) $O\ (5 \pmod{12}) = 2$

Ex. (3) $O\ (2 \pmod 7) = 3$.

Primitive Root

Definition (6.2): *'a' is called a primitive root of m, if O (a (mod m)) $= \phi(m)$.*

That is, if $k = \phi(m)$, then '*a* is a primitive root of *m*'.

In Ex. (1) $k = \phi(11) = 10$ such that $2^{\phi(m)} \equiv 1 \pmod{11}$, so 2 is a primitive root of 11.

But Ex. (2) and Ex. (3) shows that 5 and 2 are not the primitive roots of 12 and 7 respectively.

The order of an integer 1 modulo any integer $m(> 1)$ is always 1, since $1^1 \equiv 1 \pmod{m}$.

So we shall consider $a > 1$.

Also $a^x \equiv 1 \pmod{m}$ is solvable iff $(a, m) = 1$.

Thus, there exists a (+)ve integer k satisfying $a^k \equiv 1 \pmod{m}$, iff $(a, m) = 1$.

[Go through the Example.]

Let $a = 2,\ m = 12,\ (a, m) = (2, 12) = 2 \neq 1$.

$2^1 \equiv 2 \pmod{12}$

$2^2 \equiv 4 \pmod{12}$

$2^3 \equiv 8 \pmod{12}$

$2^4 \equiv 2^{\phi(m)} \not\equiv 1 \pmod{12}$ and so on.

Thus, we see that there does not exist x such that $2^x \equiv 1 \pmod{12}$.

So we conclude that $O\ (a \pmod{m}) = k$ has an implication that

(i) $(a, m) = 1$

(ii) $a^k \equiv 1 \pmod{m}$,

(iii) $a^h \not\equiv 1 \pmod{m}$, $1 < h < k$.

Again, 'a is a primitive root of m' has an implication that

(i) $(a, m) = 1$

(ii) $a^{\phi(m)} \equiv 1 \pmod{m}$

(iii) $a^k \not\equiv 1 \pmod{m}$ $1 \leq k < \phi(m)$.

Example: 3 is a primitive root of 7, since

(i) $(3, 7) = 1$

(ii) $3^{\phi(7)} = 3^6 \equiv 1 \pmod{7}$

(iii) $\nexists\, k(1 < k < 6)$ such that $3^k \equiv 1 \pmod{7}$.

We observe in Ex. (2) and in Ex. (3) that the order of 5 (mod 12) and the order of 2 (mod 7) is a divisor of $\phi(12) = 4$ and $\phi(7) = 6$ respectively.

Lemma (6.1): If $xy \equiv 1 \pmod{m}$ and $x \equiv 1 \pmod{m}$, then $y \equiv 1 \pmod{m}$, if $(x, m) = 1$.

Proof: $xy \equiv 1 \pmod{m} \iff m | xy - 1$

and $\quad x \equiv 1 \pmod{m} \iff m | x - 1$

$$m | xy - 1 \text{ and } m | x - 1 \Rightarrow m | (xy - 1) - (x - 1) \Rightarrow m | x(y - 1)$$

$$\Rightarrow \quad m | y - 1 \text{ as } (x, m) = 1 \Rightarrow y \equiv 1 \pmod{m}.$$

Theorem (6.1): *If* $(a, m) = 1,\ k = O\ (a \pmod{m})$ *and if* $a^h \equiv 1 \pmod{m}$, *then* $k | h$ *and conversely.*

Also in particular, $k | \phi(m)$.

Proof: $(a, m) = 1 \Rightarrow a^{\phi(m)} \equiv 1 \pmod{m}$, by Euler-Fermat's theorem.

Let $\quad a^h \equiv 1 \pmod{m}$.

Then by divisional algorithm, $\exists\ q, r \in Z$

such that $h = kq + r, \quad 0 \le r < k$.

Then $\quad a^h \equiv 1 \pmod{m} \Rightarrow a^{kq+r} \equiv 1 \pmod{m}$

$\Rightarrow \quad (a^k)^q \cdot a^r \equiv 1 \pmod{m}$. (1)

Also $\quad O(a \pmod{m}) = k \Rightarrow a^k \equiv 1 \pmod{m}$

$\Rightarrow \quad (a^k)^q \equiv 1 \pmod{m}$. (2)

From (1) and (2) by Lemma (6.1) we have $a^r \equiv 1 \pmod{m}$, which implies that $r = 0$, otherwise it contradicts that k is the order of a modulo m, as k is the smallest (+)ve integer.

Therefore, $h = kq$ and hence $k|h$.

Conversely, let $k|h$.

Then $h = kq$ for some $q \in Z^+$.

Hence, $\quad a^h = a^{kq} = (a^k)^q \equiv 1 \pmod{m}$ [by (2)]

In particular, $h = \phi(m) \Rightarrow k|\phi(m)$.

Example: Let $m = 17,\ a = 2 \cdot (a, m) = 1$

$2^8 \equiv 1 \pmod{17} \Rightarrow O(2 \pmod{17}) = 8\ (= k)$

$\phi(m) = \phi(17) = 16$. So $8|16$.

Hence, $k|\phi(17)$.

Theorem (6.2): *Let* $k = O(a \pmod{m}),\ (a, m) = 1,\ m > 1$. *Then*

(i) $a^p \equiv a^q \pmod{m} \Longleftrightarrow p - q \equiv 0 \pmod{k}$.

(ii) The set of numbers $1, a, a^2, \ldots, a^{k-1}$ forms an incongruent system, modulo m.

(iii) $O\ (a^h \pmod{m})$ is $\dfrac{k}{(h, k)}$.

Proof: (i) Let $a^p \equiv a^q \pmod{m}$, $p \ge q$.

Then $\quad a^{p-q} \equiv 1 \pmod{m}$ [$\because\ (a, m) = 1 \Rightarrow (a^q, m) = 1$] (1)

Also $\quad a^k \equiv 1 \pmod{m}$ (2)

as $O\ (a \pmod{m}) = k$.

From (1) and (2), it follows by Th. (6.1) that $k|p - q$.

Therefore, $k|p - q \Rightarrow p - q \equiv 0 \pmod{k}$.

Conversely, let $p - q \equiv 0 \pmod{k}$.

Then $\quad p - q = kt$ for some $t \in Z^+$.

Therefore, $a^{p-q} = a^{kt} = (a^k)^t \equiv 1 \pmod{m}$ [by (2)]

$\Rightarrow \quad a^p \equiv a^q \pmod{m}$.

(ii) The nos. $1, 2, \ldots, k - 1$ are incongruent modulo m.

If possible, let $a^i \equiv a^j \pmod{m}$ for some i, j such that $0 \le j < i < k$

$\Rightarrow \quad a^{i-j} \equiv 1 \pmod{m}$ $[\because (a^j, m) = 1]$

Now $\quad 0 < i - j < k$.

So $a^{i-j} \equiv 1 \pmod{m}$ is not possible as k is order of a (mod m).

Hence, $\quad a^i \not\equiv a^j \pmod{m}$.

Therefore, $1, a, a^2, \ldots, a^{k-1}$ are incongruent modulo m.

(iii) Given that $a^k \equiv 1 \pmod{m}$, $1 \le k \le \phi(m)$. (1)

Let $\quad (h, k) = d$.

Then $\quad d|h,\ d|k \Rightarrow h = dp,\ k = dq$ for some $p, q \in Z^+$, $(p, q) = 1$.

Now $\quad (a^h)^q = (a^{dp})^{\frac{k}{d}} = (a^p)^k = (a^k)^p \equiv 1 \pmod{m}$ [by (1)]

$\therefore \quad (a^h)^q \equiv 1 \pmod{m}$. (2)

Let $\quad x = 0(a^h \pmod{m})$, $1 \le x \le q$.

Then $\quad (a^h)^x \equiv 1 \pmod{m}$, $1 \le x < q$.

So from (2) it follows that $x|q$. (3)

Again, $(a^h)^x \equiv 1 \pmod{m} \Rightarrow a^{hx} \equiv 1 \pmod{m}$.

Since $k = 0(a \pmod{m})$, it follows $k|hx$.

Now $\quad k|hx \Rightarrow dq|dpx \Rightarrow q|px \Rightarrow q|x,\ (p, q) = 1$. (4)

So from (3) and (4), $x = q$

$$\Rightarrow \quad 0(a^h \pmod{m}) = q = \frac{k}{d} = \frac{k}{(h, k)}.$$

Hence the theorem.

Corollary (6.1): If $x \equiv a \pmod{m}$, then $O(a^h \pmod{m}) = k$, if and only if $(h, k) = 1$.

Proof: Follows from above theorem (Part iii)

Problem (6.1): Find the order of (i) 5^6 (mod 12), (ii) a (mod 12), where $a < 12$ and $(a, 12) = 1$, (iii) a (mod 15) when $a = 2$, 4, 7, 8, 11, 13 and 14.

Solution: (i) $5^2 \equiv 1 \pmod{12}$

$\Rightarrow \quad 0(5 \pmod{12}) = 2\ (= k)$.

$$\text{Therefore, } 0(5^6 \pmod{12}) = \frac{k}{(h, k)} = \frac{2}{(6, 2)} = \frac{2}{2} = 1.$$

(ii) $a < 12$ and $(a, 12) = 1$

$\Rightarrow \quad a = 1, 5, 7, 11,\ \phi(12) = 4$.

$a = 1,\ 0(1 \pmod{12}) = 1$

$a = 5,\ 5^2 \equiv 1 \pmod{12} \Rightarrow 0(5 \pmod{12}) = 2$

$a = 7,\ 7^2 \equiv 1 \pmod{12} \Rightarrow 0(7 \pmod{12}) = 2$

$a = 11,\ 11^2 \equiv 1 \pmod{12} \Rightarrow 0(11 \pmod{12}) = 2$.

(iii) $a = 2,\ \phi(15) = 8,\ 2^4 \equiv 1 \pmod{15} \Rightarrow 0(2 \pmod{15}) = 4$

$a = 7,\ 7^2 \equiv 4 \pmod{15}$

$\Rightarrow \quad 7^4 \equiv 16 \pmod{15} \equiv 1 \pmod{15} \Rightarrow 0(7 \pmod{15}) = 4$

$a = 8,\ 8^2 \equiv 4 \pmod{15}$

$\Rightarrow\ 8^4 \equiv 1 \pmod{15} \Rightarrow\ 0(8 \pmod{15}) = 4.$

$a = 11,\ 11^2 \equiv 1 \pmod{15} \Rightarrow\ 0(11 \pmod{15}) = 2$

$a = 13,\ 13^2 \equiv 4 \pmod{15}$

$\Rightarrow\ 13^4 \equiv 1 \pmod{15} \Rightarrow\ 0(13 \pmod{15}) = 4.$

$a = 14,\ 14^2 \equiv 1 \pmod{15} \Rightarrow\ 0(14 \pmod{15}) = 2.$

Problem (6.2): If $0(a \pmod n) = hk$, then $0(a^h \pmod n) = k$.

Solution: By (iii) of Th. (6.2),

$$0(a^h \pmod n) = \frac{0(a(\bmod n))}{(h, 0(a(\bmod n)))} = \frac{hk}{(h, hk)} = \frac{hk}{h} = k.$$

Alternative way

$0(a \pmod n) = hk\ \Rightarrow\ a^{hk} \equiv 1 \pmod n.$

Keeping h fixed, hk is the smallest (+)ve integer $\leq \phi(n)$, when k is smallest.

So $a^{hk} \equiv 1 \pmod n \Rightarrow (a^h)^k \equiv 1 \pmod n$

$\Rightarrow\ 0(a^h \pmod n) = k.$

Problem (6.3): Find $O(a \pmod{19})$, when $a = 3$.

Solution: $3^3 \equiv 8 \pmod{19}$

$\Rightarrow 3^6 \equiv 64 \pmod{19} \Rightarrow\ 3^6 \equiv 7 \pmod{19}\quad [\because\ 64 - 3 \cdot 19 = 7]$

$\Rightarrow 3^{18} \equiv 7^3 \pmod{19} \Rightarrow\ 3^{18} \equiv 343 \pmod{19} \equiv 343 - 18 \cdot 19 \pmod{19}$

$\Rightarrow\ 3^{18} \equiv 1 \pmod{19}.$

So $0(3 \pmod{19}) = 18 = \phi(19).$

Problem (6.4): Prove that $\phi(2^n - 1)$ is multiple of n for any prime $n > 1$.

Solution: We have $2^n \equiv 1 \pmod{2^n - 1}\ \forall\ n > 1$

$\Rightarrow\quad 0(2(\bmod (2^n - 1)) = n$ [since n is the smallest (+)ve integer $< \phi(2^n - 1)\ \forall n > 1$] (1)

Also $2^{\phi(2^n-1)} \equiv 1(\bmod (2^n - 1)),$ (2)

since $(2, 2^n - 1) = 1$ as $2^n - 1$ is odd $\forall\ n > 1$.

So from (1) and (2), it follows [by Th. (6.1)]

that $n|\phi(2^n - 1)$ and so $\phi(2^n - 1) = n.t,\ t \in Z^+$

$\Rightarrow\quad \phi(2^n - 1)$ is multiple of n.

Problem (6.5): If $O(a \pmod p)$ is 3, where p is a prime, then show that $O((a + 1)(\bmod p)) = 6$.

Solution: Given that $0(a \bmod p)) = 3$

$\Rightarrow\quad a^3 \equiv 1 \pmod p$ (1)

$\Rightarrow\quad a^3 - 1 \equiv 0 \pmod p\ \Rightarrow\ (a - 1)(a^2 + a + 1) \equiv 0 \pmod p$

$\Rightarrow\quad a^2 + a + 1 \equiv 0 \pmod p$ [$a \not\equiv 1 \pmod p$ as the order of $a \pmod p = 3$]

$\Rightarrow\quad (a + 1)^2 \equiv a \pmod p\ \Rightarrow\ (a + 1)^6 \equiv a^3 \pmod p \equiv 1 \pmod p.$ (2)

Again, (1) $\Rightarrow$ $3|\phi(p)$,

p is an odd prime, so $\phi(p)$ is even.

Therefore, $3|\phi(p) \Rightarrow \phi(p) = 3k,\ k \in Z$.

Now $\phi(p)$ is even when k is even.

$\therefore \quad \phi(p) = 3 \cdot 2m = 6m,\ m \in Z^+$.

So 6 is the smallest (+)ve integer which divides $\phi(p)$.

Hence from (2), it follows that $0((a+1)(\text{mod } p)) = 6$.

Problem (6.6): If $O(a \ (\text{mod } p)) = 2k$, p is an odd prime, then show that $a^k \equiv -1$ (mod p).

Solution: $0(a \ (\text{mod } p)) = 2k$

$\Rightarrow \quad a^{2k} \equiv 1 \ (\text{mod } p) \Rightarrow (a^k - 1)(a^k + 1) \equiv 1 \ (\text{mod } p)$

$\Rightarrow \quad a^k + 1 \equiv 0 \ (\text{mod } p) \quad [\text{But } a^k - 1 \not\equiv 0 \ (\text{mod } p)]$

$\Rightarrow \quad a^k \equiv -1 \ (\text{mod } p)$.

Theorem (6.3): *If* $(a, m) = (b, m) = 1$ *and* $0(a(\text{mod } m)),\ 0(b(\text{mod } m)) = 1$, *then* $0(ab(\text{mod } m)) = O(a(\text{mod } m)) \times O(b(\text{mod } m))$.

***Proof*:** Let $O(a(\text{mod } m)) = k_1$ and $O(b(\text{mod } m)) = k_2$

and $\quad O(ab(\text{mod } m)) = k$.

Then $\quad a^{k_1} \equiv 1 \ (\text{mod } m))$ (i)

$\quad b^{k_2} \equiv 1 \ (\text{mod } m)$, (ii)

where $\quad (k_1, k_2) = 1$

and $\quad (ab)^k \equiv 1 \ (\text{mod } m)$. (iii)

(i) and (ii) $\Rightarrow a^{k_1 k_2} \equiv 1 \ (\text{mod } m)$ and $b^{k_1 k_2} \equiv 1 \ (\text{mod } m) \Rightarrow (ab)^{k_1 k_2} \equiv 1 \ (\text{mod } m)$ (iv)

(iii) $\Rightarrow \quad [(ab)^k]^{k_2} \equiv 1 \ (\text{mod } m)$

$\Rightarrow \quad a^{kk_2} \cdot b^{kk_2} \equiv 1 \ (\text{mod } m)$

$\Rightarrow \quad a^{kk_2} \equiv 1 \ (\text{mod } m)$

[since from (ii) $b^{kk_2} \equiv 1 \ (\text{mod } m)$ and so by Lemma (6.1)]

$\Rightarrow \quad k_1 | kk_2 \Rightarrow k_1 | k$, since $(k_1, k_2) = 1$.

Similarly, we can show that $k_2 | k$.

Hence, $k_1|k,\ k_2|k \Rightarrow (k_1 k_2 | k)$. (v)

From (iv), $(a, b)^{k_1 k_2} \equiv 1 \ (\text{mod } m)$.

So by (iii), $k | k_1 k_2$. (vi)

From (v) and (vi), $k = k_1 k_2$.

That is, $O(ab(\text{mod } m)) = O(a(\text{mod } m)) \cdot O(b(\text{mod } m))$.

Example: $0(7(\text{mod } 18) = 3,\ 0(17(\text{mod } 18)) = 2$

$k_1 = 3,\ k_2 = 2,\ (k_1, k_2) = (3, 2) = 1$.

Then $\quad O(119(\text{mod } 18) = O(7 \cdot 17(\text{mod } 18)) = O(7(\text{mod } 18)) \cdot O(17(\text{mod } 18))$

$= k_1 \cdot k_2 = 2 \cdot 3 = 6$.

Theorem (6.4): *If* $O(a(\text{mod } m_1)) = k_1$, $O(b(\text{mod } m_2)) = k_2$, $(a, m_1) = 1$, $(b, m_2) = 1$, *then* $O(a(\text{mod } [m_1, m_2]) = [k_1, k_2]$.

Proof: Let $[k_1, k_2] = k$.

Given that $a^{k_1} \equiv 1(\text{mod } m_1)$ (i)

$a^{k_2} \equiv 1(\text{mod } m_2)$. (ii)

$[k_1, k_2] = k \Rightarrow k_1|k, k_2|k \Rightarrow k = k_1p, k = k_2q.\ p, q \in Z^+$.

Then from (i), $(a^{k_1})^p \equiv 1 \text{ (mod } m_1) \Rightarrow a^{k_1p} \equiv 1 \text{ (mod } m_1)$

$\Rightarrow \quad a^k \equiv 1 \text{ (mod } m_1)$. (iii)

Similarly, $a^k \equiv 1 \text{ (mod } m_2)$ (iv)

From (iii) and (iv), $m_1|a^k - 1$, $m_2|a^k - 1$

$\Rightarrow \quad [m_1, m_2]|a^k - 1 \Rightarrow a^k - 1 \equiv 0 \text{ (mod } [m_1, m_2])$

$\Rightarrow \quad a^k \equiv 1 \text{ (mod } [m_1, m_2]) \Rightarrow a^{[k_1,k_2]} \equiv 1 \text{ (mod } [m_1, m_2])$

$\Rightarrow \quad O(a(\text{mod } [m_1, m_2])) = [k_1, k_2]$.

Problem (6.7): (i) Find $O(3 \text{ (mod } 455))$.

Solution: $455 = 5 \times 7 \times 13$

Now $3^4 \equiv 1 \text{ (mod } 5)$ (i)

$3^6 \equiv 1 \text{ (mod } 7)$ (ii)

$3^3 \equiv 1 \text{ (mod } 13)$ (iii)

4, 6 and 3 are smallest (+)ve integers satisfying (i), (ii) and (iii) respectively.

Thus, $O(3(\text{mod } 5)) = 4$, $O(3(\text{mod } 7)) = 6$ and $O(3(\text{mod } 13)) = 3$.

Therefore, by Th. (6.4), $0(3(\text{mod } [5, 7.13]) = [4, 6, 3]$.

That is, $0(3(\text{mod } 455)) = 12$.

(ii) If $0(a(\text{mod } n)) = n - 1$, then prove that n is prime.

Given that $0(a(\text{mod } n)) = n - 1$.

So $a^{n-1} \equiv 1 \text{ (mod } n)$.

Also $a^{\phi(n)} \equiv 1 \text{ (mod } n)$.

Then $n-1|\phi(n)$ and this is true if n is prime, since for a prime number n, $\phi(n) = n-1$.

Theorem (6.5): *If* $O(a(\text{mod } p^k)) = h$, *then* $O(a(\text{mod } p^{k+1}))$ *is either* h *or* ph, *when* p *is prime.*

Proof: Let $O(a(\text{mod } p^{k+1})) = r$.

Then $a^r \equiv 1(\text{mod } p^{k+1})$ (i)

$\Rightarrow \quad a^r = 1 + p^{k+1}t$ for same $t \in Z$

$= 1 + p^k(pt),\ pt \in Z$

$\Rightarrow \quad a^r \equiv 1(\text{mod } p^k)$ (ii)

Also by the given condition, $a^h \equiv 1(\text{mod } p^k)$ (iii)

(ii) and (iii) $\Rightarrow h|r$ (iv)

Again, $a^h \equiv 1(\text{mod } p^k) \Rightarrow a^h = 1 + p^kt,\ t \in Z$

$$\Rightarrow \quad (a^h)^p = (1+p^k t)^p = 1 + p_{c_1} p^k \cdot t + p_{c_2} p^{2k} \cdot t^2 + \cdots + p^{kp} \cdot t^p$$

$$= 1 + p^{k+1} \cdot t + p^{2k+1} \frac{(p-1)}{2} \cdot t^2 + \cdots$$

$$= 1 + \text{ terms divisible by } p^{k+1}$$

$$\Rightarrow \quad a^{hp} \equiv 1 \pmod{p^{k+1}}. \quad \text{(v)}$$

From (i) and (v), $r|hp$.

From (iv), $h|r \Rightarrow r = hc$ for some $c \in Z$

and $\quad r|hp \Rightarrow hp = r \cdot l,\ l \in Z$

$\Rightarrow \quad hp = hc \cdot l \Rightarrow p = c \cdot l \Rightarrow c|p.$

Since p is prime, $c = 1$ or p.

$c = 1 \Rightarrow r = h,$

$c = p \Rightarrow r = hp,$

Thus r is either h or hp.

Theorem (6.6): *The necessary and sufficient condition that an integer 'a' to be a primitive root of m is* $a^{\frac{\phi(m)}{p}} \not\equiv 1 \pmod{m}$ *for every prime divisor p of* $\phi(m)$, $(a, m) = 1$.

Proof: Let 'a' be a primitive root of m. Then from definition (6.2), $a^{\phi(m)} \equiv 1 \pmod{m}$ and

$$a^k \not\equiv 1 \pmod{m},\ 1 \le k < \phi(m) \quad (6.1)$$

Let $\phi(m) = p_1^{a_1}, p_2^{a_2}, \ldots, p_n^{a_n}$, $a_i \ge 0$, p_i are primes. Then each $p_i|\phi(m)$ and so $\frac{\phi(m)}{p_i}$ is an integer lying between 1 and $\phi(m)$ and consequently by (1), $a^{\frac{\phi(m)}{p_i}} \not\equiv 1 \pmod{m}$ for every prime divisor p_i of $\phi(m)$.

Hence, $a^{\frac{\phi(m)}{p}} \not\equiv 1 \pmod{m}$ for every prime divisor of $\phi(m)$.

Next, let the given condition hold.

If possible, let 'a' be not a primitive root of m.

Then there exists a (+)ve integer $k < \phi(m)$ such that

$$a^k \equiv 1 \pmod{m}. \quad (1)$$

Again, $a^{\phi(m)} \equiv 1 \pmod{m}$ by Fermat's Theorem.

Then $\quad k|\phi(m)$.

Also p being a prime divisor of $\phi(m)$, $p|\phi(m)$.

Therefore, $k|\phi(m),\ p|\phi(m) \Rightarrow kp|\phi(m)$.

In other words, $\dfrac{\phi(m)}{k.p}$ is an integer.

Therefore, $a^{\frac{\phi(m)}{p}} = (a^k)^{\frac{\phi(m)}{kp}} = 1 \pmod{m}$ by (1),

which contradicts our given condition.

Hence the theorem.

Example: Let $m = 19,\ a = 2,\ (a, m) = (2, 19) = 1$

$\phi(m) = 18 = 2 \cdot 3^2.$

Case(1): 3 is prime divisor of $\phi(m)$.

Now $a^{\frac{\phi(m)}{p}} = 2^{\frac{18}{3}} = 2^6 \not\equiv 1 \pmod{19}$.

Case(2): 2 is a prime divisor of $\phi(m)$.

$$2^{\frac{\phi(m)}{p}} = 2^{\frac{18}{2}} = 2^9 \not\equiv 1 \pmod{19}.$$

So from Case (1) and (2), it follows that 2 is a primitive root of 19.

But 5 is not a primitive root of 19, though $(5, 19) = 1$,

since $5^9 \equiv 1 \pmod{19}$, $5^6 \equiv 7 \pmod{19}$.

Lemma (6.2): If $m = 2^h,\ h > 2$ and $(a, m) = 1$, then $a^{\frac{\phi(m)}{2}} \equiv 1 \pmod{m}$.

Proof: $m = 2^h,\ h > 2$ and $(a, m) = 1 \Rightarrow a$ is an odd integer.

Also square of an odd integer is of the form $8k + 1,\ k \in Z$. (1)

For $h = 3(> 2),\ m = 2^3 = 8,\ \phi(m) = 4$ and $(a, 8) = 1$.

Then $a^{\frac{\phi(m)}{2}} = a^{\frac{4}{2}} = a^2 \equiv 1 \pmod{2^3}$ by (1).

So the result is true for $h = 3$.

Let the result be true for $h = r(> 2)$.

Then $a^{\frac{\phi(2^r)}{2}} \equiv 1 \pmod{2^r}$ (2)

$\Rightarrow$ $a^{\frac{\phi(2^r)}{2}} = 1 + 2^r \cdot t$, for some $t \in Z$

$\Rightarrow$ $a^{\phi(2^r)} = (1 + 2^r t)^2$.

$\therefore$ $a^{\phi(2^r)} = (1 + 2^r t)^2 = 1 + 2^{r+1} t + 2^{2r} \cdot t^2$. (3)

Again, $\dfrac{\phi(2^{r+1})}{2} = \dfrac{2^{r+1} \cdot \frac{1}{2}}{2} = \dfrac{2^r}{2} = 2^{r-1} = \phi(2^r)$.

Therefore, $a^{\frac{\phi(2^{r+1})}{2}} = a^{\phi(2^r)} = 1 + 2^{r+1} \cdot t(1 + 2^{r-1} \cdot t)$ by (3)

$\Rightarrow$ $a^{\frac{\phi(2^{r+1})}{2}} \equiv 1 \pmod{2^{r+1}},\ r > 2 \Rightarrow 1 + 2^{r-1} \cdot t \in Z$.

So the result is true for $h = r + 1,\ r > 2$.

Hence, it is true $\forall\ h(> 2)$.

Therefore, $a^{\frac{\phi(m)}{2}} \equiv 1 \pmod{m}$.

Theorem (6.7): *If $m = 2^h,\ h > 2$, then m has no primitive root.*

Proof: If possible, let b be a primitive root of m.

Then $(b, m) = 1$.

Therefore, by above Lemma (6.2),

$$b^{\phi(m)/2} \equiv 1 \pmod{m},\ \frac{\phi(m)}{2} < \phi(m),$$

which shows that b is not a primitive root of m. Hence, m has no primitive root.

Example: Let $m = 2^3$. Then 3 is not a primitive root of 8. Also m has no primitive root.

$$m = 2^3,\ h = 3 > 2$$

$$\phi(m) = 4,\ b = 3,\ (b, m) = (3, 8) = 1.$$

Now $3^{\frac{\phi(m)}{2}} = 3^{\frac{4}{2}} = 3^2 \equiv 1 \pmod 8$

$\Rightarrow$ 3 is not a primitive root of 8.

$(5, 8) = 1,\ 5^{\frac{\phi(m)}{2}} = 5^2 \equiv 1 \pmod 8$

$(7, 8) = 1,\ 7^{\frac{\phi(m)}{2}} = 7^2 \equiv 1 \pmod 8$,

which shows that 5, 7 are not primitive root of 8. Hence 8 has no primitive root.

Theorem (6.8): *If 'a' is a primitive root of m, then* $a^{\frac{\phi(m)}{2}} \equiv -1 \pmod m$.

Proof: a is a primitive root of m

$\Rightarrow\quad a^{\phi(m)} \equiv 1 \pmod m,\ (a, m) = 1$

and $\quad \nexists$ any $k < \phi(m)$ such that $a^k \equiv 1 \pmod m$.

Now $\quad a^{\phi(m)} \equiv 1 \pmod m \Rightarrow a^{\phi(m)} - 1 \equiv 0 \pmod m$

$\Rightarrow\quad [a^{\frac{\phi(m)}{2}} - 1][a^{\frac{\phi(m)}{2}} + 1] \equiv 0 \pmod m$

$\Rightarrow\quad a^{\frac{\phi(m)}{2}} + 1 \equiv 0 \pmod m \quad \left[\text{since } a^{\frac{\phi(m)}{2}} \not\equiv 1 \pmod m,\ \frac{\phi(m)}{2} < \phi(m)\right]$

$\Rightarrow\quad a^{\frac{\phi(m)}{2}} \equiv -1 \pmod m$.

Problem (6.8): Find four primitive roots of 26 and 25.

Solution: Primitive roots of 26: $(3, 26), (5, 26), (7, 26), (9, 26), (11, 26), (15, 26) = 1$

$\phi(m) = \phi(26) = \phi(2.13) = \phi(2) \cdot \phi(13) = 12.$

$p =$ prime divisors of 12 are 2, 3.

$\frac{\phi(m)}{2} = \frac{12}{2} = 6$ and $\frac{\phi(m)}{3} = 4.$

(i) $3^3 \equiv 1 \pmod{26}$, $3^4 \equiv 3 \pmod{26}$

$\Rightarrow\quad 3^6 \equiv 1 \pmod{26}$.

Thus, $3^4 \not\equiv 1 \pmod{26}$, but $3^6 \equiv 1 \pmod{26}$.

So 3 is not a primitive root of 26.

(ii) $5^2 \equiv -1 \pmod{26}$

$\Rightarrow\quad 5^2 \equiv 25 \pmod{26}$ (i)

$\Rightarrow\quad 5^3 \equiv 125 \pmod{26} \equiv 21 \pmod{26}$

$\Rightarrow\quad 5^6 \equiv 441 \pmod{26} \equiv 441 - 17 \times 26 \pmod{26}$

$\equiv -1 \pmod{26}$. (ii)

Also (i) $\Rightarrow$ $5^4 \equiv 625 \pmod{26} \equiv 625 - 24 \times 26 \pmod{26}$

$\equiv 1 \pmod{26}$. (iii)

(ii) and (iii) shows that 5 is not a prime root of 26.

(iii) Now $7^2 \equiv 23 \pmod{26}$

$\Rightarrow\quad 7^3 \equiv 161 \pmod{26} \equiv 161 - 26 \times 6 \pmod{26}$

$\equiv 5 \pmod{26}$

$\Rightarrow\quad 7^6 \equiv 25 \pmod{26} \equiv -1 \pmod{26}$.

Also $7^2 \equiv 23 \pmod{26}$

$\Rightarrow \quad 7^3 \equiv 161 \pmod{26} \equiv 5 \pmod{26}$

$\Rightarrow \quad 7^4 \equiv 35 \pmod{26} \equiv 9 \pmod{26}$.

$\therefore \quad 7^6, 7^4 \not\equiv 1 \pmod{26} \Rightarrow$ 7 is a primitive root of 26.

(iv) $9^2 \equiv 3 \pmod{26} \Rightarrow 9^4 \equiv 9 \pmod{26}$.

Also $9^6 \equiv 27 \pmod{26} \Rightarrow 9^6 \equiv 1 \pmod{26}$. So 9 is not a primitive root of 26.

(v) $11^2 \equiv 17 \pmod{26}$

$\Rightarrow \quad 11^3 \equiv 187 \pmod{26} \equiv 187 - 7 \times 26 \pmod{26}$

$\equiv 5 \pmod{26}$

$\Rightarrow \quad 11^6 \equiv -1 \pmod{26}$

$11^3 \equiv 5 \pmod{26} \Rightarrow 11^4 \equiv 55 \pmod{26}$

$\equiv 3 \pmod{26}$.

$\therefore \quad 11^6, 11^4 \not\equiv 1 \pmod{26} \Rightarrow$ 11 is a primitive root of 26.

(vi) $15^2 \equiv 17 \pmod{26}$

$15^4 \equiv 289 \pmod{26}$

or, $\quad 15^4 \equiv 289 - 11 \times 26 \pmod{26}$

or, $\quad 15^4 \equiv 3 \pmod{26}$.

Also $\quad 15^6 \equiv 51 \pmod{26} \equiv -1 \pmod{26}$.

Therefore, $15^6, 15^4 \not\equiv 1 \pmod{26}$.

So 15 is a primitive root of 26.

(vii) $(17, 26) = 1$

$17^2 \equiv 3 \pmod{26}$

$\Rightarrow \quad 17^4 \equiv 9 \pmod{26}$

$\Rightarrow \quad 17^6 \equiv 27 \pmod{26} \equiv 1 \pmod{26}$.

So 17 is not a primitive root of 26.

(viii) $(19, 26) = 1$

Now $\quad 19^2 \equiv 23 \pmod{26} \equiv -3 \pmod{26}$

$19^3 \equiv -57 \pmod{26} \equiv -5 \pmod{26}$

$\Rightarrow \quad 19^4 \equiv -95 \pmod{26} \equiv -95 + 104 \pmod{26}$

$\equiv 9 \pmod{26}$.

Also $\quad 19^6 \equiv 23 \times 9 \pmod{26} \equiv 207 \pmod{26}$

$\equiv -1 \pmod{26}$.

$\therefore \quad 19^6, 19^4 \not\equiv 1 \pmod{26}$.

So 19 is a primitive root of 26.

Therefore, four primitive roots of 26 are 7, 11, 15, 19.

Primitive roots of 25:

$$(2, 25), (3, 25), (4, 25), (6, 25), (7, 25), (8, 25), (9, 25) = 1.$$

Now $m = 25,\ \phi(m) = \phi(25) = \phi(5^2) = 20.$

$p =$ prime divisor of $20 = 2, 5.$

$$\frac{\phi(m)}{2} = \frac{20}{2} = 10 \text{ and } \frac{\phi(m)}{5} = \frac{20}{5} = 4.$$

(i) $2^5 \equiv 7 \pmod{25}$, $2^4 \equiv -9 \pmod{25}$

$2^{10} \equiv 49 \pmod{25} \equiv -1 \pmod{25}.$

$\therefore\quad 2^{10}, 2^4 \not\equiv 1 \pmod{25}.$

2 is a primitive root of 25.

(ii) $3^3 \equiv 2 \pmod{25}$, $\Rightarrow 3^4 \equiv 6 \pmod{25}$

Also $3^5 \equiv 18 \pmod{25} \Rightarrow 3^{10} \equiv 324 \pmod{25} \equiv -1 \pmod{25}$

$\therefore\quad 3^{10}, 3^4 \not\equiv 1 \pmod{25}.$

3 is a primitive root of 25.

(iii) Now, $4^3 \equiv 14 \pmod{25}$

$\Rightarrow\quad 4^4 \equiv 56 \pmod{25}$

$\Rightarrow\quad 4^4 \equiv 6 \pmod{25}$ (i)

$\Rightarrow\quad 4^5 \equiv 24 \pmod{25}$

$\Rightarrow\quad 4^{10} = 576 \pmod{25}$

$\Rightarrow\quad 4^{10} \equiv 576 - 23 \times 25 \pmod{25}$

$\Rightarrow\quad 4^{10} \equiv 1 \pmod{25}$ (ii)

$\because$ from (i) & (ii) $4^4 \equiv 6 \pmod{25}$ and $4^{10} \equiv 1 \pmod{25}$

$\Rightarrow$ 4 is not a primitive root of 25.

(iv) Since $6^{10} \equiv 1 \pmod{25}$, 6 is not a primitive root.

(v) $8^2 \equiv 14 \pmod{25}$

$\Rightarrow\quad 8^4 \equiv 196 \pmod{25} \equiv 21 \pmod{25} \equiv -4 \pmod{25}$

$\Rightarrow\quad 8^5 \equiv -32 \pmod{25} = 18 \pmod{25}$

$\Rightarrow\quad 8^{10} \equiv 324 \pmod{25} \equiv 324 - 17 \times 25 \pmod{25}$

$\equiv -1 \pmod{25}.$

$\therefore\quad 8^4 \equiv -4 \pmod{25}$, $8^{10} \equiv -1 \pmod{25}$

$\Rightarrow$ 8 is a primitive root of 25.

(vi) $(13, 25) = 1$

$13^2 \equiv 19 \pmod{25}$ or, $13^2 \equiv -6 \pmod{25}$

$\Rightarrow\quad 13^3 \equiv -78 \pmod{25} \equiv -3 \pmod{25}$

$\Rightarrow\quad 13^4 \equiv -39 \pmod{25} \equiv 11 \pmod{25}$

$\Rightarrow\quad 13^5 \equiv 143 \pmod{25}$ or, $13^5 \equiv 18 \pmod{25}$

or, $\quad 13^{10} \equiv 324 \pmod{25} \equiv 324 - 13 \times 25 \pmod{25} \equiv -1 \pmod{25}.$

$\therefore \quad 13^4 \equiv 1 \pmod{25}$ and $13^{10} \equiv -1 \pmod{25}$

$\Rightarrow$ 13 is a primitive root of 25.

So the four primitive roots of 25 are 2, 3, 8 and 13.

Theorem (6.9): *The necessary and sufficient condition that 'a', when* $(a, m) = 1$, *to be a primitive root of m is that the numbers* $a, a^2, \ldots, a^{\phi(m)}$ *form a reduced residue system* mod m.

Proof: Let $(a, m) = 1$ and let 'a' be a primitive root of m. Then $a^{\phi(m)} \equiv 1 \pmod{m}$, $\phi(m)$ is the smallest (+)ve integer. Then the set of numbers $a, a^2, \ldots, a^{\phi(m)}$ are incongruent mod m by Th. [(ii) of 6.2]. That is, the numbers $a, a^2, \ldots, a^{\phi(m)}$ form a reduced residue system mod m.

Conversely, let $a, a^2, \ldots, a^{\phi(m)}$ form a reduced residue system mod m. Then $a^{\phi(m)} \equiv 1 \pmod{m}$ and $a^r \not\equiv 1 \pmod{m}$ for $1 \leq r < \phi(m)$. For if $a^r \equiv 1 \pmod{m}$, then $a^{\phi(m)} - a^r \equiv 0 \pmod{m} \Rightarrow a^{\phi(m)}, a^r$ are congruent mod m, which contradicts our assumption.

Hence 'a' is a primitive root of m.

Note: If 'a' is a primitive root of m, then we have seen that $a, a^2, \ldots, a^{\phi(m)}$ form a reduced residue system mod m, which is a G.P. series with common ratio 'a'. But since all m has no primitive root, we have a restriction and in this context we shall prove that m has a primitive root, if and only if $m = 1, 2, 4, p^\alpha$ and $2p^\alpha$, when p is an odd prime and $\alpha \geq 1$.

It is easy to verify the result for $m = 1, 2, 4$.

(i) $m = 1$, $\phi(m) = \phi(1) = 1 \Rightarrow 1' \equiv 1 \pmod{1} \Rightarrow 1^{\phi(1)} \equiv 1 \pmod{1}$

(ii) $m = 2$, $\phi(2) = 1 \Rightarrow 1^1 \equiv 1 \pmod{2} \Rightarrow 1^{\phi(2)} \equiv 1 \pmod{2}$

(iii) $m = 4$, $\phi(4) = 2 \Rightarrow 3^2 \equiv 1 \pmod{4} \Rightarrow 3^{\phi(4)} \equiv 1 \pmod{4}$.

(iv) Let $m = 3^2$, when $p = 3$ an odd prime, $\alpha = 2 > 1$.

Then $\quad \phi(m) = \phi(3^2) = 3^2\left(1 - \dfrac{1}{3}\right) = 3 \cdot 2 = 6.$

(a) $2^{\phi(m)} = 2^6 \equiv 1 \pmod{3^2}$.

(b) $\phi(2p^\alpha) = \phi(2 \cdot 3^2) = \phi(18) = \phi(2)\phi(3^2) = 1 \cdot 6 = 6.$

$$\Rightarrow \qquad 5^{\phi(2\cdot 3^2)} \equiv 1 \pmod{2 \cdot 3^2} \qquad \text{[See Prob. 6.11]}$$

Hence, it follows that 1, 3, 2, 5 are primitive roots of 2, 4, 3^2, $2 \cdot 3^2$ respectively by (ii), (iii), (a) & (b) of (iv). It is to be noted that 1 is also a primitive root of 1 [by (i)].

In general, 2, 4, p^α, $2p^\alpha$ where p is an odd prime, $\alpha \geq 1$ has primitive roots. [See Th. 6.13]

Theorem (6.10): *If p be an odd prime, then there is at least one primitive root of p such that* $a^{p-1} \not\equiv 1 \pmod{p^2}$.

Proof: Let 'a' be a primitive root. If the condition $a^{p-1} \not\equiv 1 \pmod{p^2}$ is satisfied, then the proof is complete.

Now if $a^{p-1} \equiv 1 \pmod{p^2}$ (1)

holds, then we shall establish that there is another primitive root $a' = a + p$ of p satisfying the condition $a'^{p-1} \not\equiv 1 \pmod{p^2}$.

$$(a')^{p-1} = (a+p)^{p-1} = a^{p-1} + (p-1)p \cdot a^{p-2} + \frac{(p-1)(p-2)}{2!} p^2 \cdot a^{p-3} + \cdots + p^{p-1}$$

$$= a^{p-1} - p \cdot a^{p-2} + p^2 \left[a^{p-2} + \frac{(p-1)(p-2)}{2!} a^{p-3} + \cdots + p^{p-3} \right]$$

$$= a^{p-1} - p \cdot a^{p-2} + p^2 t, \ t \in Z.$$

[Expression in the bracket is reduced to an integer, since p is an odd prime]

$$= 1 + p^2 k + p^2 t - pa^{p-2} \quad \text{[by (1) for some } k \in Z]$$

$$= 1 + (k+t)p^2 - pa^{p-2}. \tag{2}$$

Now $p \cdot a^{p-2} \not\equiv 0 \pmod{p^2}$.

For if $p^2 | pa^{p-2}$, then $p | a^{p-2}$ which is not true as 'a' being a primitive root of p, $(a, p) = 1$, i.e., $(a^{p-2}, p) = 1$ $(p \geq 3)$.

Therefore, from (2) it follows that $a'^{p-1} - 1$ is not divisible by p^2 and hence $a'^{p-1} \not\equiv 1 \pmod{p^2}$.

Thus, there is a primitive root a' of p which satisfies the condition $a^{p-1} \not\equiv 1$ (mod p^2).

Note: If 'a' be a primitive root of p, then $a+p$ is also its primitive root.

For, 'a' is a primitive root of $p \Rightarrow a^{\phi(p)} \equiv 1 \pmod{p}$

and $a^r \not\equiv 1 \pmod{p}$, $1 \leq r < \phi(p)$. (1)

Now $a + p \equiv a \pmod{p}$

$\Rightarrow (a+p)^{\phi(p)} \equiv a^{\phi(p)} \pmod{p} \equiv 1 \pmod{p}$,

which shows that $a+p$ is a primitive root of p, by virtue of (1).

Theorem (6.11): (1) *If 'a' be a primitive root of p and if $a^{p-1} \not\equiv 1 \pmod{p^2}$, then '$a$' is also a primitive root of p^2.* (2) *If 'a' be a primitive root of p and if $a^{p-1} \equiv 1 \pmod{p^2}$, then $a+p$ is a primitive root of p^2, p is an odd prime.*

Proof: (i) Since 'a' is a primitive root of p, $O(a \pmod{p}) = \phi(p) = p - 1$.

Then $O(a \pmod{p^2}) = p-1$ or $p(p-1)$ by [Th. (6.5)]. But since $a^{p-1} \not\equiv 1 \pmod{p^2}$, it follows that $O(a \pmod{p^2})$ is $p(p-1)$ and that is, $\phi(p^2)$. Consequently $a^{\phi(p^2)} \equiv 1 \pmod{p^2}$ shows that 'a' is a primitive root of p^2.

(ii) 'a' is a primitive root of p

$$\Rightarrow \quad a^{\phi(p)} \equiv 1 \pmod{p} \ \text{ and } \ a^r \not\equiv 1 \pmod{p}, \ 1 \leq r < \phi(p). \tag{1}$$

Now $a + p \equiv a \pmod{p}$

$\Rightarrow (a+p)^{\phi(p)} \equiv a^{\phi(p)} \pmod{p} \equiv 1 \pmod{p}$

$\Rightarrow a + p$ is a primitive root of p by (1)

$\Rightarrow O((a+p) \pmod{p}) = p - 1$.

Then $O((a+p) \pmod{p^2}) = p-1$ or $p(p-1)$. [by Th. (6.5)]

Now $$(a+p)^{p-1} = a^{p-1} + p(p-1)a^{p-2} + \frac{p^2(p-1)(p-2)}{2!} a^{p-3} + \cdots + p^{p-1}$$

$$= a^{p-1} - pa^{p-2} + p^2 \left[a^{p-2} + \frac{(p-1)(p-2)}{2!} a^{p-3} + \cdots + p^{p-3} \right]$$

$$= a^{p-1} - pa^{p-2} + p^2 \cdot t, \ t \in Z$$

[Since the expression in the bracket is reduced to an integer as p is an odd prime]

$$= 1 + p^2(k+t) - pa^{p-2} \quad \text{[by the given condition, for some } k \in Z]$$

But $\quad p^2 \nmid pa^{p-2}$.

For if $p^2 | pa^{p-2}$, then $p | a^{p-2}$ which is not possible as $(a^{p-2}, p) = 1$.

So $\quad (a+p)^{p-1} \not\equiv 1 \pmod{p^2}$ which contradicts the given condition.

Therefore, $O((a+p)(\text{mod } p^2)) = p(p-1) = \phi(p^2)$.

$\therefore \quad (a+p)^{\phi(p)} \equiv 1 \pmod{p^2} \Rightarrow (a+p)$ is a primitive root of p^2.

Lemma (6.3): If 'a' be a primitive root of p, p is an odd prime such that $a^{p-1} \not\equiv 1 \pmod{p^2}$, then for every $\alpha \geq 2$, $a^{\phi(p^{\alpha-1})} \not\equiv 1 \pmod{p^\alpha}$ holds.

***Proof*:** By virtue of the given condition $a^{p-1} \not\equiv 1 \pmod{p^2}$ when 'a' is a primitive root and p is an odd prime, we have $a^{\phi(p)} \not\equiv 1 \pmod{p^2}$ which shows that the result is true for $\alpha = 2$. We shall prove the lemma by method of induction.

Let the result be true for $\alpha = k\ (\geq 2)$.

Then $\quad a^{\phi(p^{k-1})} \not\equiv 1 \pmod{p^k}$. (1)

We shall establish that it is true for $\alpha = k+1$.

By Euler-Fermat's Theorem, $a^{\phi(p^{k-1})} \equiv 1 \pmod{p^{k-1}}$

$\Rightarrow \quad a^{\phi(p^{k-1})} = 1 + p^{k-1} \cdot t,$

$t \in Z$, when $p \nmid t$. (2)

For if $p|t$, then $t = qp \cdot q \in Z$

$\Rightarrow \quad a^{\phi(p^{k-1})} = 1 + p^k \cdot q$, form (2)

$\Rightarrow \quad a^{\phi(p^{k-1})} \equiv 1 \pmod{p^k}$ which contradicts (1).

From (2), $a^{p\phi(p^{k-1})} = (1 + p^{k-1}t)^p = 1 + p^k t + p_{c_2} p^{2k-2} t^2$

$$+ p_{c_3} p^{3k-3} t^3 + \cdots + p^{p(k-1)} t^p$$

$$= 1 + p^k t + p^{2k-1} \frac{(p-1)}{2!} \cdot t^2 + p^{3k-2} \cdot \frac{(p-1)(p-2)}{3!} t^3 + \cdots, k \geq 2 \quad (3)$$

Again, $2k - 1 = k + (k-1) \geq k+1 \quad (k \geq 2)$

$3k - 2 = k + 2(k-1) \geq k+2 > k+1$

....

$rk - (r-1) = k + (r-1)(k-1) \geq k + (r-1) > k+1,\ r > 1.$

Consequently, the terms after the first two terms in (3) are divisible by p^{k+1}.

Hence, $a^{p\phi(p^{k-1})} = 1 + p^k t + p^{k+1} \cdot t',\ t' \in Z$.

Here it is also to be noted that $p^{k+1} \nmid p^k t$. Otherwise, $p|t$, a contradiction arises.

So $\quad a^{p\phi(p^{k-1})} \equiv 1 + p^k t (\text{mod } p^{k+1}) \Rightarrow a^{\phi(p^k)} \not\equiv 1 (\text{mod } p^{k+1})$.

So the result is true, when $\alpha = k+1$.

Therefore, it is true $\forall\ \alpha \geq 2$.

Existence of Primitive Roots for Composite Number

Theorem (6.12): *If 'a' be a primitive root of p such that* $a^{p-1} \not\equiv 1 (\text{mod } p^2)$, *p is an odd prime, then 'a' is also a primitive root of* p^α, $\alpha \geq 2$.

***Proof*:** Let $0(a(\text{mod } p^\alpha)) = t$.

Then $a^t \equiv 1 (\text{mod } p^\alpha)$. (1)

Our intention is to show that $t = \phi(p^\alpha)$.

(1) $\Rightarrow$ $p^\alpha | a^t - 1 \Rightarrow p | a^t - 1 \Rightarrow a^t \equiv 1 \ (\text{mod } p)$. (2)

Also $a^{\phi(p)} \equiv 1 \ (\text{mod } p)$. (3)

as 'a' is a primitive root of p, $0(a(\text{mod } p)) = \phi(p)$, $(a, p) = 1$.

So from (2) and (3), we have $\phi(p) | t$.

Then $t = k\phi(p)$ for some $k \in Z$. (4)

Now $t | \phi(p^\alpha) \Rightarrow k\phi(p) | \phi(p^\alpha) \Rightarrow k(p-1) | p^{\alpha-1}(p-1) \Rightarrow k | p^{\alpha-1}$

$\Rightarrow$ $k = p^m$, $m \leq \alpha - 1$.

Then from (4), $t = p^m(p-1)$.

Now if $m = \alpha - 1$, then $t = p^{\alpha-1}(p-1) = \phi(p^\alpha)$ and hence we arrive our achievement.

Now let $m < \alpha - 1$. Then $m < \alpha - 1 \Rightarrow m \leq \alpha - 2$ and we have

$$p^m(p-1) | p^{\alpha-2}(p-1)$$

$$\Rightarrow t | p^{\alpha-2}(p-1), \text{ i.e., } t | \phi(p^{\alpha-1})$$

and so $\phi(p^{\alpha-1}) = t \cdot c$, $c \in Z$ (5)

From (1), $(a^t)^c \equiv 1 \ (\text{mod } p^\alpha)$

$$\Rightarrow a^{\phi(p^{\alpha-1})} \equiv 1 \ (\text{mod } p^\alpha) \text{ by (5),}$$

which arises a contradiction to the result proved in Lemma (6.3).

Hence, $m \not< \alpha - 1$ and consequently $m = \alpha - 1$.

Therefore, from (4), $t = k\phi(p) = p^{\alpha-1}(p-1)$ $[\because k = p^m]$

$$= \phi(p^\alpha)$$

and subsequently from (1), $a^{\phi(p^\alpha)} \equiv 1 \ (\text{mod } p^\alpha)$, $(a, p^\alpha) = 1$

$\Rightarrow$ 'a' is a primitive root of p^α, $\forall\ \alpha \geq 2$.

Theorem (6.13): ***If p be an odd prime, then there exists an odd primitive root of p^k $\forall\ k \geq 1$. Also each such primitive root of p^k is a primitive root of $2p^k$.***

***Proof*: Let 'a' be a primitive root of p^k.**

Then $O(a(\text{mod } p^k)) = \phi(p^k)$, $(a, p^k) = 1$.

Therefore, $a^{\phi(p^k)} \equiv 1 (\text{mod } p^k)$. (1)

Now $a + p^k \equiv a \ (\text{mod } p^k)$. (2)

Then $O(a + p^k(\text{mod } p^k)) = \phi(p^k)$.

For if $O(a + p^k(\text{mod } p^k) = r < \phi(p^k)$,

Then $(a + p^k)^r \equiv a^r (\text{mod } p^k)$ [by (2)]

$$\not\equiv 1 (\text{mod } p^k) \text{ [by (1)]}$$

Hence $\nexists\ r$ such that $r < \phi(p^k)$ and $(a + p^k)^r \equiv 1 (\text{mod } p^k)$.

Hence $O(a + p^k(\text{mod } p^k)) = \phi(p^k)$, $(a + p^k, p^k) = 1$

and hence $a + p^k$ is a primitive root of p^k.

Now if 'a' be an odd integer, then there exists an odd primitive root of p^k.

But if 'a' be even, then $a+p^k$ being odd (as p is an odd prime), a primitive root of p^k.

So there always exists an odd primitive root of p^k, $k \geq 1$.

Next, let $0(a(\text{mod } 2p^k)) = t$. Also $(a, 2p^k) = 1$, a being odd.

Then $\quad a^t \equiv 1(\text{mod } 2p^k).$ (3)

Our aim is to show that $t = \phi(2p^k)$.

Now, $\quad a^t \equiv 1(\text{mod } 2p^k) \Rightarrow a^t = 1 + 2p^k \cdot t$, for some $t \in Z$

$\Rightarrow a^t = 1 + p^k \cdot t'$, $t' = 2 \cdot t \in Z$

$\Rightarrow a^t \equiv 1(\text{mod } p^k).$

So $\quad \phi(p^k)|t$ by (1) (4)

$\phi(2p^k) = \phi(2) \cdot \phi(2^k) = \phi(p^k).$

Now $\quad t|\phi(2p^k)$ and so $t|\phi(p^k)$. (5)

Therefore, from (4) and (5), $t = \phi(p^k) = \phi(2 \cdot p^k)$.

Hence 'a' is a primitive root of $2p^k$, $k \geq 1$.

Example: Let $a = 2$ and $p = 13$ (odd prime).

Then $\quad \phi(p) = 12; (2, p) = 1.$

Now $\quad 2^4 \equiv 3 \text{ (mod 13)} \Rightarrow 2^{12} \equiv 27 \text{ (mod 13)}$

$\Rightarrow \quad 2^{12} \equiv 1 \text{ (mod 13)} \Rightarrow 2^{\phi(13)} \equiv 1 \text{ (mod 13)}.$

Also $\quad 2^r \not\equiv 1 \text{ (mod 13) } 1 \leq r < \phi(13).$

So 2 is a primitive root of 13.

Now $\quad p^2 = 13^2 = 169.$

We shall verify that 2 is a primitive root of 13^2. For this we have to show that $2^{p-1} \not\equiv 1 \text{ (mod } 13^2)$.

$2^6 \equiv 64 \text{ (mod } 13^2) \Rightarrow 2^{12} \equiv 64 \times 64 \text{ (mod } 13^2)$

$\Rightarrow \quad 2^{12} \equiv 40 \text{ (mod } 13^2) \Rightarrow 2^{12} \not\equiv 1 \text{ (mod } 13^2).$

Therefore, 2 is an even primitive root of 13^2.

Again, $2 + 13^2 = 171$ (odd integer).

Now $\quad 2 + 13^2 = 171 \equiv 2 \text{ (mod } 13^2)$

$\Rightarrow \quad (2 + 13^2)^{12} \equiv 2^{12} \text{ (mod } 13^2) \not\equiv 1 \text{ (mod } 13^2)$

$\Rightarrow$ $2 + 13^2$ is an odd primitive root of 13^2.

Thus, we conclude that there exists an odd primitive root of $13^2 (= p^2)$.

Therefore, by above theorem, $2 + 13^2$ is a primitive root of $2 \cdot 13^2$.

Theorem (6.14): *If 'a' is a primitive root of m, then there are exactly $\phi(\phi(m))$ incongruent primitive roots given by $S = \{a^n, 1 \leq n \leq \phi(m)\}$ if and only if $(n, \phi(m)) = 1$.*

Proof: Let 'a' be a primitive root of m and let $(n, \phi(m)) = 1$.

Then $\quad O(a(\text{mod } m)) = \phi(m)$ and hence $a^{\phi(m)} \equiv 1 \text{ (mod } m)$.

Therefore, by theorem [(6.2), (iii)],

$$O(a^n(\text{mod } m)) = \frac{O(a(\text{mod } m))}{(n, \phi(m))} = \frac{\phi(m)}{1} = \phi(m)$$

$\Rightarrow$ $a^n, 1 \leq n < \phi(m)$ is a primitive root of $m \Rightarrow$ every member of S is a primitive root of m. Let every primitive root of m be of the form a^r for some r, $1 \leq r \leq \phi(m)$.
Let 'b' be a primitive root of m.

Then $b \equiv a^r$ (mod m) for some $r = 1, 2, \ldots, \phi(m)$.

Hence, $O(a^r(\text{mod } m)) = O(b(\text{mod } m)) = \phi(m)$. (1)

$$\text{Again, } O(a^r(\text{mod } m)) = \frac{O(a(\text{mod } m))}{(r, \phi(m))} = \frac{\phi(m)}{(r, \phi(m))}$$

$$\Rightarrow \quad \phi(m) = \frac{\phi(m)}{(r, \phi(m))} \Rightarrow (r, \phi(m)) = 1, \quad 1 \leq r \leq \phi(m) \quad \text{[by (1)]}$$

Thus every primitive root of m is a member of S.

$S = \{a, a^2, \ldots, a^{\phi(m)}\}$ form a reduced system mod m. Also $(a, m) = 1$. Hence, all the incongruent primitive roots of m belong to S. a^r is a primitive root of m and $0(a^r(\text{mod } m)) = \phi(m))$.

Hence, the exact number of primitive roots of m = numbers of elements of S = numbers less than and prime to $\phi(m) = \phi(\phi(m))$.

Example: Let $m = 11$. Then $\phi(m) = \phi(11) = 10$,

p = prime divisor of $\phi(m) = 2, 5$.

Then $2^{\frac{\phi(m)}{p}} = 2^5 \equiv -1$ (mod 11).

Also $2^{\frac{\phi(m)}{5}} = 2^2 \equiv 4$ (mod 11).

$\therefore$ $2^2, 2^5 \not\equiv 1$ (mod 11).

So 2 is a primitive root of 11.
There are exactly $\phi(\phi(11)) = \phi(10) = \phi(2 \cdot 5) = \phi(2) \cdot \phi(5) = 4$ primitive roots.

The (+)ve integers less than and prime to $\phi(11)$ are 1, 3, 7, 9.

So the primitive roots are $2, 2^3, 2^7, 2^9$ which are exactly four in nos.

This example helps us to determine other primitive roots of 11, if one (smallest) primitive root can be determined.

Problem (6.9): Show that the primitive roots of 13 are given by $S = \{2^n, 1 \leq n < \phi(m)\}, (n, \phi(m)) = 1$ and 2 is a primitive root of 13. Also find the exact no. of primitive roots of 13.

Solution: $m = 13$, $\phi(m) = \phi(13) = 12$.

$$p = \text{prime divisor of } 12 = 2, 3, \frac{\phi(m)}{p} = \frac{\phi(m)}{2} = 6 \text{ and } \frac{\phi(m)}{3} = 4$$

(i) $2^6 \equiv -1$ (mod 13)

$2^4 \equiv 3$ (mod 13)

$\therefore$ $2^6, 2^4 \not\equiv 1$ (mod 13) $\Rightarrow$ 2 is a primitive root of 13.

(ii) $6^4 \equiv 9$ (mod 13)

$6^6 \equiv -1$ (mod 13)

$\therefore \quad 6^6, 6^4 \not\equiv 1 \pmod{13}$

$\Rightarrow$ 6 is a primitive root of 13.

Now n, where $1 \leq n < \phi(m) = 1, 5, 7, 11$.

$S = \{2, 2^5, 2^7, 2^{11}\}$ is a set of primitive roots,

where 2 is a primitive root of 13 (1)

$S = \{6, 6^5, 6^7, 6^{11}\}$ is another set of primitive roots while 6 is a primitive roots of 13.

Here $b = 6$ (a primitive root of 13) $\equiv 2^5 \pmod{13}$.

Similarly, we can show that 11 and 7 are primitive roots of 13.

Also $b = 11 \equiv 2^7 \pmod{13}$

$b = 7 \equiv 2^{11} \pmod{13}$.

Thus, any other primitive roots of 13 belongs to S. That is, every primitive root of 13 is of the form $2^r, 1 \leq r < \phi(m)$. So all the primitive roots of S are given by (1). That is, $S = \{2^n, 1 \leq n < \phi(m)\}$.

Exact number of primitive roots are $4 \equiv \phi(\phi(m))$.

For $\phi(m) = 12$, $\phi(12) = \phi(2^2 \cdot 3) = 2 \cdot 2 = 4$.

Problem (6.10): If 'a be a primitive root of m, then show that a^k is also a primitive root of m, if and only if $(k, \phi(m)) = 1$.

Solution: Let 'a' be a primitive root of m and let $(k, \phi(m)) = 1$.

Then $\quad O(a \pmod{m}) = \phi(m)$.

$$\text{Therefore}, O(a^k \pmod{m}) = \frac{O(a \pmod{m})}{(k, \phi(m))} = \frac{\phi(m)}{1} = \phi(m).$$

Hence a^k is a primitive root of m.

Conversely, let a^k is a primitive root of m.

Then $O(a^k \pmod{m}) = \phi(m)$.

$\Rightarrow \quad a^{k\phi(m)} \equiv 1 \pmod{m}$. (i)

Also $\quad a^{\phi(m)} \equiv 1 \pmod{m}$. (ii)

(i) and (ii) $\Rightarrow$ $\phi(m) | k\phi(m)$ $\Rightarrow$ $(\phi(m), k) = 1$.

Theory of Indices

German mathematician K. Gauss introduced the concept of indices in the light of primitive roots of an integer m. If 'a' be a primitive root of m, then the set $S = \{a^n, 1 \leq n \leq \phi(m),\ (n, \phi(m)) = 1\}$ forms the set of primitive roots with the base a.

Let b be an integer such that $(b, m) = 1$, $b < m$.

Then we have a relation $b \equiv a^r \pmod{m}$ for some $r, 1 \leq r \leq \phi(m)$.

Example: (1) Let us go through the following example:

Let $m = 14$. Then $\phi(m) = \phi(2 \cdot 7) = 6$.

$p =$ prime divisor of $6 = 2, 3$

$(3, 14) = 1$.

Also $3^2 \equiv -5 \pmod{14}$

$3^3 \equiv -1 \pmod{14}$

$\therefore$ $3^2, 3^3 \not\equiv 1 \pmod{14}$.

So 3 is a primitive root of 14.

$S = \{3, 3^5\}$ $1 \le n \le \phi(m) = 6,\ (n, 6) = 1 \ \Rightarrow\ n = 1, 5.$

Again, $(5, 14) = 1$

$5^2 \equiv -3 \pmod{14}$

$5^3 \equiv -15 \pmod{14} \equiv -1 \pmod{14}$

$5^2, 5^3 \not\equiv 1 \pmod{14}$.

$\therefore$ 5 is also a primitive root of 14.

$S = \{5, 5^5\}$.

We now consider an integer $b, (b, m) = 1, b < m$.

Hence $(b, 14) = 1,\ b < 14 \ \Rightarrow\ b = 1, 3, 5, 9, 11, 13.$

Now $3^1 \equiv 3 \pmod{14}$

$3^4 \equiv 11 \pmod{14} \Rightarrow 3^5 \equiv 5 \pmod{14}$

For $b = 3,\ 3^1 \equiv 3 \pmod{14}$

$b = 5,\ 3^5 \equiv 5 \pmod{14}$.

Thus $\nexists\, r,\ 1 \le r \le \phi(m) = 6,\ (r, 6) = 1$

such that $3^1 \equiv b \pmod{14}$, $b = 3$

$3^5 \equiv b \pmod{14}$ $b = 5$.

That is, $3^r \equiv b \pmod{14}$, $r = 1, 5 < 6$ and prime to 6 is true for other values of b with the base 'a primitive root of 14'.

Definition (6.3): *If 'a' be a primitive root of m and b is an integer, $(b, m) = 1$ and if there exists a smallest positive integer r such that $a^r \equiv b(\text{mod } m)$, then r is called 'the index of b to the base a' modulo m and is written as*

$r \equiv \text{ind}_a^b \pmod{m},\ 1 \le r \le \phi(m).$

i.e., $\text{ind}_a^b \equiv r \pmod{m}$

i.e., $a^{\text{ind}_a^b} \equiv b \pmod{m}$ $[\because\ a^r \equiv b \pmod{m}]$

We usually write it as $r = \text{ind}_a^b \pmod{m}$ or simply ind $b = r$, when base and m is clear to us.

Example (2): 3 is a primitive root of 14. [Ex 1]

Now we have the following results

$3^1 \equiv 3 \pmod{14}$ $\quad 3^4 \equiv 11 \pmod{14}$

$3^2 \equiv 9 \pmod{14}$ $\quad 3^5 \equiv 5 \pmod{14}$

$3^3 \equiv 13 \pmod{14}$ $\quad 3^6 \equiv 1 \pmod{14}$

$r = 1, 2, 3, 4, 5, 6 \cdot 1 \le r \le \phi(14) = 6.$

$1 = \text{ind}_3^3 \pmod{14}$ $\quad 4 = \text{ind}_3^{11} \pmod{14}$

$2 = \text{ind}_3^9 \pmod{14}$ $\quad 5 = \text{ind}_3^5 \pmod{14}$

$3 = \text{ind}_3^{13} \pmod{14}$ $\quad 6 = \text{ind}_3^1 \pmod{14}$

b	1	3	5	9	11	13
ind_3^b	6	1	5	2	4	3

Index Table mod (14) $(b, 14) = 1,\ b < 14$

Problem (6.11): Find the index table for modulo 18.

Solution: $m = 18,\ \phi(m) = \phi(18) = \phi(2 \cdot 3^2) = 6$

$(5, 18) = 1$

$p =$ prime divisor of $\phi(m)$ are 2 and 3

$5^2 \equiv 7 \pmod{18},\ 5^3 \equiv -1 \pmod{18}$

$\Rightarrow \quad 5^2, 5^3 \not\equiv 1 \pmod{18} \Rightarrow 5$ is a primitive root of 18.

$(11, 18) = 1$

$11^2 \equiv 2 \pmod{18},\ 11^3 \equiv 4 \pmod{18}$

$\Rightarrow \quad 11^2, 11^3 \not\equiv 1 \pmod{18} \Rightarrow 11$ is a primitive root of 18.

7, 13, 17 are not primitive roots of 18, though $(7, 18)(13, 18), (17, 18) = 1$.

Now $\quad 5^1 \equiv 5 \pmod{18},\quad 5^4 \equiv 13 \pmod{18}$

$5^2 \equiv 7 \pmod{18},\quad 5^5 \equiv 11 \pmod{18}$

$5^3 \equiv 17 \pmod{18},\quad 5^6 \equiv 1 \pmod{18}$.

$$\left.\begin{array}{ll} \text{ind}_5^5 = 1, & \text{ind}_5^{13} = 4 \\ \text{ind}_5^7 = 2, & \text{ind}_5^{11} = 5 \\ 2\text{nd}_5^{17} = 3, & \text{ind}_5^1 = 6 \end{array}\right\} \quad \text{(A)}$$

Again, $\quad 11^1 \equiv 11 \pmod{18},\quad 11^4 \equiv 7 \pmod{18}$

$11^2 \equiv 13 \pmod{18},\quad 11^5 \equiv 5 \pmod{18}$

$11^3 \equiv 17 \pmod{18},\quad 11^6 \equiv 1 \pmod{18}$.

Then
$$\left.\begin{array}{lll} \text{ind}_{11}^{11} = 1, & \text{ind}_{11}^{17} = 3, & \text{ind}_{11}^5 = 5 \\ \text{ind}_{11}^{13} = 2, & \text{ind}_{11}^7 = 4, & \text{ind}_{11}^1 = 6 \end{array}\right\} \quad \text{(B)}$$

Index Table modulo 18

Here $b = 1, 5, 7, 11, 13, 17$

b	1	5	7	11	13	17
ind_5^b	6	1	2	5	4	3
ind_{11}^b	6	5	4	1	2	3

Theorem (6.15): *If 'a' be a primitive root of m and $(x, m), (y, m) = 1$, Then*

(i) $\text{ind}_a^x + \text{ind}_a^y \equiv \text{ind}_a^{xy} \pmod{\phi(m)}$

(ii) $\text{ind}_a^{x^n} \equiv n\ \text{ind}_a^x \pmod{\phi(m)},\ n \geq 1$

(iii) $\text{ind}_a^1 = 0$ and $\text{ind}_a^a = 1$

(iv) $\text{ind}_a^{-1} = \dfrac{\phi(m)}{2}$ of $m > 2$

(v) $\text{ind}_a^{xy^n} \equiv \text{ind}_b^x + n\ \text{ind}_a^y \pmod{\phi(m)}$

(vi) $\text{ind}_a^x \equiv \text{ind}_b^x \ \text{ind}_a^b \pmod{\phi(m)}$ when b is also a primitive root of m.

(vii) $\text{ind}_a^x = \text{ind}_a^y \pmod m \iff x \equiv y \pmod m$.

Proof: (i) Let $\text{ind}_a^x \pmod m = r_1$ and $\text{ind}_a^y \pmod m = r_2$.

Then $a^{r_1} \equiv x \pmod m$ and $a^{r_2} \equiv y \pmod m$.

Then $a^{r_1+r_2} \equiv xy \pmod m \Rightarrow r_1 + r_2 \equiv \text{ind}_a^{xy} \pmod m$

$\Rightarrow \quad \text{ind}_a^x + \text{ind}_a^y \equiv \text{ind}_a^{xy} \pmod m$

$\Rightarrow \quad \text{ind}_a^x + \text{ind}_a^y \equiv \text{ind}_a^{xy} \pmod{\phi(m)}$ [By Th. 6.2 (i), here $k = \phi(m)$]

(ii) Let $a^{x^n} \pmod m = r$.

Then $a^r \equiv x^n \pmod m$.

Also let $\text{ind}_a^x \pmod m = s$. (1)

Then $a^s \equiv x \pmod m \Rightarrow a^{ns} \equiv x^n \pmod m$

$\Rightarrow \quad \text{ind}_a^{x^n} \equiv ns \pmod m \Rightarrow \text{ind}_a^{x^n} \equiv n \ \text{ind}_a^x \pmod m$ [by (1)]

$\Rightarrow \quad \text{ind}_a^{x^n} \equiv n \ \text{ind}_a^x \pmod{\phi(m)}$

(iii) Let $\text{ind}_a^1 \pmod m = r$.

Then $a^r \equiv 1 \pmod m \Rightarrow r = 0$.

$\therefore \quad \text{ind}_a^1 = 0$. That is $\text{ind}_a^1 \equiv 0 \pmod{\phi(m)}$.

Next let $\text{ind}_a^a \equiv p$.

Then $a^p \equiv a \pmod m \Rightarrow p = 1$

$\Rightarrow \quad \text{ind}_a^a = 1$. That is $\text{ind}_a^a \equiv 1 \pmod{\phi(m)}$.

(iv) Let $\text{ind}_a^{-1} \pmod m = r$.

Then $a^r \equiv -1 \pmod m$. (1)

Also 'a' being a primitive root of m,

$a^{\phi(m)/2} \equiv -1 \pmod m$. (2)

r being the smallest (+)ve integer, it follows

from (1) and (2), $r = \dfrac{\phi(m)}{2}$, $m > 2$.

$\therefore \quad \text{ind}_a^{-1} \pmod m = \dfrac{\phi(m)}{2}$.

(v) Let $\text{ind}_a^{xy^n} \pmod m = r$.

Then $a^r \equiv xy^n \pmod m$.

Again, let $\text{ind}_a^x \pmod m = p$ and $\text{ind}_a^y \pmod m = q$.

Then $a^p \equiv x \pmod m$ and $a^q \equiv y \pmod m \Rightarrow a^{nq} \equiv y^n \pmod m$.

Therefore, $a^{p+nq} \equiv xy^n \pmod m$

$\Rightarrow \quad \text{ind}_a^{xy^n} \equiv (p + nq) \pmod m \equiv \text{ind}_a^x + n \ \text{ind}_a^y \pmod{\phi(m)}$.

(vi) Let $\text{ind}_b^x \pmod m = r_1$

and $\text{ind}_a^b \pmod m = r_2$.

Then $b^{r_1} \equiv x \pmod m$ and $a^{r_2} \equiv b \pmod m$

$\Rightarrow \quad a^{r_1 r_2} \equiv b^{r_1} \pmod{m} \equiv x \pmod{m}$

$\Rightarrow \quad \text{ind}_a^x \equiv r_1 r_2 \pmod{m}$

$\Rightarrow \quad \text{ind}_a^x \equiv \text{ind}_b^x \, \text{ind}_a^b \pmod{\phi(m)}$.

(vii) Let $\text{ind}_a^x \equiv \text{ind}_a^y \pmod{m} = r$ (say).

Then $\quad a^r \equiv x \pmod{m}$ and $a^r \equiv y \pmod{m}$

$\Rightarrow \quad x \equiv y \pmod{m}$.

Now let $x \equiv y \pmod{m}$ and let $\text{ind}_a^x \pmod{m} = r$.

Then $\quad a^r \equiv x \pmod{m}$.

That is $a^{\text{ind}_a^x} \equiv x \pmod{m} \equiv y \pmod{m}$

$\Rightarrow \quad \text{ind}_a^y \equiv \text{ind}_a^x \pmod{m}$

$\Rightarrow \quad \text{ind}_a^y \equiv \text{ind}_a^x \pmod{\phi(m)}$.

Theorem (6.16): *If m has a primitive root h, and $(a, m) = (b, m) = 1$, then*

$$ax \equiv b(\mod m) \iff \text{ind}_h^a + \text{ind}_h^x \equiv \text{ind}_h^b(\mod \phi(m))$$

[or simply : $\text{ind } a + \text{ind } x \equiv \text{ind } b \pmod{\phi(m)}$].

***Proof*:** Let $ax \equiv b \pmod{m}$.

Then $\quad ax \equiv b \pmod{m}$

$\Rightarrow \quad \text{ind}_h^{ax} \equiv \text{ind}_h^b \pmod{\phi(m)}$ [by (vii) of Th. 6.15]

$\Rightarrow \quad \text{ind}_h^a + \text{ind}_h^x \equiv \text{ind}_h^b \pmod{\phi(m)}$ [by (i) of Th. 6.15]

i.e., $\quad \text{ind } a + \text{ind } x \equiv \text{ind } b \pmod{\phi(m)}$.

Conversely, let $\text{ind } a + \text{ind } x \equiv \text{ind } b \pmod{\phi(m)}$.

Then $\quad \text{ind}_h^{ax} \equiv \text{ind}_h^b \pmod{\phi(m)}$ [by (i) Th. 6.15]

$\Rightarrow \quad ax \equiv b \pmod{m}$ [by (vii) of Th. 6.15].

Corollary (6.2): If a and b are primitive roots of the odd prime m and $(x, m) = 1$, then $\text{ind}_b^x \equiv \text{ind}_a^x \cdot \text{ind}_b^a \pmod{m-1}$.

***Proof*:** Follows from (vi) of Th. 6.15, since $\phi(m) = m - 1$ as m is an odd prime.

Existence of the solution

We have discussed that $ax \equiv b \pmod{m}$ is solvable, if and only if $d|b$, where $d = (a, b)$.

If $d \nmid b$, there does not exist any solution [See Th. (3.13), Ch-3].

Now $\quad ax \equiv b \pmod{m} \Rightarrow \text{ind}\, a + \text{ind} x \equiv \text{ind } b \pmod{\phi(m)}$

$\Rightarrow \quad 1 \cdot \text{ind } x \equiv \text{ind } b - \text{ind } a \pmod{\phi(m)}$

is solvable, since $(1, \phi(m)) = 1$.

Again, if we consider an equation $ax^n \equiv b \pmod{m}$,

then $\quad \text{ind } a + n \text{ ind } x \equiv \text{ind } b \pmod{\phi(m)}$

$\Rightarrow \quad n \text{ ind } x \equiv \text{ind } b - \text{ind } a \pmod{\phi(m)}$,

which is solvable, if and only if $(n, \phi(m)) | (\text{ind}\, b - \text{ind}\, a)$. Also if $(n, \phi(m)) = d$, then there exists exactly d-solutions.

Problem (6.12): Solve: $7x^3 \equiv 3 \pmod{11}$.

Solution: Index table: modulo 11.

$(2, 11) = 1$. 2 is a primitive root of 11.

$\phi(11) = 10$.

$2^1 \equiv 2 \pmod{11}$ $2^6 \equiv 9 \pmod{11}$

$2^2 \equiv 4 \pmod{11}$ $2^7 \equiv 7 \pmod{11}$

$2^3 \equiv 8 \pmod{11}$ $2^8 \equiv 3 \pmod{11}$

$2^4 \equiv 5 \pmod{11}$ $2^9 \equiv 6 \pmod{11}$

$2^5 \equiv 10 \pmod{11}$ $2^{10} \equiv 1 \pmod{11}$

b	1	2	3	4	5	6	7	8	9	10
ind_2^b	10	1	8	2	4	9	7	3	6	5

Given equation can be written as

$\text{ind}_2^7 + 3\ \text{ind}_2^x \equiv \text{ind}_2^3 \pmod{\phi(11)}$

i.e., $\text{ind } 7 + 3 \text{ ind } x \equiv \text{ind } 3 \pmod{\phi(11)}$

or, $3 \text{ ind } x \equiv \text{ind } 3 - \text{ind } 7 \pmod{10}$

or, $3 \text{ ind } x \equiv 8 - 7 \pmod{10}$ [from the index table]

or, $3 \text{ ind } x \equiv 1 \pmod{10} \equiv 21 \pmod{10}$

or, $\text{ind } x \equiv 7 \pmod{10}$ or, $x \equiv 7 \pmod{10}$ [as $\text{ind}_2^7 = 7$]

So, $x \equiv 7 \pmod{\phi(11)}$, is a solution.

Problem (6.13): Solve, if possible,

(i) $x^8 \equiv 10 \pmod{11}$ **(ii)** $3x^4 \equiv 5 \pmod{11}$.

Solution: (i) $x^8 \equiv 10 \pmod{11}$

$\Rightarrow$ $8\ \text{ind}_2^x \equiv \text{ind}_2^{10} \pmod{\phi(11)}$ [2 is a primitive root of 11]

i.e., $8 \text{ ind } x \equiv \text{ind } 10 \pmod{10}$.

Now $(8, 10) = 2$, $\text{ind}_2^{10} = 5$ [from index table in Problem(6.12)]

$2 \nmid 5$

So the equation is not solvable.

(ii) $3x^4 \equiv 5 \pmod{11}$

$\Rightarrow$ $\text{ind}_2^3 + 4\ \text{ind}_2^x \equiv \text{ind}_2^5 \pmod{\phi(11)}$

i.e., $\text{ind } 3 + 4 \text{ ind } x \equiv \text{ind } 5 \pmod{10}$

or, $4 \text{ ind } x \equiv \text{ind } 5 - \text{ind } 3 \pmod{10}$

or, $4 \text{ ind } x \equiv 4 - 8 \pmod{10}$ [from index table in Problem(6.12)]

or, $4 \text{ ind } x \equiv -4 \pmod{10} \equiv 6 \pmod{10}$. (1)

Now $(4, 10) = 2$ and $2 | 6$.

Equation is solvable and $\exists$ 2-solutions only.

$\therefore$ $4 \text{ ind } x \equiv 16 \pmod{10}$ or, $\text{ind } x \equiv 4 \pmod 5$

which is solvable for ind $x = 4$ and ind $x = 9$. That is, for $x = 5$ and $x = 6$, follows from the index table given above

Hence, ind $x \equiv$ ind 5 (mod 10) [from index table in Problem(6.12)]

$\Rightarrow \quad x \equiv 5 \pmod{\phi(11)}$.

Again, from (1), 4 ind $x \equiv 36 \pmod{\phi(11)}$

$\Rightarrow \quad$ ind $x \equiv 9 \pmod{\phi(11)}$

$\Rightarrow \quad$ ind $x \equiv$ ind 6 $\pmod{\phi(11)}$ [from index table in Problem (6.12)]

$\Rightarrow \quad x \equiv 6 \pmod{\phi(11)}$.

Thus the solutions are given by $x \equiv 5, 6 \pmod{11}$.

Problem (6.14): If 'a' be a primitive root of an odd prime p, then show that $\text{ind}_a^{-1} = \text{ind}_a^{p-1} = \frac{1}{2}(p-1)$.

Solution: 'a' being a primitive root of p,

$$O(a \pmod p) = \phi(p).$$

Therefore, $a^{\phi(p)} \equiv 1 \pmod p \Rightarrow a^{p-1} \equiv 1 \pmod p$

$$\Rightarrow \quad (a^{\frac{p-1}{2}} - 1)(a^{\frac{p-1}{2}} + 1) \equiv O \pmod p$$

$$\Rightarrow \quad a^{\frac{p-1}{2}} + 1 \equiv O \pmod p,$$

since $a^{\frac{p-1}{2}} \not\equiv 1 \pmod p$ as 'a' is a primitive root.

$$\Rightarrow \quad a^{\frac{p-1}{2}} \equiv -1 \pmod p \qquad (1)$$

$$\equiv p - 1 \pmod p$$

$$\Rightarrow \quad \text{ind}_a^{p-1} \equiv \frac{p-1}{2} \pmod{\phi(p)}.$$

Also from (1), $\text{ind}_a^{-1} \equiv \dfrac{p-1}{2} \pmod{\phi(p)}$.

Index Table

Index table of $\text{ind}_a^b \pmod m$, $(b, m) = 1$ and 'a' is the primitive root of m.

Here 'a' is taken as the smallest primitive root of m for forming the table. (a is kept fixed).

Index tables become different for different primitive roots of m. However, the table given here will solve our purpose in solving the given congruence.

We have already submitted some index tables for $m = 11$, 14, 18

Index Table (1)
(for prime numbers only)

Ind_a^b / b	1	2	3	4	5	6	7	8	9	10	11	12	13	14	15	16	17	18	19	20	21	22
m = 3	2	1																				
m = 5	4	1	3	2																		
m = 7	6	2	1	4	5	3																
m = 11	10	1	8	2	4	9	7	3	6	5												
m = 13	12	1	4	2	9	5	11	3	8	10	7	6										
m = 17	16	14	1	12	5	15	11	10	2	3	7	13	4	9	6	8						
m = 19	18	1	13	2	16	14	6	3	8	17	12	15	5	7	11	4	10	9				
m = 23	22	2	16	4	1	18	19	6	10	3	9	20	14	21	17	8	7	12	15	5	13	11

Continuing Table (2)

Ind_a^b / b →	1	2	3	4	5	6	7	8	9	10	11	12	13	14	15	16	17	18	19	20	21	22	23
m = 29	28	1	5	2	22	6	12	3	10	23	25	7	18	13	17	4	21	11	9	24	17	16	20
m = 31	30	24	1	18	20	25	28	12	2	14	23	19	11	22	21	6	7	26	4	8	29	17	27
m = 37	36	1	26	2	23	27	32	3	16	24	30	28	11	33	13	4	7	17	35	25	22	31	15
m = 41	40	26	15	12	22	1	39	38	30	8	3	27		31	25	37	24	33	16	9	34	14	29
m = 43	42	27	1	12	25	28	35	39	2	10	30	13	32	20	26	24	38	29	19	37	36	15	16
m = 47	46	18	20	36	1	38	32	8	40	19	7	10	11	4	21	26	16	12	45	37	6	25	5

Ind_a^b / b →	24	25	26	27	28	29	30	31	32	33	34	35	36	37	38	39	40	41	42	43	44	45	46
m = 29	8	16	19	15	14																		
m = 31	13	10	5	3	16	9	15																
m = 37	29	10	12	6	34	21	14	9	5	20	8	19	18										
m = 41	36	13	4	17	5	11	7	23	28	10	18	19	21	2	32	35	6	20					
m = 43	40	8	17	3	5	41	11	34	9	31	23	18	14	7	4	33	22	6	21				
m = 47	28	2	29	14	22	35	39	3	44	27	34	33	30	42	17	31	9	15	24	13	43	41	23

Exercise 6

1. Prove that the sum of the primitive roots of m is congruent to $\mu(m-1)$ (mod m).

2. If m is an odd prime > 3, prove that the product of primitive roots of m is congruent to 1 (mod m).

3. Let $d|\phi(m)$ and $a^r \equiv d$ (mod m), where r is the smallest (+)ve integer, i.e., a is a primitive root of $x^d \equiv 1$ (mod m). Prove that if $x^{\phi(m)} \equiv 1$ (mod m) has a primitive root, then it has $\phi(\phi(m))$ incongruent (mod m) primitive roots.

4. The prime $m = 71$ has 7 as a primitive root. Find all primitive roots of 71 and also find a primitive roots of m^2 and $2m^2$.

5. Prove that 3 is a primitive root of m, if m is a prime of the form $2^n + 1$, $n > 1$.

6. Prove that 2 is a primitive root of m, if m is a prime of the form $4k + 1$ where k is an odd prime.

7. If $m(> 2)$ has a primitive root and $(a, m) = 1$, then

   (a) $x^2 \equiv a$ (mod m) has a solution if and only if $a^{\phi(m)/2} \equiv 1$ (mod m).

   (b) If aRm, then $x^2 \equiv a$ (mod m) has exactly two solutions.

8. If m is any prime except 2 or 5, then $t = m - 1$ or t is a divisor of $m - 1$ according as 10 is or is not a primitive root of m.

9. If m is an odd prime, then show that the congruence $x^{m-2} + \cdots + x^2 + x + 1 \equiv 0$ (mod m) has exactly $(m-2)$ incongruent solutions and those are $2, 3, \ldots, (m-1)$.

10. For an odd prime m, verify the sum $1^n + 2^n + 3^n + \cdots + (m-1)^n \equiv 0$ (mod m), if $(m-1) \not| n$ and $\equiv -\phi$ (mod m), if $(m-1)|n$.

Chapter 7

Quadratic Congruence

Solution of Quadratic Congruence

Here we deal with the solution of a quadratic congruence of the form $Ax^2 + Bx + C \equiv 0 \pmod{m}$, where m is an odd prime and $(2A, m) = 1$.

In this light, Eulerean criterion, Legender's symbol, Gauss' lemma play an important role.

We discuss the solution of $Ax^2 + Bx + C \equiv 0 \pmod{m}$ when m is an odd prime and $(2A, m) = 1$.

$$Ax^2 + Bx + C \equiv 0 \pmod{m}$$

$$\Rightarrow \quad 4A^2x^2 + 4ABx + 4AC \equiv 0 \pmod{m}$$

$$\Rightarrow \quad (2Ax + B)^2 \equiv B^2 - 4AC \pmod{m}.$$

Writing $2Ax + B \equiv y \pmod{m}$ and $r \equiv B^2 - 4AC \pmod{m}$,

we have, $y^2 \equiv r \pmod{m}$. (1)

Let $y = y_0$ be a solution of (1).

Then $2Ax + B \equiv y_0 \pmod{m}$, (2)

which is a linear congruence and $(2A, m) = 1$. Therefore, it is solvable and hence, $Ax^2 + Bx + C \equiv 0 \pmod{m}$ can be solved.

Problem (7.1): Solve $x^2 + 7x + 10 \equiv 0 \pmod{11}$.

Solution: $x^2 + 7x + 10 \equiv 0 \pmod{11}$ (1)

$\Rightarrow \quad 4x^2 + 28x + 40 \equiv 0 \pmod{11}$

$\Rightarrow \quad (2x + 7)^2 \equiv 49 - 40 \pmod{11}$

$\Rightarrow \quad (2x + 7)^2 \equiv 9 \pmod{11}$.

Let $\quad 2x + 7 \equiv y \pmod{11}$.

Then $\quad y^2 \equiv 9 \pmod{11}$. (2)

Solutions of (2) are $y \equiv 3 \pmod{11}$ and $y \equiv -3 \pmod{11}$.

$$y \equiv 3 \pmod{11} \Rightarrow 2x + 7 \equiv 3 \pmod{11} \Rightarrow 2x \equiv -4 \pmod{11},$$

which is solvable since $(2, 11) = 1$.

$\therefore \quad 2x \equiv 18 \pmod{11} \Rightarrow x \equiv 9 \pmod{11}$.

Again, $\quad 2x + 7 \equiv -3 \pmod{11} \Rightarrow 2x + 7 \equiv 8 \pmod{11}$

$\Rightarrow \quad 2x \equiv 1 \pmod{11}$, which is solvable since $(2, 11) = 1$

$\Rightarrow \quad 2x \equiv 12 \pmod{11} \Rightarrow x \equiv 6 \pmod{11}$.

Hence, $x = 6,\ 9 \pmod{11}$ are the solutions of (1).

Problem (7.2) (i) Solve: $3x^2 + 9x + 7 \equiv 0 \pmod{13}$

Solution: $3x^2 + 9x + 7 \equiv 0 \pmod{13}$ (1)

$\Rightarrow \quad 36x^2 + 108x + 84 \equiv 0 \pmod{13}$

$\Rightarrow \quad (6x+9)^2 \equiv 81 - 84 \pmod{13}$

$\Rightarrow \quad (6x+9)^2 \equiv -3 \pmod{13} \equiv 10 \pmod{13}.$

Let $\quad 6x + 9 \equiv y \pmod{13}.$

Then $\quad y^2 \equiv 10 \pmod{13} \Rightarrow y^2 \equiv 36 \pmod{13}$

$\Rightarrow \quad y \equiv 6 \pmod{13}$ and $y \equiv -6 \pmod{13}.$

Now $\quad y \equiv 6 \pmod{13} \Rightarrow 6x + 9 \equiv 6 \pmod{13}$

$\Rightarrow \quad 6x \equiv -3 \pmod{13} \equiv 36 \pmod{13},$

which is solvable as $(6, 13) = 1$.

$\therefore \quad x \equiv 6 \pmod{13}.$

Again, $\quad y \equiv -6 \pmod{13} \Rightarrow 6x + 9 \equiv -6 \pmod{13}$

$\Rightarrow \quad 6x \equiv -15 \pmod{13} \Rightarrow 6x \equiv 24 \pmod{13},$

which is solvable as $(6, 13) = 1$

$\Rightarrow \quad x \equiv 4 \pmod{13}.$

$\therefore x \equiv 4, 6 \pmod{13}$ are the solutions of (1).

(ii) Solve: $x^2 + 6x + 5 \equiv 0 \pmod{15}$

Solution: Here $m = 15$ is not prime.

$x^2 + 6x + 5 \equiv 0 \pmod{15}$ (1)

$\Rightarrow \quad (2x+6)^2 \equiv 36 - 20 \pmod{15} \Rightarrow (2x+6)^2 \equiv 16 \pmod{15}.$

Let $\quad 2x + 6 \equiv y \pmod{15}.$

Then $\quad y^2 \equiv 16 \pmod{15} \equiv 1 \pmod{15}.$ (2)

$y \equiv 4 \pmod{15}$ and $y \equiv -4 \pmod{15}$, $y \equiv 1 \pmod{15}$ and $y \equiv -1 \pmod{15}$ are the solutions of (2).

Then $\quad 2x + 6 \equiv 4 \pmod{15} \Rightarrow 2x \equiv -2 \pmod{15},$

which is solvable as $(2, 15) = 1$

and $\quad 2x \equiv 28 \pmod{15} \Rightarrow x \equiv 14 \pmod{15}.$

$2x + 6 \equiv -4 \pmod{15} \Rightarrow 2x \equiv -10 \pmod{15}$

$\Rightarrow \quad 2x \equiv 20 \pmod{15} \Rightarrow x \equiv 10 \pmod{15}.$

Again, $\quad 2x + 6 \equiv 1 \pmod{15} \Rightarrow 2x \equiv -5 \pmod{15} \equiv 10 \pmod{15}$

$\Rightarrow \quad x \equiv 5 \pmod{15}.$

Also $\quad 2x + 6 \equiv -1 \pmod{15} \Rightarrow 2x \equiv -7 \pmod{15} \equiv 8 \pmod{15}$

$\Rightarrow \quad x \equiv 4 \pmod{15}.$

Thus the solutions of (1) are given by

$$x \equiv 4, 5, 10, 14 \pmod{15}.$$

In this case the quadratic has more than two solutions.

Thus, we observe from the Examples [7.1], [7.2–i] and [7.2–ii] that if m is an odd prime, then the quadratic has exactly two solutions and if m is not prime (i.e., a composite number), the quadratic has more than two solutions. Therefore, the quadratic congruence $x^2 \equiv a \pmod{m}$, m is an odd prime and $a \not\equiv 0 \pmod{m}$ has at most two solutions.

If x_0 is a solution, then the other solution is $m - x_0$, of course x_0 is the least solution of it. Also x_0 and $m - x_0$ are incongruent (mod m). Thus, there exist two incongruent solutions, $x \equiv x_0 \pmod{m}$ and $x \equiv m - x_0 \pmod{m}$ of $x^2 \equiv a \pmod{m}$.

Also we can observe that a quadratic congruence $x^2 \equiv a \pmod{m}$, m is an odd prime, $a \not\equiv 0 \pmod{m}$ has no solution.

The above observations follow from the examples.

Example: (i) $x^2 \equiv 4 \pmod{11}$ has two solutions, one is $x \equiv 2 \pmod{11}$ and the other is $-2 + 11 \pmod{11}$.

(ii) $x^2 \equiv 2 \pmod{11}$ has no solution.

Theorem (7.1): *The quadratic congruence $x^2 \equiv a \pmod{m}$, where m is an odd prime and $a \not\equiv 0 \pmod{m}$, has exactly two incongruent solutions x_0 and $m - x_0$, provided the solutions exist.*

Proof: Let the solution of the given congruence exist and let x_0 be its solution.

Then $x_0^2 \equiv a \pmod{m}$.

Now $(m - x_0)^2 = m^2 - 2mx_0 + x_0^2 \equiv m^2 - 2mx_0 + a \pmod{m}$

or, $(m - x_0)^2 \equiv -2mx_0 + a \pmod{m}$

or, $(m - x_0)^2 \equiv a \pmod{m}$

$\Rightarrow$ $m - x_0$ is a solution of the given congruence.

Also $x_0 \not\equiv m - x_0 \pmod{m}$, since $m - 2x_0 = m - x_0 - x_0$ is not divisible by m, as $m \nmid 2$. So x_0 and $m - x_0$ are two incongruent solutions.

Also if possible, let x_1 be a third solution of it other than x_0 and $m - x_0$.

Therefore, $x_1^2 \equiv a \pmod{m}$, and $x_0^2 \equiv a \pmod{m}$

$\Rightarrow x_1^2 - x_0^2 \equiv 0 \pmod{m}$

$\Rightarrow$ $(x_1 - x_0)(x_1 + x_0) \equiv 0 \pmod{m}$

$\Rightarrow$ either $m | x_1 - x_0$ or, $m | x_1 + x_0$.

When $m | x_1 - x_0$, $x_1 \equiv x_0 \pmod{m}$ and $m | x_1 + x_0$, $x_1 \equiv -x_0 \pmod{m}$

$\Rightarrow$ $x_1 \equiv m - x_0 \pmod{m}$.

Hence, x_1 is not a distinct solution of the given congruence.

Hence the theorem.

Quadratic Residue and Quadratic Non-residue

Definition (7.1): *If a quadratic congruence $x^2 \equiv a \pmod{m}$,* (1)
where m is an odd prime and $a \not\equiv 0 \pmod{m}$ [i.e., $(a, m) = 1$] has a solution, then 'a' is called a quadratic residue (mod m) and is denoted by $a R m$. If (1) has no solution then 'a' is called a quadratic non-residue mod m and is denoted by $a \not{R} m$.

Example: (1) We consider the quadratic congruence $x^2 \equiv a \pmod{7}$, 7 is an odd prime and $(a, 7) = 1$.

We shall determine the values of 'a' for which $a\,R\,7$ and $a \not R\, 7$.

Here 'x' runs over from 1 to 6.

$$1^2 \equiv 1 \pmod 7 \quad 4^2 \equiv 2 \pmod 7$$
$$2^2 \equiv 4 \pmod 7 \quad 5^2 \equiv 4 \pmod 7$$
$$3^2 \equiv 2 \pmod 7 \quad 6^2 \equiv 1 \pmod 7$$

Then the given congruence has the solution when 'a' $= 1, 4, 2$, while the congruence has no solution when 'a' $= 3, 5, 6$.

Therefore, 'a' $= 1, 4, 2\, R\, 7$

and 'a' $= 3, 5, 6 \not R\, 7$.

(2) Let us consider the congruence $x^2 \equiv a \pmod{17}$, $(a, 17) = 1$

$$1^2 \equiv 1 \pmod{17} \quad 9^2 \equiv 13 \pmod{17}$$
$$2^2 \equiv 4 \pmod{17} \quad 10^2 \equiv 15 \pmod{17}$$
$$3^2 \equiv 9 \pmod{17} \quad 11^2 \equiv 2 \pmod{17}$$
$$4^2 \equiv 16 \pmod{17} \quad 12^2 \equiv 8 \pmod{17}$$
$$5^2 \equiv 8 \pmod{17} \quad 13^2 \equiv 16 \pmod{17}$$
$$6^2 \equiv 2 \pmod{17} \quad 14^2 \equiv 9 \pmod{17}$$
$$7^2 \equiv 15 \pmod{17} \quad 15^2 \equiv 4 \pmod{17}$$
$$8^2 \equiv 13 \pmod{17} \quad 16^2 \equiv 1 \pmod{17}$$

Here 'x' runs over from 1 to 16.

'a' $= 1, 2, 4, 8, 9, 13, 15, 16,\ R\ 17$

'a' $= 3, 5, 6, 7, 10, 11, 12, 14 \not R\, 17$.

Theorem (7.2): *The congruence* $x^{\frac{m-1}{2}} \equiv 1 \pmod m$ *where* m *is an odd prime has exactly* $\frac{m-1}{2}$ *quadratic residues belonging to the set* $S = \left\{1^2, 2^2, 3^2, \ldots, \left(\frac{m-1}{2}\right)^2\right\}$ *of residue classes mod* m.

***Proof*:** We consider the set $S = \left\{1^2, 2^2, 3^2, \ldots, \left(\frac{m-1}{2}\right)^2\right\}$ containing $\frac{m-1}{2}$ members.

Every member of S is of the form t^2 for $t = 1, 2, \ldots, \frac{m-1}{2}$ and $(t, m) = 1$.

Now $(t^2)^{\frac{m-1}{2}} = t^{m-1} \equiv 1 \pmod m$, by Fermat's theorem. Therefore, $(t^2)^{\frac{m-1}{2}} \equiv 1 \pmod m \Rightarrow x^{\frac{m-1}{2}} \equiv 1 \pmod m$ has a solution t^2, a member of S. Also the members of S are incongruent modulo m.

For, $r^2 \equiv s^2 \pmod m$, $1 \le r, s \le \dfrac{m-1}{2}$

$\Rightarrow \quad (r-s)(r+s) \equiv 0 \pmod m$

$\Rightarrow \quad m | (r-s)(r+s) \Rightarrow m | r-s$ or $m | r+s$.

But $\quad 1 \le r + s < m - 1 < m \Rightarrow m \nmid (r+s)$.

Hence, $m | r - s$ and so $r \equiv s \pmod m$, which is not true as $\{0, 1, 2, \ldots, m-1\}$ is a set of residue classes mod m.

Hence, S forms a set of residue classes mod m.

Therefore, $1^2, 2^2, 3^2, \ldots, \left(\frac{m-1}{2}\right)^2 \; R m$.

Thus, the congruence $x^{\frac{m-1}{2}} \equiv 1 \pmod{m}$ has exactly $\frac{m-1}{2}$ quadratic residues.

Corollary: The congruence $x^{\frac{m-1}{2}} \equiv 1 \pmod{m}$, where m is an odd prime has exactly $\frac{m-1}{2}$ quadratic non-residues mod m.

Proof: Let $S = \{1, 2, 3, \ldots, m-1\}$ and let $x \in S$ such that $x^{\frac{m-1}{2}} \not\equiv 1 \pmod{m}$. That is, $x \not\equiv r^2 \pmod{m}$, $1 \le r \le m-1$.

Then the numbers of such x are $(m-1) - \frac{m-1}{2} = \frac{m-1}{2}$, as $\frac{m-1}{2}$ members of S are quadratic residues mod m.

Therefore, $x^{\frac{m-1}{2}} \equiv 1 \pmod{m}$ has exactly $\frac{m-1}{2}$ quadratic non-residues mod m.

Problem (7.3): Determine the quadratic residues and quadratic non-residues of $m = 7, 17$ with the help of Th. (7.2) and also show that the quadratic residues mod (m) belongs to $S = \left\{1^2, 2^2, \ldots, \left(\frac{m-1}{2}\right)^2\right\}$.

Solution: (i) $m = 7 \Rightarrow x^{\frac{m-1}{2}} = x^3 \equiv 1 \pmod{7}$ (1)

$$S = \{1, 2, 3, 4, 5, 6\}.$$

x assumes the values of S.

$x = 1,\quad 1^3 \equiv 1 \pmod{7}$

$x = 2,\quad 2^3 \equiv 1 \pmod{7}$

$x = 4,\quad 4^3 \equiv 1 \pmod{7}$

But $x = 3,\quad 3^3 \not\equiv 1 \pmod{7}$

$x = 5,\quad 5^3 \not\equiv 1 \pmod{7}$

$x = 6,\quad 6^3 \not\equiv 1 \pmod{7}$

Therefore, $1, 2, 4 \, R 7$ and $3, 5, 6 \, \not R 7$.

Also $2 \equiv 3^2 \pmod{7} \Rightarrow 1^2, 3^2, 2^2 \, R 7$, i.e., $1^2, 2^2, 3^2 \, R 7$.

Thus, the quadratic residues mod 7 belong to $\{1^2, 2^2, 3^2\}$ having exactly $\frac{7-1}{2} = 3$ numbers.

Again, $3, 5, 6 \not\equiv r^2 \pmod{7}$, $1 \le r \le 6 \Rightarrow 3, 5, 6 \, \not R \, 7 \Rightarrow x^3 \equiv 1 \pmod{7}$ has exactly $3 = \frac{7-1}{2}$ quadratic non-residues mod 7.

(ii) $m = 17$.

$x^{\frac{m-1}{2}} = x^{\frac{17-1}{2}} = x^8 \equiv 1 \pmod{17}$ (1)

$$S = \{1, 2, 3, 4, 5, 6, 7, 8, 9, 10, 11, 12, 13, 14, 15, 16\}.$$

x assumes the values of S.

It can be verified that $x = 1, 2, 4, 8, 9, 13, 15, 16$ satisfies (1).

Again, $2 \equiv 6^2 \pmod{17}$, $13 \equiv 8^2 \pmod{17}$

$8 \equiv 5^2 \pmod{17}$, $15 \equiv 7^2 \pmod{17}$

$1, 2, 4, 8, 9, 13, 15, 16 \, R 17 \Rightarrow 1^2, 6^2, 2^2, 5^2, 3^2, 8^2, 7^2, 4^2 \, R 17$.

$\therefore\quad x \, R 17 \Rightarrow x \in \{1^2, 2^2, 3^2, 4^2, 5^2, 6^2, 7^2, 8^2\}$, which are exactly $\frac{17-1}{2} = 8$ in numbers

$$3, 5, 6, 7, 10, 11, 12, 14 \not R\, 17.$$

Also $\quad x \not R\, 17 \Rightarrow x \not\equiv r^2 \pmod{17},\ 1 \le r \le 16.$

A short table of $a\,R\,m$ and $a \not R\, m$ for a quadratic $x^2 \equiv a \pmod m$, $(a, m) = 1$ and m is an odd prime.

m	3	5	7	11	13	17
$a\,R\,m$	1	1, 4	1, 2, 4	1, 3, 4, 5, 9	1, 3, 4, 9, 10, 12	1, 2, 4, 8, 9, 13, 15, 16
$a \not R\, m$	2	2, 3	3, 5, 6	2, 6, 7, 8, 10	2, 5, 6, 7, 8, 11	3, 5, 6, 7, 10, 11, 12, 14

Theorem (7.3): *If g be a primitive root of an odd prime m, then $g^2, g^4, g^6, \ldots, g^{m-1}\, R\, m$ and $g^1, g^3, g^5, \ldots, g^{m-2} \not R\, m$.*

Proof: Since g is a primitive root of m, $(g, m) = 1$. We consider the set $S = \{g^2, g^4, \ldots, g^{m-1}\}$ in which each power of g is even.

Thus we have

(i) $g^{2k} \in S,\ k = 1, 2, \ldots, \dfrac{m-1}{2}$

(ii) $(g^{2k}, m) = 1,\ \forall\, k = 1, 2, \ldots, \dfrac{m-1}{2}$, since $(g, m) = 1$.

(iii) $g^{2p} \not\equiv g^{2q} \pmod m,\ 1 \le p,\ q \le \dfrac{m-1}{2}$ as $S \subseteq \{1, g, g^2, \ldots, g^{m-1}\}$ which is a set of reduced system mod m.

Therefore, S forms a reduced system of mod m, containing $\frac{m-1}{2}$ members. They form a solution of $x^2 \equiv a \pmod m$, proved in Th.[7.2]. Every solution of the quadratic is of form g^n, $n = 2k$.

$$g^n = g^{2k} = (g^k)^2$$

$\therefore\quad g^n \equiv x^2 \pmod m \Rightarrow x^2 \equiv g^n \pmod m$

$\Rightarrow\quad a \equiv g^n \pmod m$ for each primitive root g of m.

$\Rightarrow\quad g^n\ R\ m,\ n = 2k,\ k = 1, 2, \ldots, \dfrac{m-1}{2}.$

Consequently, $g^n \not R\, m,\ n = 2k - 1,\ k = 1, 2, \ldots, \dfrac{m-1}{2}.$

Theorem (7.4) (Euler's criterion): *The necessary and sufficient condition that $a\ R\ m$ when m is an odd prime, $(a, m) = 1$ is $a^{\frac{m-1}{2}} \equiv 1 \pmod m$.*

Proof: **Condition is necessary**

Let $a\,R\,m$.

Then $x^2 \equiv a \pmod m$ has a solution.

Let $x = t$ be a solution of it.

Then $(t^2)^{\frac{m-1}{2}} = t^{m-1} \equiv 1 \pmod m$ by Fermat's theorem, as $(a, m) = 1$ leads to $(t, m) = 1$ (1)

Hence, $\quad a \equiv t^2 \pmod m \Rightarrow a^{\frac{m-1}{2}} \equiv (t^2)^{\frac{m-1}{2}} \pmod m$

$\Rightarrow\quad a^{\frac{m-1}{2}} \equiv 1 \pmod m$, by (1).

Condition is sufficient

Let the given condition $a^{\frac{m-1}{2}} \equiv 1 \pmod m$ hold.

We have to show that $a\,R\,m$.

In other words, $x^2 \equiv a \pmod{m}$ has a solution.

Let g be a primitive root of m.

Therefore, $a \equiv g^k \pmod{m}$, for some k, $1 \le k \le m-1$ [by Th. (7.3)].

$\Rightarrow \quad g^k \equiv a \pmod{m}$ (1)

$\Rightarrow \quad g^{k\left(\frac{m-1}{2}\right)} \equiv a^{\frac{m-1}{2}} \pmod{m} \equiv 1 \pmod{m}$ [by the given condition] (2)

Again, g being a primitive root of m,

$$O(g \pmod{m}) = \phi(m).$$

$\Rightarrow \quad g^{\phi(m)} \equiv 1 \pmod{m}$. (3)

From (2) and (3), it follows that $\phi(m) | k\left(\frac{m-1}{2}\right)$,

i.e., $\quad (m-1) | k\left(\frac{m-1}{2}\right) \Rightarrow k = 2q,\ q \in Z^+$.

Therefore, $\quad (g^q)^2 = g^{2q} = g^k \equiv a \pmod{m}$

$\Rightarrow \quad g^q$ is a solution of $x^2 \equiv a \pmod{m}$

$\Rightarrow \quad a\,R\,m$.

Legendre's Symbol and its Properties

The great mathematician Legendre introduced a symbol $\left(\frac{a}{m}\right)$ against a function $f(a, m)$ of two variables a, m where $(a, m) = 1$, having two values $+1$ and -1 according as $a\,R\,m$ and $a\,\not R\,m$.

Definition (7.2): *We now define the symbol* $\left(\frac{a}{m}\right)$, *known as Legendre's symbol, when m is an odd prime and $(a, m) = 1$ by*

$$\left(\frac{a}{m}\right) = 1 \Leftrightarrow a\,R\,m \quad \text{and} \quad \left(\frac{a}{m}\right) = -1 \Leftrightarrow a(\not R)m.$$

If $a \equiv 0 \pmod{m}$, then $\left(\frac{a}{m}\right) = 0$.

This symbol is very much useful in developing the theory of quadratic congruences.

Example(3): For a congruence $x^2 \equiv a \pmod{17}$. We have already shown in [Ex 1,2 Ch-7] that

$1, 4, 9, 16, 8, 2, 15, 13\ R\,m$ (i)

and $3, 5, 6, 7, 10, 11, 12, 14\ \not R\ m$. (ii)

Then (i) $\Rightarrow \left(\frac{1}{17}\right), \left(\frac{4}{17}\right), \left(\frac{9}{17}\right), \left(\frac{16}{17}\right), \left(\frac{8}{17}\right), \left(\frac{2}{17}\right), \left(\frac{15}{17}\right), \left(\frac{13}{17}\right) = 1.$

(ii) $\Rightarrow \left(\frac{3}{17}\right), \left(\frac{5}{17}\right), \left(\frac{6}{17}\right), \left(\frac{7}{17}\right), \left(\frac{10}{17}\right), \left(\frac{11}{17}\right), \left(\frac{12}{17}\right), \left(\frac{14}{17}\right) = -1.$

Theorem (7.5): *If m is an odd prime and a, b two integers such that $(a, m) = 1$, $(b, m) = 1$, then*

(i) $\left(\frac{a}{m}\right) \equiv a^{\frac{m-1}{2}} \pmod{m}$.

(ii) $a \equiv b \pmod{m} \Rightarrow \left(\frac{a}{m}\right) = \left(\frac{b}{m}\right)$.

(iii) $\left(\frac{a}{m}\right)$ is completely a multiplicative function of a.

(iv) $\left(\frac{a}{m}\right) = 1 \Rightarrow \left(\frac{a^2}{m}\right) = 1$

(v) $\left(-\frac{1}{m}\right) \equiv (-1)^{\frac{m-1}{2}} \pmod{m} = \begin{cases} 1, & \text{when } m \equiv 1 \pmod 4. \\ -1, & \text{when } m \equiv 3 \pmod 4, \end{cases}$

Proof: (i) Let $\left(\frac{a}{m}\right) = 1$. Then $a\,R\,m$.

Therefore, by Euler's criterion [Th. 7.4],

$$a^{\frac{m-1}{2}} \equiv 1 \pmod{m} \Rightarrow 1 \equiv a^{\frac{m-1}{2}} \pmod{m}$$

$\Rightarrow \quad \left(\frac{a}{m}\right) \equiv a^{\frac{m-1}{2}} \pmod{m}$.

Let $\quad \left(\frac{a}{m}\right) = -1$. Then $a \not R\, m$.

Therefore, $\quad a^{\frac{m-1}{2}} \equiv -1 \pmod{m} \Rightarrow -1 \equiv a^{\frac{m-1}{2}} \pmod{m}$

$\Rightarrow \quad \left(\frac{a}{m}\right) \equiv a^{\frac{m-1}{2}} \pmod{m}$

[**Remarks:** $(a^{\frac{m-1}{2}} - 1)(a^{\frac{m-1}{2}} + 1) = a^{m-1} - 1 \equiv 0 \pmod{m}$. (1)

Since m is an odd prime and $(a, m) = 1$, (1) is true by Fermat's theorem.

Hence, $\quad$ either $a^{\frac{m-1}{2}} - 1 \equiv 0 \pmod{m}$ or, $a^{\frac{m-1}{2}} + 1 \equiv 0 \pmod{m}$,

i.e., $\quad$ either $a^{\frac{m-1}{2}} \equiv 1 \pmod{m}$ or, $a^{\frac{m-1}{2}} \equiv -1 \pmod{m}$.

Both the congruences do not hold simultaneously. If so, then $1 \equiv -1 \pmod{m} \Rightarrow m|2$, which is not possible, as m is an odd prime.

Therefore, if $a \not R\, m$, then 'a' cannot satisfy $a^{\frac{m-1}{2}} \equiv 1 \pmod{m}$.

Hence, it must satisfy $a^{\frac{m-1}{2}} \equiv -1 \pmod{m}$, which is also taken as an alternative formula of Euler's criterian, when an integer $a \not R\, m$].

Proof: (ii) Let $\left(\frac{a}{m}\right) = 1$. Then $a\,R\,m$.

Therefore, $x^2 \equiv a \pmod{m}$ has a solution.

$\Rightarrow \quad x^2 \equiv b \pmod{m}$ has the same solution as $a \equiv b \pmod{m}$

$\Rightarrow \quad b\,R\,m \Rightarrow \left(\frac{b}{m}\right) = 1$.

$\therefore \quad \left(\frac{a}{m}\right) = \left(\frac{b}{m}\right)$.

Again, if $\left(\frac{a}{m}\right) = -1$, then $a \not R\, m$.

Hence, $x^2 \equiv a \pmod{m}$ has no solution

$\Rightarrow \quad x^2 \equiv b \pmod{m}$ has no solution

$\Rightarrow \quad b \not R\, m \Rightarrow \left(\frac{b}{m}\right) = -1$.

$\therefore \quad \left(\frac{a}{m}\right) = \left(\frac{b}{m}\right)$.

[Go through the example.]

Example(4): $x^2 \equiv 3 \pmod{11}$ and $3 \equiv 14 \pmod{11}$

Then $x^2 \equiv 25 \pmod{11}$ has the solutions $x = \pm 5$.

For, $3 \equiv 14 \pmod{11} \Rightarrow x^2 \equiv 14 \pmod{11} \Rightarrow x^2 \equiv 25 \pmod{11}$

$\Rightarrow$ has the same solution ± 5, as that of $x^2 \equiv 3 \pmod{11}$.

So $\left(\frac{3}{11}\right) = \left(\frac{14}{11}\right)$ $[(3, 11), (14, 11) = 1]$

Again, $x^2 \equiv 6 \pmod{11}$, has no solution.

Now, $6 \equiv 28 \pmod{11}$.

Then $x^2 \equiv 28 \pmod{11} \Rightarrow x^2 \equiv 6 \pmod{11}$ has no solution.

$\therefore$ $\left(\frac{6}{11}\right) = -1, \left(\frac{28}{11}\right) = -1 \Rightarrow \left(\frac{6}{m}\right) = \left(\frac{28}{m}\right)$.

Proof: (iii) **Case(1):** Let $a \equiv 0 \pmod{m}$ and $b \equiv 0 \pmod{m}$.

Then $m|a$ and $m|b \Rightarrow m|ab$.

$\therefore$ $\left(\frac{a}{m}\right) = 0, \left(\frac{b}{m}\right) = 0, \left(\frac{ab}{m}\right) = 0$. [by Def. (7.2)]

Hence, $\left(\frac{ab}{m}\right) = 0 = \left(\frac{a}{m}\right) \cdot \left(\frac{b}{m}\right)$.

Conversely, let $\left(\frac{ab}{m}\right) = 0$.

Then $\left(\frac{ab}{m}\right) = 0 \Rightarrow m|ab \Rightarrow m|a$ or $m|b$.

$m|a \Rightarrow \left(\frac{a}{m}\right) = 0$ or $m|b \Rightarrow \left(\frac{b}{m}\right) = 0$.

Then $\left(\frac{a}{m}\right)\left(\frac{b}{m}\right) = 0 = \left(\frac{ab}{m}\right)$.

Case(2): Let $m \nmid a$ and $m \nmid b$. Them $m \nmid ab$.

Therefore, we have

$$\left(\frac{ab}{m}\right) \equiv (ab)^{\frac{m-1}{2}} \pmod{m}$$

$$\Rightarrow \left(\frac{a \cdot b}{m}\right) \equiv a^{\frac{m-1}{2}} b^{\frac{m-1}{2}} \pmod{m}$$

$$\equiv \left(\frac{a}{m}\right)\left(\frac{b}{m}\right) \pmod{m}. \qquad (1)$$

Now the following possibilities arise, if $\left(\frac{a}{m}\right) = 1, \left(\frac{b}{m}\right) = 1$, then $\left(\frac{a \cdot b}{m}\right) = 1$, because $\left(\frac{ab}{m}\right) = -1 \Rightarrow -1 \equiv 1 \pmod{m} \Rightarrow m| - 2$, which is not possible as m is an odd prime.

Therefore, $\left(\frac{a \cdot b}{m}\right) = 1 = \left(\frac{a}{m}\right) \cdot \left(\frac{b}{m}\right)$.

Again, if $\left(\frac{a}{m}\right) = -1$, $\left(\frac{b}{m}\right) = -1$, then $\left(\frac{a \cdot b}{m}\right) = 1$, because $\left(\frac{ab}{m}\right) = -1$ will arise $m|2$, which is not possible.

Therefore, $\left(\frac{ab}{m}\right) = 1 = (-1)(-1) = \left(\frac{a}{m}\right)\left(\frac{b}{m}\right)$.

Also, if $\left(\frac{a}{m}\right) = 1$, $\left(\frac{b}{m}\right) = -1$ [vice versa], then $\left(\frac{a \cdot b}{m}\right) = -1$, because $\left(\frac{ab}{m}\right) = 1$ will arise $m|2$, not possible.

Therefore, $\left(\frac{ab}{m}\right) = -1 = 1(-1) = \left(\frac{a}{m}\right)\left(\frac{b}{m}\right)$.

Hence, we have $\left(\frac{ab}{m}\right) = \left(\frac{a}{m}\right)\left(\frac{b}{m}\right)$.

[Go through the example]

Example (5): Let $m = 17$ [See Ex. (3), Ch.7].

Case (i): Let $a = 4$, $b = 15$.

Then $\left(\frac{4}{17}\right) = 1$, $\left(\frac{15}{17}\right) = 1$.

So $\left(\frac{ab}{17}\right) = \left(\frac{4 \cdot 15}{17}\right) = \left(\frac{9}{17}\right) = 1$. [$\because$ $60 \equiv 9 \pmod{17}$]

Therefore, $\left(\frac{60}{17}\right) = 1 = \left(\frac{4}{17}\right) \cdot \left(\frac{5}{17}\right)$.

Case (ii): Let $a = 4$, $b = 11$.

Then $\left(\frac{4}{17}\right) = 1$, $\left(\frac{11}{17}\right) = -1$.

Therefore, $\left(\frac{44}{17}\right) = \left(\frac{4 \cdot 11}{17}\right) = \left(\frac{10}{17}\right) = -1$ [$\because$ $44 \equiv 10 \pmod{17}$].

Thus, $\left(\frac{44}{17}\right) = (+1)(-1) = \left(\frac{4}{17}\right) \cdot \left(\frac{11}{17}\right)$.

Case (iii): Let $a = 10$, $b = 6$.

Then $\left(\frac{10}{17}\right) = -1$, $\left(\frac{6}{17}\right) = -1$

But $\left(\frac{10 \cdot 6}{17}\right) = \left(\frac{60}{17}\right) = \left(\frac{9}{17}\right) = 1$

$= (-1)(-1) = \left(\frac{10}{17}\right) \cdot \left(\frac{6}{17}\right)$.

***Proof* (iv):** $\left(\frac{a}{m}\right) = 1 \Rightarrow a\ R\ m \Rightarrow a^{\frac{m-1}{2}} \equiv 1 \pmod{m}$ [by Th. (7.4)].

Then $(a^2)^{\frac{m-1}{2}} = a^{m-1} \equiv 1 \pmod{m}$, [by Fermat's theorem, as $(a, m) = 1$]

$\Rightarrow$ $a^2\ R\ m \Rightarrow \left(\frac{a^2}{m}\right) = 1$.

Proof **(v):** From (i) $\left(\frac{a}{m}\right) \equiv a^{\frac{m-1}{2}} \pmod{m}$.

Putting $a = -1$,

$$\left(-\frac{1}{m}\right) \equiv (-1)^{\frac{m-1}{2}} \pmod{m}$$

$$\equiv 1, \text{ if } \quad \frac{m-1}{2} \text{ is even.}$$

$$\text{i.e., if } \frac{m-1}{2} = 2k, \ k \in z$$

$$\text{i.e., if } m = 4k+1$$

$$\text{i.e., if } m \equiv 1 \pmod{4}.$$

Thus, $\left(-\frac{1}{m}\right) = 1$, when $m \equiv 1 \pmod{4}$.

Now $\left(-\frac{1}{m}\right) = -1$, if $\quad \frac{m-1}{2}$ is odd.

$$\text{i.e., if } \frac{m-1}{2} = (2k+1), \ k \in z.$$

$$\text{i.e., if } m = 2(2k+1)+1 = 4k+3$$

$$\text{i.e., if } m \equiv 3 \pmod{4}.$$

Thus, $\left(-\frac{1}{m}\right) = -1$, if $m \equiv 3 \pmod{4}$.

Theorem (7.6): Gauss' Lemma *If m be an odd prime and $(a, m) = 1$ and if l denotes the number of those least (+)ve residues of $a, 2a, 3a, \ldots, \frac{m-1}{2}a$, mod m exceeding $\frac{m}{2}$, then $\left(\frac{a}{m}\right) \equiv (-1)^l$ (mod m).*

***Proof*:** The set $S = \left\{a, 2a, 3a, \ldots, \frac{m-1}{2}a\right\}$ is a reduced system mod m. We get two sets of distinct least residues mod m as follows :

$$S_1 = \{r_1, r_2, \ldots, r_k\}, \ 0 \le r_i < \frac{m}{2}, \ i = 1, 2, \ldots, k.$$

$$S_2 = \{s_1, s_2, \ldots, s_l\}, \ \frac{m}{2} < s_j < m, \ j = 1, 2, \ldots, l.$$

such that $S_1 \cap S_2 = \phi$. Hence, $k + l = \dfrac{m-1}{2}$.

$$\left.\begin{array}{ll} \text{Also} & r_i \equiv ta \pmod{m} \\ \text{and} & s_j \equiv t'a \pmod{m} \end{array}\right\} \quad (1)$$

for some t, t' satisfying $1 \le t, t' \le \dfrac{m-1}{2}$.

We consider $S_3 = \{x_1, x_2, \ldots, x_l\}$, where $x_j = m - s_j$, $j = 1, 2, \ldots, l$; also $0 < x_j < \dfrac{m}{2}$.

Therefore, $S_3 \subset \left[0, \dfrac{m}{2}\right]$

Then $\quad S_1 \cap S_3 = \phi$.

If not, then $x_j \equiv r_i \pmod{m} \Rightarrow m - s_j \equiv r_i \pmod{m}$

$\Rightarrow \quad r_i + s_j \equiv 0 \pmod{m} \Rightarrow ta + t'a \equiv 0 \pmod{m}$

$\Rightarrow \quad (t + t')a \equiv 0 \pmod{m}$. $\quad (2)$

But from (1), $0 < 1 < t + t' < m$ and $(a, m) = 1$. So (2) is not possible.

Hence, $S_1 \cap S_3 = \phi$.

Therefore, $n(S_1 \cup S_3) = \dfrac{m-1}{2}$ and $S_1 \cup S_3 \subset \left[1, \dfrac{m-1}{2}\right]$.

$\therefore \quad S_1 \cup S_3 = \{r_1, r_2, r_3, \cdots, r_k, x_1, x_2, \ldots, x_l\} = \left\{1, 2, 3, \ldots, \dfrac{m-1}{2}\right\}$.

So $\quad r_1 r_2 \cdots r_k x_1 x_2 \cdots x_l = \left(\dfrac{m-1}{2}\right)!$.

Therefore,

$$\begin{aligned}
\left(\frac{m-1}{2}\right)! &= r_1 \cdot r_2 \cdots r_k x_1 \cdot x_2 \cdots x_l \\
&= r_1 \cdot r_2 \cdots r_k (m - s_1) \cdots (m - s_l) \\
&= r_1 r_2 \cdots r_k (-s_1)(-s_2) \cdots (-s_l) \pmod m \\
&\qquad [m - s_j \equiv -s_j \pmod m] \\
&\equiv (-1)^l r_1 r_2 \cdots r_k s_1 \cdot s_2 \cdots s_l \pmod m \\
&\equiv (-1)^l a \cdot 2a \cdots \frac{m-1}{2} a \pmod m
\end{aligned}$$

$\left[\text{since } r_i, s_j \equiv ta \pmod m,\ t = 1, 2, \ldots, \dfrac{m-1}{2} \text{ in some order}\right]$

$$\equiv (-1)^l a^{\frac{m-1}{2}} \left(\frac{m-1}{2}\right)! \pmod m$$

$\Rightarrow \quad 1 \equiv (-1)^l a^{\frac{m-1}{2}} \pmod m \quad \left[\left(\left(\dfrac{m-1}{2}\right)!, m\right) = 1\right]$

$\Rightarrow \quad a^{\frac{m-1}{2}} \equiv (-1)^l \pmod m$

$\Rightarrow \quad \left(\dfrac{a}{m}\right) \equiv (-1)^l \pmod m$ [Eulerean criterion]

[Go through the Example.]

Example: Let $m = 17$ and $a = 3$, $\dfrac{m-1}{2} = 8$.

$$S = \{3,\ 2 \cdot 3,\ 3 \cdot 3,\ 4 \cdot 3,\ 5 \cdot 3,\ 6 \cdot 3,\ 7 \cdot 3,\ 8 \cdot 3\}$$

Least (+)ve residues are 3, 6, 9, 12, 15, 1, 4, 7

$S_1 = \{1,\ 3,\ 4,\ 6,\ 7\}, \quad \therefore \quad r_i < \dfrac{m}{2} = 8$

$S_2 = \{9,\ 12,\ 15\},\ s_j > \dfrac{m}{2} = 8.$

Then $S_1 \cap S_2 = \phi$.

Also $S_3 = \{17 - 9,\ 17 - 12,\ 17 - 15\}$, $x_i \equiv 17 - s_j \pmod{17}$.

Here $l = 3$, number of residues exceeding 8.

Also $S_1 \cap S_3 = \phi$.

Then $S_1 \cup S_3 = \{1,\ 3,\ 4,\ 6,\ 7,\ 17 - 9,\ 17 - 12,\ 17 - 15\} \subset [1, 8]$.

$n(S_1 \cup S_3) = 8 \Rightarrow S_1 \cup S_3 = \{1, 2, 3, 4, 5, 6, 7, 8\}$.

Then the product of elements of $S_1 \cup S_3$

$$1\cdot 2\cdot 3\cdot 4\cdot 5\cdot 6\cdot 7\cdot 8 \equiv 1\cdot 3\cdot 4\cdot 6\cdot 7\cdot (17-9)(17-15)(17-12) \pmod{17}$$

$\Rightarrow \quad 8! \equiv 1\cdot 3\cdot 4\cdot 6\cdot 7\cdot(-9)(-12)(-15) \pmod{17}$

$\equiv (-1)^3 1\cdot 3\cdot 4\cdot 6\cdot 7\cdot 9\cdot 12\cdot 15 \pmod{17}$

$\equiv (-1)^3(1\cdot 3)(2\cdot 3)(3\cdot 3)(4\cdot 3)(5\cdot 3)(6\cdot 3)(7\cdot 3)(8\cdot 3) \pmod{17}$ by (1)

$\Rightarrow \quad 8! \equiv (-1)^3 3^8\cdot 8! \pmod{17}$

$\Rightarrow \quad 1 \equiv (-1)^3 3^{\frac{17-1}{2}} \pmod{17}$

$\Rightarrow \quad 3^{\frac{17-1}{2}} \equiv (-1)^3 \pmod{17}$

$\Rightarrow \quad \left(\frac{3}{17}\right) \equiv (-1)^3 \pmod{17}.$

Lattice point

Definition (7.3): *A point (x, y) in XY-plane is said to be a lattice point if $x, y \in Z$. $P(2,3) \to$ a point on X-Y plane is a lattice point, since 2,3 are integers.*

$P(-3,-5)$ is a lattice point, but $P\left(-\frac{3}{4}, 5/2\right)$ is not a lattice point.

Lemma (7.1): If $P(m, n)$ be a lattice point in the 1st quadrant and $(m, n) = 1$, then there does not exist any lattice point on the line OP, excluding the points O and P.

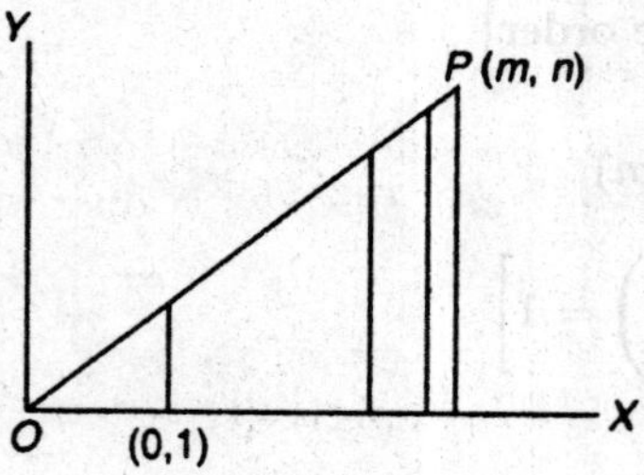

Proof: Equation of OP is $y = \frac{nx}{m}$, $0 < x < m$, $0 < y < n$. Then (x, y) is a lattice point on OP, iff $\frac{nx}{m}$ is an integer, for some (+)ve integer $x < m$. But this is not possible as $(n, m) = 1$ and $m \nmid x$ as $x < m$. Hence the lemma.

Ex: Let $P = (5, 7)$, where $(5, 7) = 1$

Equation of OP is $y = \dfrac{7x}{5}$, $x < 5$.

Then $\dfrac{7x}{5}$ is not an integer as $(7, 5) = 1$ and $5 \nmid x$ as $x < 5$.

So y is not an integer.

$\therefore$ (x, y) is not a lattice point on OP excluding O and P.

Lemma (7.2) If $P(m, n)$ be a lattice point, then the number of lattice points on the vertical line DE drawn through a point D $(r, 0)$ meeting the line OP at a point E is $\left[\dfrac{rn}{m}\right]$, excluding the point D.

Proof: DE is the vertical line drawn through D $(r, 0)$, $1 \le r < m$.

Equation of OP is $y = \dfrac{nx}{m}$.

Length of $DE = \dfrac{rn}{m}$.

Then the number of lattice points on DE (excluding the point D), is the largest integer but not exceeding $\dfrac{rn}{m}$.

Hence, the number of lattice point on DE is $\left[\dfrac{rn}{m}\right]$, here it is to be noted that the point E is excluded by Lemma (7.1).

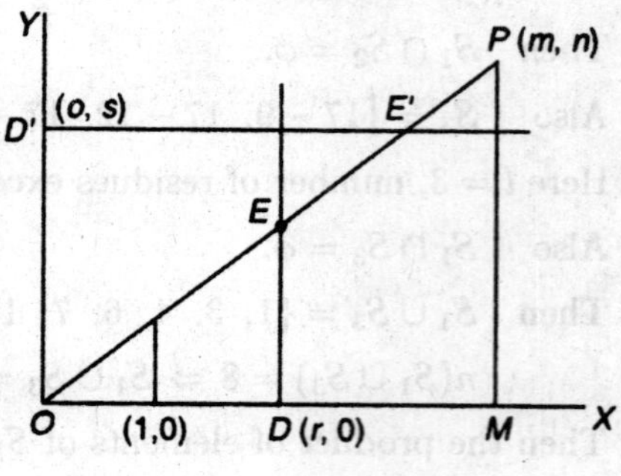

Lemma (7.3): If $P(m,n)$ be a lattice point, then the number of lattice points on the horizontal line $D'E'$ drawn through a point $D'(0,s)$ meeting the line OP at E' is $\left[\frac{sm}{n}\right]$, excluding the point D'.

Result follows from Lemma (7.2).

Theorem [Einestein] (7.7): *If m,n be two odd (+)ve integers such that $(m,n)=1$, then*

$$\sum_{r=1}^{\frac{m-1}{2}}\left[\frac{rn}{m}\right]+\sum_{s=1}^{\frac{n-1}{2}}\left[\frac{sm}{n}\right]=\frac{m-1}{2}\times\frac{n-1}{2}.$$

Proof:

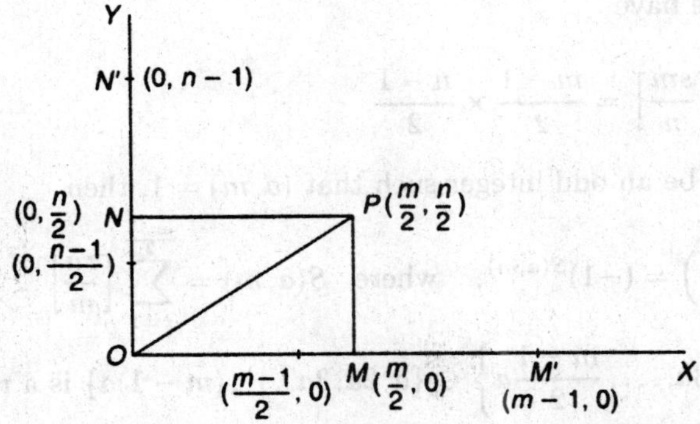

We construct a rectangle □ $OMPN$ in the 1st quadrant with the vertices $O(0,0)$, $M\left(\frac{m}{2},0\right)$, $P\left(\frac{m}{2},\frac{n}{2}\right)$, $N\left(0,\frac{n}{2}\right)$, where OP is a diagonal of □ $OMPN$ and it divides the rectangle into two equal triangles △ OMP and △ ONP.

Let R denote the region within the rectangle □ $OMPN$.

R_1 and R_2 denote the region within △ OMP and △ ONP respectively.

Then the number of lattice points in R is equal to the sum of the number of lattice points in R_1 and the number of lattice points in R_2.

Let $S_1=\left\{x, 1\le x\le\frac{m-1}{2}\right\}$, where x, y are integers.

$$S_2=\left\{y, 1\le y\le\frac{n-1}{2}\right\},$$

$$S_1\times S_2=\left\{(x,y),\ 1\le x\le\frac{m-1}{2},\ 1\le y\le\frac{n-1}{2}\right\}.$$

All the points (x,y) of $S_1\times S_2$ are the lattice points in R.

$$n(S_1\times S_2)=\frac{m-1}{2}\times\frac{n-1}{2}.$$

Therefore, the number of lattice points in $R=\frac{m-1}{2}\times\frac{n-1}{2}$. (1)

Lattice points in R do not lie on OP, the diagonal of □ $OMPN$.

Now, the number of lattice points in R_1 is equal to the sum of the number of lattice points lying on the vertical lines drawn through the points $(r,0)$, $1\le r\le\frac{m-1}{2}$

and this is equal to $\sum_{r=1}^{\frac{m-1}{2}}\left[\frac{rm}{n}\right]$ [by Lemma (7.2)].

Also the number of lattice points in R_2 is equal to the sum of the number of lattice points lying on the horizontal lines drawn through the points $(0, s)$ $1 \leq s \leq \frac{n-1}{2}$ and is equal to $\sum_{s=1}^{\frac{n-1}{2}} \left[\frac{sm}{n}\right]$ [by Lemma (7.3)].

Therefore, the number of lattice points in R

= the number of lattice points in R_1 + the number of lattice points in R_2

$$= \sum_{r=1}^{\frac{m-1}{2}} \left[\frac{rn}{m}\right] + \sum_{s=1}^{\frac{n-1}{2}} \left[\frac{sm}{n}\right]. \tag{2}$$

From (1) and (2), we have

$$\sum_{r=1}^{\frac{m-1}{2}} \left[\frac{rn}{m}\right] + \sum_{s=1}^{\frac{n-1}{2}} \left[\frac{sm}{n}\right] = \frac{m-1}{2} \times \frac{n-1}{2}.$$

Lemma (7.4): If a be an odd integer such that $(a, m) = 1$, then

$$\left(\frac{a}{m}\right) = (-1)^{S(a,m)}, \quad \text{where} \quad S(a, m) = \sum_{r=1}^{\frac{m-1}{2}} \left[\frac{ra}{m}\right].$$

Proof: $S = \left\{a, 2a, 3a, \ldots, \frac{m-1}{2}a\right\} \subset \{a, 2a, 3a, \ldots, (m-1)a\}$ is a reduced system mod m.

Let the distinct least residues mod m of S belong to $S_1 = \{r_1, r_2, \ldots, r_k\}$ and $S_2 = \{s_1, s_2, \ldots, s_l\}$ where $0 < r_1 < \frac{m}{2}$ and $s_j > \frac{m}{2}$.

Then $S_1 \cup S_2 = \{r_1, r_2, \ldots, r_k, s_1, s_2, \ldots, s_l\}$ and also $S_1 \cap S_2 = \phi$ and hence, $k + l = \frac{m-1}{2}$.

If $\quad S_3 = \{x_1, x_2, \ldots, x_l\}, \quad$ where $\quad x_j = m - s_j,\ j = 1, 2, \ldots, l,\ x_j < \frac{m}{2}$.

Then $\quad S_1 \cap S_3 = \phi \quad$ and $\quad S_1 \cup S_3 = \left[1, \frac{m-1}{2}\right]$.

Therefore, $r_1 + r_2 + \cdots + r_k + m - s_1 + \cdots + m - s_j$

$$= 1 + 2 + \cdots + \frac{m-1}{2} = \left(\frac{m-1}{2}\right)\left(\frac{m+1}{2}\right)/2$$

$$= \frac{m^2-1}{8}. \tag{1}$$

Then the sum of the least residues of S

$$= r_1 + r_2 + \cdots + r_k + s_1 + s_2 + \cdots + s_l$$

$$= r_1 + \cdots + r_k + (m - s_1) + (m - s_2) + \cdots + (m - s_l) + 2(s_1 + s_2 + \cdots + s_l) - lm$$

$$= \frac{m^2-1}{8} + 2\sum_{j=1}^{l} s_j - lm \text{ by (1).} \tag{2}$$

Let R_j be the least residues of ja, $j = 1, 2, \ldots, \frac{m-1}{2}$,

then $\quad ja = \left[\frac{ja}{m}\right] m + Rj \Rightarrow R_j = ja - \left[\frac{ja}{m}\right] \cdot m.$

$$\sum_{j=1}^{\frac{m-1}{2}} R_j = \text{sum of the least residues of } S$$

$$= a + 2a + \cdots + \frac{m-1}{2}a - m\sum_{j=1}^{\frac{m-1}{2}}\left[\frac{ja}{m}\right]$$

$$= a\frac{m^2-1}{8} - mS(a,m). \tag{3}$$

Then from (2) and (3),

$$\frac{m^2-1}{8}a - mS(a,m) = \frac{m^2-1}{8} + 2\sum_{j=1}^{l} s_j - lm$$

$$\Rightarrow \qquad lm = mS(a,m) - (a-1)\frac{m^2-1}{8} + 2\sum_{j=1}^{l} s_j.$$

$a-1$ is even, $\dfrac{m^2-1}{8}$ = integer as m is an odd prime.

Therefore, $lm \equiv mS(a,m) \pmod 2$

$\Rightarrow \qquad l \equiv S(a,m) \pmod 2$ $[\because\ (m,2)=1]$

$$\Rightarrow \qquad l \equiv S(a,m) = \sum_{j=1}^{\frac{m-1}{2}}\left[\frac{ja}{m}\right].$$

Therefore, by Th.(7.6),

$$\left(\frac{a}{m}\right) = (-1)^l = (-1)^{S(a,m)}.$$

Problem (7.4): Evaluate $\left[\frac{7}{17}\right]$, when $m = 17$, $a = 7$ with the help of the above Lemma [7.4].

Solution: $m = 17$, $a = 7$, $\dfrac{m-1}{2} = 8$.

We find out $l = S(a,m) = \displaystyle\sum_{j=1}^{m-1}\left[\frac{ja}{m}\right]$.

$$\text{Here } \ l = S(7,17) = \sum_{j=1}^{8}\left[\frac{j7}{17}\right]$$

$$= \left[\frac{7}{17}\right] + \left[\frac{14}{17}\right] + \left[\frac{21}{17}\right] + \left[\frac{28}{17}\right] + \left[\frac{35}{17}\right] + \left[\frac{42}{17}\right] + \left[\frac{49}{17}\right] + \left[\frac{56}{17}\right]$$

$$= 0+0+1+1+2+2+2+3 = 11.$$

Therefore, by Lemma (7.4),

$$\left(\frac{7}{17}\right) = (-1)^l = (-1)^{11} = -1.$$

Theorem (7.8): *If m be an odd prime, then*

$$\left(\frac{2}{m}\right) = (-1)^{\left[\frac{m+1}{4}\right]} = (-1)^{\frac{m^2-1}{8}}.$$

Proof: From Th.(7.6), for $a = 2$, we have $\left(\frac{2}{m}\right) = (-1)^l$.

Here l is the number of least residues mod m, of the set of integers $1 \cdot 2$, $2 \cdot 2, \ldots, \frac{m-1}{2} \cdot 2$, greater than $\frac{m}{2}$.

Then the number of integers which are less than $\frac{m}{2}$ is

$$1 \cdot 2,\ 2 \cdot 2,\ 3 \cdot 2, \ldots, \left(\frac{m-1}{2} - l\right) \cdot 2. \tag{1}$$

The next number of the series in (1) is

$$\left(\frac{m-1}{2} - l + 1\right) \cdot 2. \tag{2}$$

From (1) and (2), it follows that

$$\left(\frac{m-1}{2} - l\right) \cdot 2 < \frac{m}{2} < \left(\frac{m-1}{2} - l + 1\right) \cdot 2$$

$$\Rightarrow \quad (m-1) - 2l < \frac{m}{2} < m + 1 - 2l.$$

Now $\quad m - 1 - 2l < \frac{m}{2}$

$$\Rightarrow \quad 2m - 2 - 4l < m \Rightarrow \frac{m-2}{4} < l. \tag{i}$$

Also $\quad \frac{m}{2} < m + 1 - 2l \Rightarrow m < 2m + 2 - 4l \Rightarrow 4l < m + 2$

$$\Rightarrow \quad l < \frac{m+2}{4}. \tag{ii}$$

Thus from (i) and (ii), we get

$$\frac{m-2}{4} < l < \frac{m+2}{4} \Rightarrow \frac{m+2}{4} - 1 < l < \frac{m+2}{4}$$

$$\Rightarrow \quad l = \left[\frac{m+2}{4}\right] = \left[\frac{m+1}{4}\right] \forall\, m.$$

Let $\quad m = 4n + 1.$

Then $\quad \frac{m^2-1}{8} = \frac{16n^2 + 8n + 1 - 1}{8} = 2n^2 + n$

$$\left[\frac{m+1}{4}\right] = \left[\frac{4n+2}{4}\right] = \left[n + \frac{1}{2}\right] = n\ \forall\, n.$$

Therefore, $\frac{m^2-1}{8} = 2n^2 + \left[\frac{m+1}{4}\right] \Rightarrow \frac{m^2-1}{8} \equiv \left[\frac{m+1}{4}\right] \pmod 2.$

Also, if $m = 4n + 3$,

$$\frac{m^2-1}{8} = 2n^2 + 3n + 1 \quad \text{and} \quad \left[\frac{m+1}{4}\right] = \left[\frac{4n+3+1}{4}\right] = n + 1.$$

Therefore, $\frac{m^2-1}{8} = 2n(n+1) + \left[\frac{m+1}{4}\right]$

$$\Rightarrow \quad \frac{m^2-1}{8} \equiv \left[\frac{m+1}{4}\right] \pmod 2.$$

Hence, for all odd primes,

$$\left(\frac{2}{m}\right) = (-1)^{\left[\frac{m+1}{4}\right]} = (-1)^{\left(\frac{m^2-1}{8}\right)}.$$

Example: Let us consider an odd prime $m = 11$.

The set of integers is

$$1\cdot 2,\ 2\cdot 2,\ 3\cdot 2,\ 4\cdot 2,\ \left(\frac{11-1}{2}\right)\cdot 2$$

i.e., $1\cdot 2,\ 2\cdot 2,\ 3\cdot 2,\ 4\cdot 2\ 5\cdot 2.$

$$1\cdot 2,\ 2\cdot 2 < \frac{m}{2} = \frac{11}{2} \quad \text{(i)}$$

$$3\cdot 2,\ 4\cdot 2,\ 5\cdot 2 > \frac{m}{2}.$$

$l = 3$, since the number of least residues mod(11), greater than $\frac{m}{2}$ is 3.

The next number of the series (1) is

$$\left(\frac{m-1}{2} - l + 1\right) \times 2 = \left(\frac{11-1}{2} - 3 + 1\right)\cdot 2 = (5-3+1)\cdot 2. \quad \text{(ii)}$$

Again, the number of integers less than $\frac{11}{2}$

$$= 1\cdot 2,\ \left(\frac{11-1}{2} - 3\right)\cdot 2. \quad \text{(iii)}$$

From (ii) and (iii), it follows that

$$\left(\frac{11-1}{2} - 3\right)\cdot 2 < \frac{11}{2} < \left(\frac{11-1}{2} - 3 + 1\right)\cdot 2$$

$$\Rightarrow \quad 5\cdot 2 - 3\cdot 2 < \frac{11}{2} < (5-3+1)\cdot 2.$$

Now $\quad 5\cdot 2 - 3\cdot 2 < \frac{11}{2} \Rightarrow 5\cdot 4 - 3\cdot 4 < 11$

$$\Rightarrow \quad \frac{5\cdot 4 - 11}{4} < 3, \text{ i.e., } \frac{9}{4} < 3. \quad \text{(iii)}$$

Again, $\frac{11}{2} < (5-3+1)2 \Rightarrow 11 < 24 - 3\cdot 4$

$$\Rightarrow \quad 3\cdot 4 < 24 - 11 = 13 \Rightarrow 3 < \frac{13}{4}. \quad \text{(iv)}$$

Thus, from (iii) and (iv),

$$\frac{9}{4} < 3 < \frac{13}{4}, \text{ i.e., } \frac{11+2}{4} - 1 < 3 < \frac{11+2}{4}$$

i.e., $\quad 3 = \left[\frac{11+2}{4}\right] = \left[\frac{11+1}{4}\right].$

Theorem (7.9): [Quadratic Reciprocity Law] *If m, n be any two distinct odd primes, then*

$$\left(\frac{m}{n}\right)\left(\frac{n}{m}\right) = (-1)^{\frac{m-1}{2}\times\frac{n-1}{2}}.$$

***Proof*:** Since m, n are odd primes and $m \neq n$, we have $(m, n) = 1$. So $\left(\frac{m}{n}\right)$ can be defined, and m plays a role of (+)ve odd integer.

Therefore, by Lemma (7.4),

$$\left(\frac{m}{r}\right) = (-1)^{S(m,n)} = (-1)^{\sum_{r=1}^{\frac{n-1}{2}}\left[\frac{rm}{n}\right]}.$$

$\left(\frac{n}{m}\right)$ can be defined and n plays a role of (+)ve odd integer.

Therefore, $\left(\frac{n}{m}\right) = (-1)^{S(n,m)} = (-1)^{\sum_{s=1}^{\frac{m-1}{2}}\left[\frac{sn}{m}\right]}$.

Hence, $$\left(\frac{m}{n}\right)\left(\frac{n}{m}\right) = (-1)^{\sum_{r=1}^{\frac{n-1}{2}}\left[\frac{rm}{n}\right]} \cdot (-1)^{\sum_{s=1}^{\frac{m-1}{2}}\left[\frac{sm}{n}\right]} = (-1)^{\sum_{r=1}^{\frac{n-1}{2}}\left[\frac{rm}{n}\right] + \sum_{s=1}^{\frac{m-1}{2}}\left[\frac{sm}{n}\right]}$$

$$= (-1)^{\frac{n-1}{2} \times \frac{m-1}{2}} = (-1)^{\frac{m-1}{2} \times \frac{n-1}{2}} \text{ [by Theorem [7.7]}$$

Corollary: For any two distinct odd primes m, n,

(i) $\left(\frac{n}{m}\right) = (-1)^{\frac{m-1}{2} \times \frac{n-1}{2}} \left(\frac{m}{n}\right)$

(ii) $\left(\frac{n}{m}\right) = \pm\left(\frac{m}{n}\right)$

according as one of m, n is of the form $4k + 1$ and both are of the form $4k + 3$.

***Proof*:** (i) $\left(\frac{m}{n}\right) = (-1)^l$ [by Lemma (7.4)]

$\Rightarrow$ $\left(\frac{m}{n}\right) \cdot \left(\frac{m}{n}\right) = (-1)^{2l} = 1.$ (i)

Now $(-1)^{\frac{m-1}{2} \times \frac{n-1}{2}} \left(\frac{m}{n}\right) = \left(\frac{m}{n}\right)\left(\frac{n}{m}\right)\left(\frac{m}{n}\right)$ [by Th (7.9)]

$$= \left(\frac{n}{m}\right)\left(\frac{m}{n}\right)\left(\frac{m}{n}\right) = \left(\frac{n}{m}\right) \cdot 1 \text{ by (i).}$$

Hence, $\left(\frac{n}{m}\right) = (-1)^{\frac{m-1}{2} \times \frac{n-1}{2}} \cdot \left(\frac{m}{n}\right)$.

(ii) Let $m = 4k + 1$. Then $\frac{m-1}{2} = 2k =$ an even integer.

Again, since n is an odd prime, $\frac{n-1}{2}$ is even.

Therefore, $\left(\frac{n}{m}\right) = (-1)^{\frac{m-1}{2} \times \frac{n-1}{2}} \frac{m}{n} = \frac{m}{n}$.

(iii) Let $m, n = 4k + 3$.

Then each of $\frac{m-1}{2}$ and $\frac{n-1}{2}$ is of the form $2k + 1$, an odd integer.

Therefore, $\left(\frac{n}{m}\right) = (-1)^{\frac{m-1}{2} \times \frac{n-1}{2}} \cdot \left(\frac{m}{n}\right) = -1\left(\frac{m}{n}\right)$.

$\therefore$ $\left(\frac{n}{m}\right) = -\left(\frac{m}{n}\right)$.

Problem (7.5): Find the values of the following Legendre symbols:

(i) $\left(\frac{20}{31}\right)$ (ii) $\left(-\frac{23}{59}\right)$ (iii) $\left(\frac{199}{3}\right)$ (iv) $\left(\frac{5}{199}\right)$.

Solution: (i) $\left(\frac{20}{31}\right) = \left(\frac{2^2 \cdot 5}{31}\right) = \left(\frac{2^2}{31}\right) \cdot \left(\frac{5}{31}\right) \left[\text{since } \left(\frac{ab}{m}\right) = \left(\frac{a}{m}\right)\left(\frac{b}{m}\right)\right]$

$$= 1 \cdot \frac{5}{31} \left[\text{since } \left(\frac{2^2}{31}\right) = 1 \text{ as } \left(\frac{2}{31}\right) = 1\right]$$

$$= \left(\frac{31}{5}\right), \text{ [since } 5 = 4 \cdot 1 + 1, \text{ so by Cor. of Th.(7.9)]}$$

$$= \left(\frac{1}{5}\right) = 1 \left[\text{since } \left(\frac{1}{m}\right) = 1\right].$$

(ii) $\left(-\frac{23}{59}\right) = \left(-\frac{1}{59}\right)\left(\frac{23}{59}\right) = -\left(\frac{23}{59}\right). \left[\text{since } \left(-\frac{1}{m}\right) = -1\right]$

$= (-1)(-1)\left(\frac{59}{23}\right)$ [by Cor. of Th.[7.8] , as $59 = 4 \cdot 14 + 3,\ 23 = 4 \cdot 5 + 3$]

$= \left(\frac{36}{23}\right)$ [$\because\ 59 \equiv 36 \pmod{23}$]

$= \left(\frac{2^2 \cdot 3^2}{23}\right) = \left(\frac{2^2}{23}\right)\left(\frac{3^2}{23}\right) = 1 \cdot 1 = 1 \left[\because \left(\frac{2}{31}\right) = 1, \left(\frac{3}{31}\right) = 1\right].$

(iii) $\left(\frac{199}{3}\right) = \left(\frac{1}{3}\right)$ [$\because\ 199 \equiv 1 \pmod 3$]

$= 1.$

(iv) $\left(\frac{5}{199}\right) = \left(\frac{199}{5}\right)$, [since $5 = 4 \cdot 1 + 1$, so by Cor. of Th.(7.9)]

$$= \left(\frac{9}{5}\right) \quad [\because\ 199 \equiv 9 \pmod 5]$$

$$= \left(\frac{4}{5}\right) \quad [\because\ 9 \equiv 4 \pmod 5]$$

$$= \left(\frac{2^2}{5}\right) = 1 \left[\text{since } \left(\frac{a^2}{m}\right) = 1\right].$$

Problem (7.6): Examine whether the following equations (quadratic congruences) are solvable:

(i) $x^2 \equiv 219 \pmod{419}$

(ii) $3x^2 + 6x + 5 \equiv 0 \pmod{89}$.

Solution: A quadratic congruence $x^2 \equiv a \pmod m$ is solvable if aRm, i.e., if $\left(\frac{a}{m}\right) = 1$.

(i) Here $a = 219$, $m = 419$.

$$\left(\frac{a}{m}\right) = \left(\frac{219}{419}\right) = \left(\frac{3 \times 73}{419}\right) = \left(\frac{3}{419}\right)\left(\frac{73}{419}\right). \tag{1}$$

Now $\left(\frac{3}{419}\right) = -\left(\frac{419}{3}\right)$

[since $3 = 4 \cdot 0 + 3,\ 419 = 4 \times 104 + 3$; so by Cor. of Th.(7.9)]

$$= -\left(\frac{2}{3}\right) \; [\because\; 419 \equiv 2 \pmod 3]$$

$$= -1(-1)^{\left[\frac{3+1}{4}\right]} = (-1)(-1) = 1 \text{ [by Th.(7.8)]} \qquad (2)$$

$$\left(\frac{73}{419}\right) = \left(\frac{419}{73}\right), \text{ [since } 73 = 4\cdot 18 + 1;\text{ so by Cor. of Th.(7.9)]}$$

$$= \left(\frac{54}{73}\right) \; [\because\; 419 \equiv 54 \pmod{73}]$$

$$= \left(\frac{73}{54}\right) \; [\because\; 73 = 4\cdot 18 + 1]$$

$$= \left(\frac{19}{54}\right) \; [\because\; 73 \equiv 19 \pmod{54}]$$

$$= (-1)^{\left[\frac{54+1}{4}\right]} = (-1)^{13} = -1. \qquad (3)$$

Hence $\left(\frac{219}{419}\right) = 1\cdot(-1) = -1$ [from (1) by (2) & (3)]

Therefore, $219 \not{R}\, 419$ and hence, $x^2 \equiv 219 \pmod{419}$ is not solvable.

(ii) $3x^2 + 6x + 5 \equiv 0 \pmod{89}$

$\Rightarrow \quad (3x+3)^2 \equiv -6 \pmod{89} \equiv 83 \pmod{89}$

$\Rightarrow \quad y^2 \equiv 83 \pmod{89}$, where $y = 3x + 3$.

Now $\left(\frac{83}{89}\right) = \left(\frac{89}{83}\right)$, [since $89 = 4\cdot 22 + 1$]

$$= \left(\frac{6}{83}\right) \; [\because\; 89 \equiv 6 \pmod{83}]$$

$$= \left(\frac{2}{83}\right)\left(\frac{3}{83}\right) = (-1)^{\left[\frac{83+1}{4}\right]}\cdot\left(\frac{3}{83}\right)$$

$$= -\left(\frac{3}{83}\right) \; [\because\; 83 = 4\cdot 20 + 3, 3 = 4\cdot 0 + 3]$$

$$= (-1)(-1)\left(\frac{83}{3}\right),$$

$$= (-1)(-1)\left(\frac{2}{3}\right) \; [\because\; 83 \equiv 2 \pmod 3]$$

$$= (-1)^{\left[\frac{3+1}{4}\right]} = -1.$$

$83 \not{R}\, 89$.

So $y^2 \equiv 83 \pmod{89}$ is not solvable and hence, $3x^2 + 6x + 5 \equiv 0 \pmod{89}$ is not solvable.

Theorem (7.10): *3 is a quadratic residue of m, if and only if m is an odd prime* > 3, *but of the form* $12k \pm 1$.

Proof: m is an odd prime $> 3 \;\Rightarrow\; m = 5, 7, 11, 13, 17, 19, \ldots$

which can be expressed in the form $12k \pm 1$, or $12k \pm 5$.

Now $3\,R\,m \Leftrightarrow \left(\frac{3}{m}\right) = 1$.

Case(1): Let $m = 12k + 1$.

$$\left(\frac{3}{m}\right) = \left(\frac{3}{12k+1}\right) = \left(\frac{12k+1}{3}\right) \quad [\because\ 12k+1 = 4\cdot 3k+1]$$

$$= \left(\frac{1}{3}\right), \quad [\because\ 12k+1 \equiv 1 \pmod 3]$$

$$= 1 \left[\because\ \left(\frac{1}{m}\right) = 1\right].$$

Case(2): Let $m = 12k - 1$.

Then $\left(\frac{3}{m}\right) = \left(\frac{3}{12k-1}\right) = -\left(\frac{12k-1}{3}\right)$

$$[\because\ 3 = 4\cdot 0+3,\ 12k-1 = 4(3k-1)+3]$$

$$= -\left(-\frac{1}{3}\right) \quad [\because\ 12k-1 \equiv -1 \pmod 3]$$

$$= (-1)(-1) = 1 \left[\because\ \left(-\frac{1}{m}\right) = -1, \text{ when } m \equiv 3(\bmod 4)\right]$$

Case(3): Let $m = 12k + 5$.

$$\left(\frac{3}{m}\right) = \left(\frac{3}{12k+5}\right) = \left(\frac{12k+5}{3}\right) \quad [\because\ 12k+5 = 4(3k+1)+1]$$

$$= \left(\frac{5}{3}\right) \quad [\because\ 12k+5 \equiv 5 \pmod 3]$$

$$= \left(\frac{2}{3}\right) \quad [\text{as } 5 \equiv 2 \pmod 3]$$

$$= \left(-\frac{1}{3}\right) \quad [\text{as } 2 \equiv -1 \pmod 3]$$

$$= -1$$

Case (4): Let $m = 12k - 5$.

$$\left(\frac{3}{m}\right) = \left(\frac{3}{12k-5}\right) \left[\begin{array}{c}\because\ 12k-5 = 4(3k-2)+3 \\ 3 = 4\cdot 0+3\end{array}\right]$$

$$= -\left(\frac{12k-5}{3}\right)$$

$$= -\left(-\frac{5}{3}\right) \quad [\because\ 12k-5 \equiv -5 \pmod 3]$$

$$= -\left(-\frac{2}{3}\right) \quad [\because\ -5 \equiv -2 \pmod 3]$$

$$= -\left(\frac{1}{3}\right) \quad [\because\ -2 \equiv 1 \pmod 3]$$

$$= -1.$$

From Cases (1) and (2) it follows that $3\,R\,m$ when $m = 12k \pm 1$ and from Cases (3) and (4) it follows that $3\,\not R\,m$ when $m = 12k \pm 5$.

Thus, $3\,R\,m$ when $m = 12k \pm 1$.

Problem (7.7): Show that $6\,R\,m$ if and only if $m \equiv 1, 5, 19, 23 \pmod{24}$ where m is an odd prime.

Solution: $m = 24k+1,\ 24k+5,\ 24k+19,\ 24k+23$, where k assumes those (+)ve integers for which m becomes an odd primes.

Case(1): $m = 24k+1$

$$k \neq 0,1,2,5,\quad k=3,\quad m=73$$
$$k=4,\quad m=97 \quad \text{and so on.}$$

Now $\left(\frac{.6}{m}\right) = \left(\frac{2\cdot 3}{m}\right) = \left(\frac{2}{m}\right)\left(\frac{3}{m}\right) = \left(\frac{2}{24k+1}\right)\left(\frac{3}{24k+1}\right)$

$$= (-1)^{\left[\frac{24k+1+1}{4}\right]}\cdot\left(\frac{24k+1}{3}\right).$$

$$= (-1)^{6k}\left(\frac{1}{3}\right) = 1\cdot 1$$

$$= 1, \quad \text{as } 24k+1 \equiv 1 \pmod 3.$$

Case(2): $m = 24k+5$

Now $\left(\frac{6}{m}\right) = \left(\frac{2}{24k+5}\right)\left(\frac{3}{24k+5}\right)$

$$= (-1)^{\left[\frac{24k+5+1}{4}\right]}\left(\frac{24k+5}{3}\right) \quad [\because\ 24k+5 = 4(6k+1)+1]$$

$$= (-1)^{6k+1}\left(\frac{2}{3}\right) \quad [\because\ 24k+5 \equiv 2 \pmod 3]$$

$$= (-1)(-1)^{\left[\frac{3+1}{4}\right]}$$

$$= (-1)(-1) = 1.$$

Case(3): $m = 24k+19$

$$\left(\frac{6}{m}\right) = \left(\frac{2}{24k+19}\right)\left(\frac{3}{24k+19}\right)$$

$$= (-1)^{\left[\frac{24k+19+1}{4}\right]}(-1)\left(\frac{24k+19}{3}\right) \quad \left[\begin{array}{r}\because\ 24k+19 = 4(6k+4)+3\\ 3 = 4\cdot 0+3\end{array}\right]$$

$$= (-1)^{6k+5}(-1)\left(\frac{4}{3}\right) \quad [\because\ 24k+19 \equiv 4 \pmod 3]$$

$$= (-1)(-1)\left(\frac{1}{3}\right)$$

$$= 1.$$

Case(4): $m = 24k+23$.

$$\left(\frac{6}{m}\right) = \left(\frac{2}{24k+23}\right)\left(\frac{3}{24k+23}\right)$$

$$= (-1)^{\left[\frac{24k+23+1}{4}\right]}(-1)\left(\frac{24k+23}{3}\right) \quad \left[\begin{array}{r}\because\ 24k+23 = 4(6k+5)+3\\ \text{and } 3 = 0\cdot k+3\end{array}\right]$$

$$= (-1)^{6k+6}(-1)\left(\frac{2}{3}\right) \quad [\because\ 24k+23 \equiv 2 \pmod 3]$$

$= 1 \cdot (-1)(-1)^{[\frac{3+1}{4}]}$
$= (-1) \cdot (-1) = 1.$

Therefore, $6 \, R m$.

Quadratic congruence with composite module m^h, $h \geq 1$

Theorem (7.11): *If m be an odd prime and $(a, m) = 1$, then $x^2 \equiv a \pmod{m^h}$, $h \geq 1$ has a solution if and only if $\left(\frac{a}{m}\right) = 1$.*

***Proof*:** Let $x^2 \equiv a \pmod{m^h}$ has a solution.

Then $a \, R \, m^h$ and so $a \, R \, m$.

Therefore, $\left(\frac{a}{m}\right) = 1$.

Now let $\left(\frac{a}{m}\right) = 1$, we have to show that $\left(\frac{a}{m^h}\right) = 1$.

We shall prove it by method of induction.

Let $h = 1$, then $\left(\frac{a}{m^h}\right) = \left(\frac{a}{m}\right) = 1$.

So it holds for $h = 1$.

Let it be true for $h = k \; (> 1)$.

Then $\left(\frac{a}{m^k}\right) = 1$.

Therefore, $x^2 \equiv a \pmod{m^k}$ has a solution, say x_0.

Hence, $x_0^2 = a + m^k \cdot t$ for some $t \in Z$. (1)

Let us consider the linear congruence

$$2x_0 y \equiv -t \pmod{m}. \quad (2)$$

Then it has a unique solution, say y_0, as $(2x_0, m) = 1$.

By (2), it follows that $m | 2x_0 y_0 + t$.

That is, $2x_0 y_0 + t = mt', \; t' \in Z$. (3)

Now we consider an integer $x_1 = x_0 + y_0 m^k$;

Then $x_1^2 = x_0^2 + 2x_0 y_0 m^k + y_0^2 m^{2k}$

$= a + m^k t + 2x_0 y_0 m^k + y_0^2 m^{2k}$ [by (1)]

$= a + (t + 2x_0 y_0) m^k + y_0^2 m^{2k}$

$= a + mt' \cdot m^k + y_0^2 m^{2k}$ [by (3)].

$= a + m^{k+1} t' + y_0^2 m^{2k}$

$= a + m^{k+1}(t' + y_0^2 m^{k-1}), \; k > 1$

$\Rightarrow \quad x_1^2 \equiv a \pmod{m^{k+1}}$

$\Rightarrow \quad x^2 \equiv a \pmod{m^{k+1}}$ has a solution x_1.

Hence, $\left(\frac{a}{m^{k+1}}\right) = 1$.

So it is true for $h = k + 1$.

Therefore, it is true $\forall \, h, \, h \geq 1$.

Hence $\left(\frac{a}{m^h}\right) = 1, \; h \geq 1$.

Theorem (7.12): $-3\,R\,m$ *when* m *is an odd prime* > 3 *of the form* $6k+1$.

Proof: $m = 6k+1$ induces the primes $7, 13, 19, 31, 37, \ldots$ which can be further expressed in the form $12k+1$, $12k-5$.

Case(1): Let $m = 12k+1$.

$$\text{Then } \left(-\frac{3}{m}\right) = \left(-\frac{3}{12k+1}\right) = \left(-\frac{1}{12k+1}\right)\left(\frac{3}{12k+1}\right)$$

$$= (-1)^{\frac{12k+1-1}{2}} \cdot \left(\frac{12k+1}{3}\right) \quad [\because\ 12k+1 = 4\cdot 3k+1]$$

$$= 1 \cdot \left(\frac{1}{3}\right) \quad [\because\ 12k+1 \equiv 1 \pmod 3]$$

$$= 1 \quad \left[\because\ \left(\frac{1}{m}\right) = 1\right]$$

Case(2): Let $m = 12k-5$.

$$\text{Then } \left(-\frac{3}{m}\right) = \left(-\frac{1}{12k-5}\right)\left(\frac{3}{12k-5}\right)$$

$$= (-1)^{\frac{12k-5-1}{2}}(-1)\left(\frac{12k-5}{3}\right)$$

$$[\because\ 12k-5 = 4(3k-2)+3,\ 3 = 4\cdot 0+3]$$

$$= (-1)(-1)\left(\frac{7}{3}\right) \quad [\because\ 12k-5 \equiv 7 \pmod 3)]$$

$$= \left(\frac{1}{3}\right) \quad [\because\ 7 \equiv 1 \pmod 3)]$$

$$= 1.$$

Therefore, $-3\,R\,m$, when $m = 12k+1$, $12k-5$.

Corollary (7.3): $-3\,\not{R}\,m$, if m is an odd prime > 3 of the form $6k+5$.

Problem (7.8): Solve (i) $x^2 \equiv 7 (\text{mod } 3^3)$ (ii) $x^2 \equiv 14 \pmod{5^3}$.

Solution: To solve the given congruence we start with $x^2 \equiv 7 \pmod{3^2}$, where $x = 4 (= x_0)$ is a solution of it.

Then $16 = 7 + t\cdot 3^2 \Rightarrow t = 1$.

The linear congruence is of the form $2x_0y \equiv -1 \pmod 3$.

That is, $8y \equiv -1 \pmod 3$.

Then $y = 1 (= y_0)$ is a unique solution of it, as $(8,3) = 1$.

Therefore, $x_1 = x_0 + y_0 \cdot 3^2 = 4 + 1 \cdot 3^2 = 13$, is a solution of $x^2 \equiv 7 \pmod{3^3}$

Also -13 is its another solution, but $-13 \equiv 14 \pmod{3^3}$

$\Rightarrow 14$ is its another solution.

Hence, $x \equiv 13, 14 \pmod{3^3}$ are its solution.

(ii) We start with the congruence $x^2 \equiv 14 \pmod{5^2}$.

Then $x = 8\ (= x_0)$ is a solution of it.

Then $8^2 = 14 + 25t \ \Rightarrow\ t = 2$.

The linear congruence is of the form $2x_0y \equiv -2 \pmod 5$, i.e.,$16y \equiv -2 \pmod 5$, which has a unique solution $y = 3\ (= y_0)$ as $(16, 5) = 1$.

Therefore, $x_1 = x_0 + y_0 \cdot 5^2 = 8 + 3 \cdot 5^2 = 83$ is a solution of $x^2 \equiv 14 \pmod{5^3}$.

Also -83 is its another solution, but $-83 \equiv 42 \pmod{5^3}$.

So $x \equiv 42, 83 \pmod{5^3}$ are its solution.

Theorem (7.13): *The congruence $x^2 \equiv a \pmod{2^h}$, $h \geq 3$ is solvable, if and only if $a \equiv 1 \pmod 8$ $(a, 2) = 1$.*

Proof: Let $x^2 \equiv a \pmod{2^h}$ has a solution x_0 for $h \geq 3$.

Then $\quad x_0^2 \equiv a \pmod{2^h}$

$\Rightarrow \quad x_0^2 \equiv a \pmod{2^3}$, for $h = 3$. (1)

Again, $\quad (a, 2) = 1 \Rightarrow a$ is an odd integer.

$\Rightarrow x_0$ is an odd integer, by (1).

Also square of any odd integer is of the form $8k + 1$.

Therefore, $x_0^2 = 8k + 1$

$\Rightarrow \quad x_0^2 \equiv 1 \pmod 8$. (2)

So from (1) and (2), we have $a \equiv 1 \pmod 8$.

Conversely, let $a \equiv 1 \pmod 8$.

For $h = 3$, $x^2 \equiv a \pmod{2^3}$.

$\Rightarrow x^2 \equiv 1 \pmod 8$ is solvable. So the result holds for $h = 3$.

Now let it be true for $h = k$ $(k > 3)$.

Then $x^2 \equiv a \pmod{2^k}$ has a solution, say, x_0.

Then $\quad x_0^2 = a + 2^k t$ for some $t \in Z$ (3)

$\Rightarrow x_0$ is an odd integer as a is an odd integer. (3) $\Rightarrow t$ is an odd integer.

Consequently the linear congruence $x_0 y \equiv -t \pmod 2$ (4)

determines a unique solution y_0 (say).

Now we claim that the integer $x_1 = x_0 + y_0 2^{k-1}$ is a solution of $x^2 \equiv a \pmod{2^{k+1}}$.

Then
$$\begin{aligned} x_1^2 &= (x_0 + y_0 2^{k-1})^2 \\ &= x_0^2 + 2x_0 y_0 2^{k-1} + y_0^2 2^{2k-2} \\ &= a + 2^k \cdot t + x_0 y_0 2^k + y_0^2 2^{2k-2}, \text{ by (3)} \\ &= a + 2^k(t + x_0 y_0) + y_0^2 2^{2k-2}. \end{aligned}$$

Now (4) $\Rightarrow 2 | x_0 y_0 + t \Rightarrow t + x_0 y_0 = 2q$, $q \in Z$.

$$\begin{aligned} \therefore \quad x_1^2 &= a + 2^k \cdot 2q + y_0^2 2^{2k-2} \\ &= a + 2^{k+1}(q + y_0^2 \cdot 2^{k-3}), \ k > 3 \\ &= a + 2^{k+1} t', \ t' = q + y_0^2 2^{k-3} \in Z. \end{aligned}$$

$\Rightarrow \quad x_1^2 \equiv a \pmod{2^{k+1}}$

$\Rightarrow \quad x_1$ is a solution of $x^2 \equiv a \pmod{2^{k+1}}$.

Therefore, the result holds for $h = k + 1$.

Hence it holds $\forall\, h \geq 3$.

Therefore, $x^2 \equiv a \pmod{2^h}$ is solvable, iff $a \equiv 1 \pmod 8$.

Note: If x_0 be a solution of $x^2 \equiv a \pmod{2^h}$, then its all other four solutions are $x_0,\ -x_0,\ x_0+2^{h-1},\ -x_0+2^{h-1}$.

Problem (7.9): Find the value of a and then solve the congruence $x^2 \equiv a \pmod{2^5}$.

Solution: The congruence $x^2 \equiv a \pmod{2^5}$ is solvable if $a \equiv 1 \pmod 8$.

Then $a = 1, 9, 17, 25$ $(< 2^5 = 32)$.

Case(1): When $a = 1$, we start with $x^2 \equiv 1 \pmod{2^3}$.

Now $x^2 \equiv 1 \pmod{2^3}$ has a solution $x_0 = 1$.

Now $x_0^2 - a = 1 - 1 = 0$ is divisible by 2^4.

So 1 is a solution of $x^2 \equiv 1 \pmod{2^5}$ and hence the four solutions are $1,\ -1,\ 1+2^4,\ -1+2^4$, i.e., $1,\ -1,\ 17,\ 15 \pmod{2^5}$.

Case(2): When $a = 9$, we start with $x^2 \equiv 9 \pmod{2^3}$, it has a solution $x_0 = 1$.

Now $x_0^2 - a = 1 - 9 = -8 = -2^3$ is divisible by 2^3. Highest power of 2 is 3 and $t = -1$.

Then $x_1 = x_0 + 2^{3-1}y_0 = 1 + 2^{3-1} \cdot 1$ as the linear congruence $xy \equiv +1 \pmod 2$ has a solution $y_0 = 1$, for $x_0 = 1$.

$\therefore\ x_1 = 1 + 2^2 = 5$ is a solution of $x^2 \equiv 9 \pmod{2^4}$.

Again, $x_1^2 - 9 = 25 - 9 = 16 = 2^4$ is divisible by 2^4, the highest power of 2 is 4 and $t = 1$.

Linear congruence $xy \equiv -1 \pmod 2 \Rightarrow 5y \equiv -1 \pmod 2$.

$\Rightarrow y = 1 (= y_0)$ is its solution.

So $x_0 + y_0 2^{k-1} = 5 + 1 \cdot 2^{4-1} = 5 + 8 = 13$ is a solution of $x^2 \equiv 9 \pmod{2^5}$.

Hence, the other four solutions are $13, -13, 13+2^4, -13+2^4$, i.e., $13, -13, 29, 3$, i.e., $x \equiv 3, 13, 19, 29 \pmod{2^5}$ $[\because -13 \equiv 19 \pmod{2^5}]$

Case(3): When $a = 17$, we start with $x^2 \equiv 17 \pmod{2^3}$.

It has a solution $x_0 = 1$.

$x_0^2 - 17 = 1 - 17 = (-1)2^4$, is divisible by 2^4. The highest power of 2 is 4 and $t = -1$.

The linear congruence $xy \equiv 1 \pmod 2 \Rightarrow y \equiv 1 \pmod 2 \Rightarrow y = 1 = (y_0)$ is its solution.

So $x_0 + 2^{4-1} \cdot y_0 = 1 + 2^3 \cdot 1 = 9$ is a solution of $x^2 \equiv 17 \pmod{2^5}$.

The four solutions are $9, -9, 9+2^4, -9+2^4$, i.e., $9, -9, 25, 7$, i.e., $x \equiv 7, 9, 23, 25 \pmod{2^5}$ $[\because -9 \equiv 23 \pmod{2^5}]$.

Case(4): When $a = 25$, we start with $x^2 \equiv 25 \pmod{2^3}$.

It has a solution $x_0 = -1$.

Now $x_0^2 - 25 = 1 - 25 = -3 \cdot 2^3$, is divisible by 2^3. The highest power of 2 is 3. Also $t = -3$.

The linear congruence $xy \equiv 3 \pmod 2 \Rightarrow -y \equiv 3 \pmod 2 \Rightarrow y = 1 = (y_0)$ is its solution.

So $x_0 + y_0 \cdot 2^{3-1} = -1 + 1 \cdot 2^2 = 3$ is a solution of $x^2 \equiv 25 \pmod{2^4}$.

Again, $3^2 - 25 = (-1) \cdot 2^4$ is divisible by 2^4. The highest power of 2 is 4 and $t = -1$.

The linear congruence $xy = 1 \pmod 2 \Rightarrow 3y \equiv 1 \pmod 2 \Rightarrow y = 1 = y_0$ is its solution.

So $3 + 2^{4-1} \cdot 1 = 3 + 2^3 = 11$ is a solution of $x^2 \equiv 25 \pmod{2^5}$.

The four solutions are 11, -11, $11+2^4$, $-11+2^4$, i.e., 11, -11, 27, 5, i.e., $x \equiv 5, 11, 21, 27 \pmod{2^5}$ [$\because -11 \equiv 21 \pmod{2^5}$].

Problem (7.10): Prove that $2x^2+1 \equiv 0 \pmod{p}$, where p is an odd prime, has a solution, if and only if $p \equiv 1$ or $3 \pmod 8$.

Solution: $2x^2+1 \equiv 0 \pmod{p}$

$\Rightarrow \quad 2x^2 \equiv -1 \pmod{p}$

$\Rightarrow \quad 2x^2 \equiv p-1 \pmod{p}$

$\Rightarrow \quad x^2 \equiv \dfrac{p-1}{2} \pmod{p}$ [$p-1$ is even, $(2,p)=1$] (1)

Square of any integer is of the form $8q$ or $8q+1$ or $8q+4$, $q \in Z$.

(1) is solvable if and only if

$$\frac{p-1}{2} \equiv 0 \pmod 8 \text{ or } \frac{p-1}{2} \equiv 1 \pmod 8 \text{ or, } \frac{p-1}{2} \equiv 4 \pmod 8.$$

Now $\dfrac{p-1}{2} \equiv 0 \pmod 8 \Rightarrow p \equiv 1 \pmod 8$ (i)

$\dfrac{p-1}{2} \equiv 1 \pmod 8 \Rightarrow p \equiv 3 \pmod 8$ (ii)

$\dfrac{p-1}{2} \equiv 4 \pmod 8 \Rightarrow p-1 \equiv 8 \pmod 8$

$\Rightarrow p-1 \equiv 0 \pmod 8$

$\Rightarrow p \equiv 1 \pmod 8.$ (iii)

So from (i), (ii) and (iii) $\Rightarrow p \equiv 1$ or $3 \pmod 8$.

Exercise 7

1. Let m be an odd prime and $(a,m)=1$. Establish that the quadratic congruence $ax^2+bx+c \equiv 0 \pmod{m}$ is solvable, if and only if b^2-4ac is either zero or a quadratic residue of m.

2. If $m = 2^n+1$ is prime, then verify that every quadratic non-residue of m is a primitive root of m.

3. If m be an odd prime and $(a,m)=1$, $(b,m)=1$, prove that either all three of the quadratic congruences $x^2 \equiv a \pmod{m}$, $x^2 \equiv b \pmod{m}$, $x^2 \equiv ab \pmod{m}$ are solvable or exactly one of them admits a solution.

4. Verify that each of the congruences $x^2 \equiv 1 \pmod{15}$, $x^2 \equiv -1 \pmod{65}$ and $x^2 \equiv -2 \pmod{33}$ has four incongruent solutions. Hence, Lagrange's Theorem need not hold if the modulus is a composite number.

5. If m is a prime, show that $m|(a^m+(m-1)!a)$ and $m|((m-1)!a^m+a)$ for any integer a.

6. If p is an odd (+)ve integer and $p = p_1p_2\cdots p_k$, then

   (a) $\left(\dfrac{1}{p}\right) = 1.$

   (b) $\left(\dfrac{a}{p}\right) = \left(\dfrac{b}{p}\right)$, if $a \equiv b \pmod{p}$.

(c) If $a = a_1 a_2 \dots a_n$, then $\left(\frac{a}{p}\right) = \left(\frac{a_1}{p}\right)\left(\frac{a_2}{p}\right)\cdots\left(\frac{a_n}{p}\right)$.

(d) $\frac{p-1}{2} \equiv \frac{p_1 - 1}{2} + \frac{p_2 - 1}{2} + \cdots + \frac{p_k - 1}{2} \pmod 2$.

7. If b, b' are (+)ve odd integers and $(aa', bb') = 1$, then

(a) $a \equiv a' \pmod b \Rightarrow \left(\frac{a}{b}\right) = \left(\frac{a'}{b}\right)$.

(b) $\left(\frac{aa'}{b}\right) = \left(\frac{a}{b}\right)\left(\frac{a'}{b}\right)$.

(c) $\left(\frac{a}{bb'}\right) = \left(\frac{a}{b}\right)\left(\frac{a}{b'}\right)$.

(d) $\left(\frac{a^2}{b}\right) = \left(\frac{a}{b^2}\right) = 1$.

(e) $\left(\frac{2}{b}\right) = (-1)^{\frac{b^2-1}{8}}$.

8. Establish that 7 is a primitive root of any prime of the form $p = 2^{4n} + 1$.

9. Solve $x^2 \equiv 31 \pmod{11^4}$.

10. Prove that, for a fixed n, all the solvable congruences $x^2 \equiv a \pmod n$ with $(a, n) = 1$ have the same number of solutions.

Chapter 8

Non-linear Diophantine Equation

$x^n + y^n = z^n$, $n \geq 2$ is known as **non-linear Diophantine equations.**

That is, equations of the types $x^2 + y^2 = z^2$, $x^3 + y^3 = z^3$, $x^4 + y^4 = z^4 \cdots$ $x^n + y^n = z^n$ are all non-linear Diophantine equations.

Here we shall consider the solution of $x^2 + y^2 = z^2$, of course the solution consists of (+)ve integers only.

We observe that

$$3^2 + 4^2 = 5^2$$

$$5^2 + 12^2 = 13^2$$

$$7^2 + 24^2 = 25^2$$

$$9^2 + 40^2 = 41^2$$

and so on such that $a^2 + b^2 = c^2$, where a, b, c are all (+)ve integers.

So (3,4,5), (5,12,13), ... satisfy the equation $x^2 + y^2 = z^2$ and these are the solutions of $x^2 + y^2 = z^2$.

In this connection, we refer the most important theorem 'Pythagorean Theorem' for right-angled triangle.

That is, if a and b denote the length of the two sides containing the right angle and c denotes the length of the opposite side of the right angle, then $a^2 + b^2 = c^2$. So, no doubt, the solutions of $x^2 + y^2 = z^2$ satisfy 'Pythagorean Theorem'. (3,4,5), (5,12,13), ... are known as **Pythagorean triple.**

Thus, (a, b, c) is said to be a Pythagorean triple if $a^2 + b^2 = c^2$, in other wards, if a, b, c satisfy $x^2 + y^2 = z^2$.

It is to be noted that Pythagoras himself discovered the formula,

$$x = 2n + 1, \; y = 2n^2 + 2n, \; z = 2n^2 + 2n + 1,$$

where n is a (+)ve integer for infinitely many Pythagorean triples $(2n + 1,\; 2n^2 + 2n,\; 2n^2 + 2n + 1)$. Further we observe that

$$6^2 + 8^2 = 10^2$$

$$10^2 + 24^2 = 26^2$$

$$9^2 + 12^2 = 15^2 \text{ and so on.}$$

The triples (6,8,10), (10,24,26), $\cdots$ are also Pythagorean triples and infinitely many such Pythagorean triples are given by $(4n + 2,\; 4n^2 + 4n,\; 4n^2 + 4n + 2)$ when n is a (+)ve integer.

Thus the Pythagorean triples (a, b, c) can be divided into two groups:

(i) When G.C.D. $(a, b, c) = 1$

Ex: (3,4,5), (5,12,13).

(ii) When G.C.D. $(a, b, c) \geq 2$

Ex: (6,8,10), (9,12,15).

Definition (8.1): *The Pythagorean triple (a, b, c) is called Primitive if G.C.D. $(a, b, c) = 1$.*

Ex: (3,4,5), (5,12,13) are Primitive, while (6,8,10) is not primitive.

Definition (8.2): *If G.C.D. $(a, b, c) = 1$ and $a^2 + b^2 = c^2$, then (a, b, c) is called a Primitive solution of $x^2 + y^2 = z^2$*

It also follows that for every non-primitive solution (a', b', c'), there exists a primitive solution (a, b, c) such that $(a', b', c') = (ka, kb, kc)$, k is a (+)ve integer.

Ex: (6,8,10) = (2.3,2.4,2.5) where (6,8,10) is non-primitive and (3,4,5) is primitive solution, if $x^2 + y^2 = z^2$.

Theorem (8.1): *If the Diophantine equation $x^2 + y^2 = z^2$ has a primitive solution (a, b, c), then one of a and b is even and the other is odd.*

Proof: (a, b, c) is a primitive solution of $x^2 + y^2 = z^2$

$\Rightarrow$ G.C.D. $(a, b, c) = 1$.

If possible, let a and b be both even. Then $a = 2m$, $b = 2k \Rightarrow c^2 = 4(m^2 + k^2)$

$\Rightarrow c$ is even,

and hence G.C.D. $(a, b, c) \geq 2$, a contradiction. (1)

Again, if possible, let a and b be both odd. Then $a = 2m + 1$ and $b = 2k + 1$.

$\Rightarrow c^2 = 4(m^2 + k^2) + 4(k + m) + 2$

$\Rightarrow c^2 \equiv 2 \pmod 4$.

But the square of any integer x, is of the form, either $x^2 = 0 \pmod 4$ or $x^2 \equiv 1 \pmod 4$.

Therefore, $c^2 \equiv 2 \pmod 4$ is not possible, (2)

Hence, from (1) and (2), it follows that if a is even, then b is odd and vice versa and c remains odd in each case.

Theorem (8.2): *a is even and b is odd or vice versa, if and only if $a + b \equiv 1 \pmod 2$.*

Proof: $a + b \equiv 1 \pmod 2$

$\Leftrightarrow \quad a + b = 2q + 1, \ q \in z$

$\Leftrightarrow \quad a + b =$ an odd integer

$\Leftrightarrow \quad$ either a is odd, b is even

or a is even, b is odd.

Corollary: If (a, b, c) be a primitive solution of $x^2 + y^2 = z^2$, then $a + b \equiv 1 \pmod 2$.

[Follows from Theorems (8.1) and (8.2)].

Lemma (8.1): If $(a, b) = 1$ and $a \cdot b = c^n$, n is a (+)ve integer (≥ 1), then a and b are the nth power of an integer.

Proof: Let $(a, c) = d$.

Then $d|a, \ d|c \Rightarrow a = da_1, \ c = dc_1$, where $(a_1, c_1) = 1$.

So $\quad ab = c^n \Rightarrow da_1 \cdot b = (dc_1)^n = d^n c_1^n$

$\Rightarrow \quad a_1 b = d^{n-1} c_1^n \Rightarrow c_1^n | b$ (i)

as $(a_1, c_1^n) = 1$.

Again, $ab = c^n \Rightarrow ab = d^n c_1^n \Rightarrow b | c_1^n d^n$.

Now $\quad (a, b) = 1 \Rightarrow (da_1, b) = 1 \Rightarrow (d, b) = 1 \Rightarrow (d^n, b) = 1$.

So $\quad b | c_1^n d^n \Rightarrow b | c_1^n$. (ii)

Therefore, from (i) and (ii),

$$b = c_1^n. \quad \text{(iii)}$$

Also $\quad a = \dfrac{c^n}{b} = \left(\dfrac{c}{c_1}\right)^n = \left(\dfrac{dc_1}{c_1}\right)^n = d^n$. (iv)

Result follows from (iii) and (iv). So a and b are nth power of an integer.

Thus, $a = d^2$, $b = c_1^2$, squares of integers, for $n = 2$.

Theorem (8.3): *Any primitive solution of $x^2 + y^2 = z^2$ is of the form $(2ab, a^2 - b^2, a^2 + b^2)$ for some integers a and b such that $a > b > 0$, $(a, b) = 1$ and $a + b \equiv 1$ (mod 2)*

Proof: Let (p, q, r) be a primitive solution of $x^2 + y^2 = z^2$.

Then we have to show that $p = 2ab$, $q = a^2 - b^2$ and $r = a^2 + b^2$.

We have $p + q \equiv 1 \pmod 2$, by Cor. of Th. (8.2).

That is, if p is even, then q is odd and consequently r is odd.

Then $\quad 2 | r + q$ and $2 | r - q$

and so $\quad \left(\dfrac{r+q}{2}, \dfrac{r-q}{2}\right) = 1$ (i)

$\left[\begin{array}{l} (r, q) = 1 \\ r + q = 2k_1, r - q = 2k_2 \\ \text{for some } k_1, k_2 \in z \\ \frac{r+q}{2} = k, \ \frac{r-q}{2} = k_2 \\ \because (k_1, k_2) = 1 \\ \Rightarrow \left(\frac{r+q}{2}, \frac{r-q}{2}\right) = 1 \end{array}\right.$

Also $\quad r^2 - q^2 = p^2$

$\Rightarrow \quad (r - q)(r + q) = p^2$

$\Rightarrow \quad \left(\dfrac{r-q}{2}\right)\left(\dfrac{r+q}{2}\right) = \left(\dfrac{p}{2}\right)^2$ (ii)

$\Rightarrow \quad \dfrac{r-q}{2} =$ square of any integer, say, b^2. [by Lemma (8.1)]

So, $\quad \dfrac{r-q}{2} = b^2$ (iii)

Also $\quad \dfrac{r+q}{2} = a^2$, say, (iv)

(iii) + (iv) $\Rightarrow r = a^2 + b^2$

(iv) − (iii) $\Rightarrow q = a^2 - b^2$

$q > 0 \Rightarrow a > b$.

Also it follows from (ii) that

$$\left(\frac{p}{2}\right)^2 = b^2 a^2 = a^2 b^2 \Rightarrow p = 2ab.$$

Now $\quad p > 0 \Rightarrow ab > 0 \Rightarrow a > 0, \ b > 0$ (or $a < 0, \ b < 0$)

Here we take $a, b > 0$.

Thus, $a > 0$, $b > 0$. [It is to be noted that we get the same result if $a < 0$, $b < 0$]

Again from (i), $(a^2, b^2) = 1$ and hence $(a, b) = 1$.

Also for any (+)ve integers a, b,

$$a^2 \equiv a \pmod 2, \; b^2 \equiv b \pmod 2$$

$$\Rightarrow \quad a^2 + b^2 \equiv (a+b) \pmod 2$$

$$\Rightarrow \quad (a+b) \equiv (a^2+b^2) \pmod 2$$

$$\Rightarrow \quad (a+b) \equiv r \pmod 2 \equiv 1 \pmod 2, \text{ as } r \text{ is odd.}$$

Hence the theorem.

Converse of Th. (8.3)

Theorem (8.4): *If a and b be any two integers such that $a+b \equiv 1 \pmod 2$, $a > b > 0$ and $(a, b) = 1$, then $(2ab,\ a^2 - b^2,\ a^2 + b^2)$ is a primitive solution of $x^2 + y^2 = z^2$.*

***Proof*:** We have $(2ab)^2 + (a^2 - b^2)^2 = (a^2 + b^2)^2 \ \forall a, b \in z$.
and, therefore, the triple $(2ab,\ a^2 - b^2,\ a^2 + b^2)$ is a solution of $x^2 + y^2 = z^2$.

Now it is sufficient to show that G.C.D. $(2ab,\ a^2 - b^2,\ a^2 + b^2) = 1$.

If possible, let $(2ab,\ a^2 - b^2,\ a^2 + b^2) = d(> 1)$.

Let x be any prime divisor of d.

Then $d | a^2 - b^2,\ d | a^2 + b^2$

$$\Rightarrow \quad x | a^2 - b^2,\ x | a^2 + b^2. \qquad \text{(i)}$$

But $a^2 - b^2$, $a^2 + b^2$ are both odd as $a + b \equiv 1 \pmod 2$ and hence, $x \neq 2$.

Again, (i) $\Rightarrow x | 2a^2$ and $x | 2b^2$

$\Rightarrow x | a$ and $x | b$, which is impossible as $(a, b) = 1$.

Hence, considering all possibilities, we can conclude that $d = 1$.

Consequently, G.C.D. $(2ab,\ a^2 - b^2,\ a^2 + b^2) = 1$ and hence the triple $(2ab,\ a^2 - b^2,\ a^2 + b^2)$ is a primitive solution of $x^2 + y^2 = z^2$.

Definition (8.3) *The pair of integers a, b satisfying $a > b > 0$, $(a, b) = 1$ and $a+b \equiv 1 \pmod 2$ generates a primitive solution $(2ab,\ a^2 - b^2,\ a^2 + b^2)$ of $x^2 + y^2 = z^2$.*

The triple $(2ab,\ a^2 - b^2,\ a^2 + b^2)$ is called the **generator** of the primitive solution and the pair (a, b) is called the generating pair.

a	b	$x = 2ab$	$y = a^2 - b^2$	$z = a^2 + b^2$	$x^2 + y^2 = z^2$	$x + y \equiv 7 \pmod 8$ or $(x+y) \equiv 1 \pmod 8$ $x - y \equiv 1 \pmod 8$ or $(x-y) \equiv 7 \pmod 8$
2	1	4	3	5	$4^2 + 3^2 = 5^2$	$4+3 \equiv 7 \pmod 8$, $4-3 \equiv 1 \pmod 8$
3	2	12	5	13	$12^2 + 5^2 = 13^2$	$12+5 \equiv 1 \pmod 8$, $12-5 \equiv 7 \pmod 8$
4	1	8	15	17	$8^2 + 15^2 = 17^2$	$8+15 \equiv 7 \pmod 8$, $8-15 \equiv 1 \pmod 8$
12	9	216	63	225	$216^2 + 63^2 = 225^2$	$216+63 \equiv 7 \pmod 8$, $216-63 \equiv 1 \pmod 8$

Thus, we obtain the generating pair (a, b) forming the primitive solution of $x^2 + y^2 = z^2$ satisfying the relation $x + y$, $x - y \equiv 1 \pmod 8$ or $\equiv 7 \pmod 8$.

Problem (8.1): Find all primitive triples (x, y, z) in which (i) $x = 40$, (ii) $x = 60$.

Solution: $x = 40$, which can be expressed in the form $2ab$ by the following ways :

$$40 = 2 \cdot 5 \cdot 4, \ (a, b) = (5, 4) = 1$$

$$= 2 \cdot 20 \cdot 1, \ (a, b) = (20, 1) = 1$$

$$(x, y, z) = (2ab,\ a^2 - b^2,\ a^2 + b^2)$$
$$= (40,\ 9,\ 41).$$

Also $(40, 399, 401)$.

(ii) $60 = 2 \cdot 30 \cdot 1 \qquad (a, b) = (30, 1) = 1$

$= 2 \cdot 6 \cdot 5 \qquad (a, b) = (6, 5) = 1$

$= 2 \cdot 10 \cdot 3 \qquad (a, b) = (10, 3) = 1$

$= 2 \cdot 15 \cdot 2 \qquad (a, b) = (15, 2) = 1$

Therefore, $(x, y, z) = (2ab,\ a^2 - b^2,\ a^2 + b^2)$

$\Rightarrow \qquad (x, y, z) = (60, 899, 901)$

$(60, 11, 61)$

$(60, 91, 109)$

$(60, 221, 229)$.

Problem (8.2): Find Pythagorean triple (not necessarily primitive) having $n(\geq 3)$ as one of its members.

Solution: Case(1): Let n be even.

Then, $n = 2ab = 2 \cdot \frac{n}{2} \cdot 1$, $\frac{n}{2}$ may be even or odd.

So $\quad (x, y, z) = \left(n,\ \left(\frac{n}{2}\right)^2 - 1,\ \left(\frac{n}{2}\right)^2 + 1\right) = \left(n,\ \frac{n^2}{4} - 1,\ \frac{n^2}{4} + 1\right).$

Case(2): Let n be odd.

Then n is odd $\Rightarrow n^2$ is odd $\Rightarrow n^2 - 1$ is even and also $n^2 + 1$ is even.

Then, $\frac{n^2 - 1}{2}$ = even or odd.

Let $\frac{n^2 - 1}{2}$ = even. We take $x = \frac{n^2 - 1}{2}$.

Then $\quad (x, y, z) = \left(\frac{n^2 - 1}{2},\ n,\ \frac{n^2 - 1}{2} + 1\right)$

$$= \left(\frac{n^2 - 1}{2},\ n,\ \frac{n^2 + 1}{2}\right). \tag{i}$$

Next, let $\frac{n^2 - 1}{2}$ = odd.

Then $\quad \frac{n^2 - 1}{2} + 1 = \frac{n^2 + 1}{2}$. = even. We take $x = \frac{n^2 + 1}{2}$.

Then $\quad (x, y, z) = \left(\frac{n^2 + 1}{2},\ n,\ \frac{n^2 - 1}{2}\right). \tag{ii}$

Considering the same for $\frac{n^2+1}{2}$, we get the same triple as in (i) and (ii).

Therefore, Pythagorean triples are

$$\left(n,\ \frac{n^2}{4} - 1,\ \frac{n^2}{4} + 1\right) \quad \text{and} \quad \left(\frac{n^2 - 1}{2},\ n,\ \frac{n^2 + 1}{2}\right)$$

or $\left(\frac{n^2+1}{2},\ n,\ \frac{n^2-1}{2}\right)$

having n as one of the members.

Problem (8.3): Prove that in a Pythagorean primitive triple (x, y, z), $12|xy$ and hence $60|xyz$.

Solution: Let (a, b) generates the primitive triple $(x, y, z) = (2ab,\ a^2 - b^2,\ a^2 + b^2)$.

If $a \equiv 0 \pmod 3$ or, $b \equiv 0 \pmod 3$.

Then $ab \equiv 0 \pmod 3 \Rightarrow 3|x$ ($\because$ $x = 2ab$).

Let $a \not\equiv 0 \pmod 3$ and $b \not\equiv 0 \pmod 3$.

Then the possible remainders are 1,2.

So $a \equiv 1, 2 \pmod 3 \Rightarrow a \equiv -1, -2 \pmod 3$.

Similarly, $b \equiv \pm 1, \pm 2 \pmod 3$.

So $a^2 \equiv 1, 4 \pmod 3$, $b^2 \equiv 1, 4 \pmod 3$.

Therefore, $a^2 - b^2 \equiv 0 \pmod 3$, for all cases.

$\Rightarrow$ $y \equiv 0 \pmod 3$ $[\because y = a^2 - b^2]$

$\Rightarrow$ $3|y$.

Thus, we have $3|x$ or $3|y$.

Now one of a and b is even.

Then $x = 2ab \Rightarrow 4|x$.

Therefore, $3|x, 4|x \Rightarrow 12|x \Rightarrow 12|xy$ [for any integer y].

Also $3|y, 4|x \Rightarrow 12|xy$.

Thus, we have $12|xy$.

Again, if $a \equiv 0 \pmod 5$ or $b \equiv 0 \pmod 5$,

then $ab \equiv 0 \pmod 5 \Rightarrow 5|x$ $[\because x = 2ab]$.

Now if $a \not\equiv 0 \pmod 5$ and $b \not\equiv 0 \pmod 5$,

then the possible remainders are 1,2,3,4.

That is, $a \equiv 1, 2, 3, 4 \pmod 5$, i.e., $a \equiv \pm 1, \pm 2 \pmod 5$.

Similarly, $b \equiv \pm 1, \pm 2 \pmod 5$.

Then it follows that $a^4 \equiv 1 \pmod 5$ and $b^4 \equiv 1 \pmod 5 \Rightarrow a^4 - b^4 \equiv 0 \pmod 5$.

Therefore, $yz = (a^2 - b^2)(a^2 + b^2) = a^4 - b^4 \equiv 0 \pmod 5 \Rightarrow 5|yz$.

Now $4|x,\ 5|yz \Rightarrow 20|xyz$.

Also $3|x \Rightarrow 3|xyz$.

$\therefore$ $20|xyz,\ 3|xyz \Rightarrow 60|xyz$.

Hence, $60|xyz$.

Problem (8.4): If (x, y, z) be a primitive triple in which $z - y = 2$, then prove that $x = 2t$, $y = t^2 - 1$, $z = t^2 + 1$ for some $t > 1$.

Solution: Let $(x, y, z) = (2ab,\ a^2 - b^2,\ a^2 + b^2)$.

$$z - y = 2 \Rightarrow (a^2 + b^2) - (a^2 - b^2) = 2$$

$$\Rightarrow 2b^2 = 2 \Rightarrow b^2 = 1 \Rightarrow b = 1\ (b > 0).$$

For $b = 1,\ x = 2a,\ y = a^2 - 1,\ z = a^2 + 1$.

So for some t,

$$x = 2t,\ y = t^2 - 1,\ z = t^2 + 1 \quad \text{and} \quad y > 0 \Rightarrow t > 1.$$

Problem (8.5): Prove that if (x, y, z) be a primitive triple and x and z are two consecutive integers, then $x = 2t(t + 1),\ y = 2t + 1,\ z = 2t(t + 1) + 1$.

Solution: Let $(x, y, z) = (2ab,\ a^2 - b^2,\ a^2 + b^2)$.

Let $x = n$ (any (+)ve integer)

Then $z = n + 1$, since x, z are two consecutive integers.

Therefore, $z - x = 1 \Rightarrow a^2 + b^2 - 2ab = 1$

$$\Rightarrow (a - b)^2 = 1 \Rightarrow a - b = 1 \quad (\text{as } a > b)$$

$$\Rightarrow a = b + 1.$$

$$\therefore \quad x = 2b(b + 1),\ y = (b + 1)^2 - b^2 = 2b + 1,$$
$$z = a^2 + b^2 = (b + 1)^2 + b^2 = 2b(b + 1) + 1.$$

So for some $b = t\ (> 0)$,

$$x = 2t(t + 1),\ y = 2t + 1,\ z = 2t(t + 1) + 1.$$

Problem (8.6): Show that $3n, 4n, 5n$ for $n = 1, 2, 3, \ldots$ are the only Pythagorean triple whose terms are in A.P.

Solution: Let the triple be $x - d,\ x,\ x + d$ which are in A.P. with common difference d.

Then $\quad (x - d)^2 + x^2 = (x + d)^2 \Rightarrow 2x^2 - 2xd + d^2 = x^2 + 2xd + d^2$

$\Rightarrow \quad x^2 = 4xd \Rightarrow x = 4d \quad (x \neq 0)$.

Also $\quad x > 0, \Rightarrow d > 0$.

Hence, the triple is $3d,\ 4d,\ 5d,\ d > 0$.

$d = 1$, triple is, 3, 4, 5.

$d = 2$, triple is, 3.2, 4.2, 5.2 and so on.

Hence, $3n,\ 4n,\ 5n,\ n \in N$ are the only Pythagorean triples.

Theorem (8.5) (Fermat): *The Diophantine equation $x^4 + y^4 = z^2$ has no (+)ve integral solution.*

***Proof*:** If possible, let the equation $x^4 + y^4 = z^2$ have a solution $x_0,\ y_0,\ z_0$ where x_0, y_0, z_0 are (+)ve integers. Let us assume that $(x_0, y_0) = 1$.

For if $(x_0, y_0) = d (> 1)$, then

$$d | x_0,\ d | y_0 \Rightarrow x_0 = dx_1,\ y_0 = dy_1,\ (x_1, y_1) = 1.$$

Then $\quad (dx_1)^4 + (dy_1)^4 = z_0^2 \Rightarrow d^4x_1^4 + d^4y_1^4 = z_0^2$

$\Rightarrow \quad d^4 | z_0^2$.

So we take $z_0 = d^2z_1$ and hence, $(x_1^2)^2 + (y_1^2)^2 = z_1^2,\ (x_1, y_1) = 1$.

So our assumption is meaningful.

Thus, we have

$$(x_0^2)^2 + (y_0^2)^2 = z_0^2,\ (x_0^2, y_0^2) = 1 \text{ [as } (x_0, y_0) = 1].$$

Then (x_0^2, y_0^2, z_0) becomes a primitive solution of $x^4 + y^4 = z^2$.

Also if x_0^2 is even, then y_0^2 is odd (or vice versa).

Let us take x_0^2 as even.

Also there exist one or more values of x_0^2, y_0^2 and z_0, and let z_0 be the smallest value of z.

Then we have $x_0^2 = 2ab$, $y_0^2 = a^2 - b^2$ and $z_0 = a^2 + b^2$, where $a > b > 0$, $(a, b) = 1$ and $a + b \equiv 1 \pmod 2$. (i)

If possible, let a be even and b be odd.

Then $a^2 \equiv 0 \pmod 4$ and $b^2 \equiv 1 \pmod 4$.

Then $y_0^2 \equiv a^2 - b^2 \equiv -1 \pmod 4$ which is not possible as the square of an integer cannot be expressed in the form $4q - 1$.

Therefore, a is odd and b is even.

So let $b = 2c$, where c is a (+)ve integer.

Then $x_0^2 = 4ac \Rightarrow ac = \left(\frac{x_0}{2}\right)^2 \Rightarrow a$ and c are both squares of an integer, [by Lemma (8.1)]

So let $a = z_1^2$ and $c = t^2$, $z_1, t \in Z^+$. (ii)

Also $a = z_1^2 \Rightarrow z_1$ is odd as a is odd.

Therefore, $b = 2c = 2t^2 \Rightarrow\ y_0^2 = a^2 - b^2 = z_1^4 - 4t^4 = (z_1^2)^2 - 4(t^2)^2$

$\Rightarrow \qquad (2t^2)^2 + y_0^2 = (z_1^2)^2,\ \text{ where } (z_1^2, 2t^2) = 1,$

which shows that $(2t^2, y_0, z_1^2)$ is a primitive solution of $x^2 + y^2 = z^2$.

Then we have

$$2t^2 = 2uv,$$
$$y_0 = u^2 - v^2,$$
$$z_1^2 = u^2 + v^2,$$

where u, v are integers satisfying $u > v > 0$, $(u, v) = 1$ and $u + v \equiv 1 \pmod 2$.

Now $\quad 2t^2 = 2uv \Rightarrow t^2 = uv, \Rightarrow u$ and v are squares of an integer.

Let $\quad u = p^2$ and $v = q^2$, $p, q \in Z^+$.

Hence, $\quad p^4 + q^4 = u^2 + v^2 = z_1^2$.

Now for all $z_1 (z_1 \neq 1)$, $z_1 < z_1^2 = a < a^2 + b^2 = z_0$. [from (i) and (ii)]

That is, $z_1^2 < z_0$, which contradicts that z_0 is the smallest value of z. Hence, $x^4 + y^4 = z^2$ has no solution.

Theorem (8.6) *The equation $x^4 + y^4 = z^4$ has no positive integral solution.*

Proof: If possible, let (x_0, y_0, z_0) be a (+)ve integral solution of $x^4 + y^4 = z^4$.

Then $x_0^4 + y_0^4 = z_0^4 = (z_0^2)^2$,

which shows that (x_0, y_0, z_0^2) is an integral solution of $x^4 + y^4 = z^2$, but this is not true as we have proved it in Th. (8.5).

Hence, $x^4 + y^4 = z^4$ has no integral solution.

Theorem (8.7): [Fermat's Last Theorem]* *The Diophantine equation $x^n + y^n = z^n$ has no positive integral solution except trivial solutions $(0, y, z)$ or (x, θ, z) for $n > 2$.*

***Proof*:** We proved in Th. (8.6), that the equation $x^4 + y^4 = z^4$ has no (+)ve integral solution.

Now it remains to be proved, when $n > 4$ and n is an odd prime.

So we consider (i) $n = 2^m$, $m > 2$ and $n = pk$, $m, k \in Z^+$, $k \geq 1$, p is an odd prime.

Case(1): Let $n = 2^m$. Then we can write, $n = 4t$, for some $t \geq 1$.

Then $x^n + y^n = z^n$ is reduced to the form

$$x^{4t} + y^{4t} = z^{4t} \;\Rightarrow\; (x^t)^4 + (y^t)^4 = (z^t)^4 \;\Rightarrow\; X^4 + Y^4 = Z^4,$$

which has no (+)ve integral solution [Th. (8.6)].

Case(2): Let $n = pk$, p is an odd prime, $k \geq 1$, $k \in Z^+$. Then $x^n + y^n = z^n$ is reduced to the form,

$$x^{pk} + y^{pk} = z^{pk} \;\Rightarrow\; (x^k)^p + (y^k)^p = (z^k)^p \;\Rightarrow\; X^p + Y^p = Z^p,$$

where p is an odd prime. But the proof of the solution of this equation has remained unsolved.

---

*The theorem was proved in 1995 and was named after the French mathematician Pierre de Fermat.

Exercise 8

1. If z_0 be a given integer such that $z_0 = u^2 + v^2$, $(u, v) = 1$ for some integers u and v, then $(2uv, u^2 - v^2, z_0)$ is a primitive solution of $x^2 + y^2 = z^2$.
2. If y_0 be a given odd integer and if $y_0 = u \cdot v$ where $(u, v) = 1$ and u, v are (+)ve integers such that $u > v$, then $(\frac{u^2-v^2}{2}, y_0, \frac{u^2+v^2}{2})$ is a primitive solution of $x^2 + y^2 = z^2$.
3. Show that the system of simultaneous equations $x^2 + y^2 = z^2$ and $x^2 - y^2 = \omega^2$ admits no solution in (+)ve integers x, y, z, ω.
4. Prove that $x^4 - y^4 = 2z^2$ has no solutions in (+)ve integers x, y, z.
5. Show that there exist infinitely many primitive Pythagorean triples (x, y, z) whose even member x is a perfect square.
6. Verify that only solution in relatively prime (+)ve integers of the equation $x^4 + y^4 = 2z^2$ is $x = y = z = 1$.
7. Find all the primitive solutions of $x^2 + y^2 = z^2$ having $0 < z < 30$.

Chapter 9

Continued Fraction

One of the greatest mathematician Lenardo of Pisa introduced continued fraction in the field of mathematics. He wrote it in the name of Fibonacce. But it has been found that the idea of continued fraction was already been developed in finding the square root by means of infinite continued fraction by the great algebraist Rafael Bombelli of Renaisance, Italy, in 1572.

Definition (9.1): *Fraction of the form*

$$a_0 \pm \cfrac{b_1}{a_1 \pm \cfrac{b_2}{a_2 \pm \cfrac{b_3}{a_3 + \cdots}}},$$

where a_i, b_i are real numbers is called a continued fraction.

The above expression can be written in the form, $a_0 \pm \frac{b_1}{a_1 \pm}\ \frac{b_2}{a_2 \pm}\ \frac{b_3}{a_3 \pm} \cdots$.

Let $F = a_0 + \frac{b_1}{a_1 +}\ \frac{b_2}{a_2 +} \cdots$.

Then $a_0, \frac{b_1}{a_1}, \frac{b_2}{a_2}, \cdots$ are called the **elements** of F.

The fraction $a_0, a + \frac{b_1}{a_1}, a_0 + \frac{b_1}{a_1 + \frac{b_2}{a_2}}, \cdots$ terminated at some particular stages, i.e., terminated at 1st, 2nd, 3rd,... stages respectively which are called **convergents** as 1st, 2nd, 3rd and so on.

It is to be noted that the continued fraction F remains unaffected, if b_n, a_n, $b_{n+1} (n = 1, 2, \dots)$ be multiplied by the same constant k (say), or if the signs of b_n, a_n, b_{n+1} be changed.

Example: (1) $\frac{1}{1-}\ \frac{1}{2-}\ \frac{3}{3-}\ \frac{4}{4}$ $\left[\text{i.e., of the form } \cfrac{1}{1 - \cfrac{1}{2 - \cfrac{3}{3 - \frac{4}{4}}}}\right]$

(2) $2 + \frac{1}{1+}\ \frac{1}{5+}\ \frac{1}{7+}\ \frac{1}{11}$

(3) $-3 + \cfrac{1}{2 + \cfrac{1}{1 + \frac{1}{7}}}$

(4) $1 + \cfrac{1}{\frac{1}{2x+} - \frac{9}{2x+}\ \frac{1}{3+\cdots}}$

(5) Let us consider a finite continued fraction, $F = a_0 + \cfrac{b_1}{a_1 + \frac{b_2}{a_2}}$.

If b_1, a_1, b_2 be multiplied by the same constant k, then we see that F remains unchanged.

$$a_0 + \frac{kb_1}{ka_1 + \frac{kb_2}{a_2}} = a_0 + \frac{kb_1a_2}{ka_1a_2 + kb_2} = a_0 + \frac{b_1a_2}{a_1a_2 + b_2}$$

$$= a_0 + \frac{b_1}{a_1 + \frac{b_2}{a_2}} = F.$$

Also if b_1, a_1, b_2 be replaced by $-b_1$, $-a_1$, $-b_2$, then

$$a_0 + \frac{-b_1}{-a_1 - \frac{b_2}{a_2}} = a_0 + \frac{b_1a_2}{a_1a_2 + b_2} = a_0 + \frac{b_1}{a_1 + \frac{b_2}{a_2}} = F.$$

So F remains unchanged.

The same can be established if for a continued fraction,

$$F = a_0 + \frac{b_1}{a_1+}\ \frac{b_2}{a_2+\cdots+} \cdots \frac{b_{n-1}}{a_{n-1}+}\ \frac{b_n}{a_n+}\ \frac{b_{n+1}}{a_{n+1}},$$

b_n, a_n, b_{n+1} be multiplied by the same constant k or replaced by $-b_n, -a_n, -b_{n+1}$, then F remains unchanged.

It is to the noted that F remains unchanged if there be any constant k common to b_n, a_n, b_{n+1}, then k can be cancelled from b_n, a_n, b_{n+1}, for $n = 1, 2, \ldots$

That is, if $b_n = 15$, $a_n = 5$, $b_{n+1} = 20$, then we can replace them by 3, 1, 4 by removing the common factor 5 from them.

Problem (9.1): Show that $\frac{2}{4+}\ \frac{2}{4+}\ \frac{2}{4+}\ \frac{2}{4+} = \frac{1}{2+}\ \frac{1}{4+}\ \frac{1}{2+}\ \frac{1}{4}$

Solution: Here $b_1 = 2$, $a_1 = 4$, $b_2 = 2$.

So cancelling 2, the continued fraction becomes $\frac{1}{2+}\ \frac{1}{4+}\ \frac{2}{4+}\ \frac{2}{4+}$.

Again, $b_3 = 2$, $a_3 = 4$, $b_4 = 2$.

So cancelling 2, we have $\frac{2}{4+}\ \frac{2}{4+}\ \frac{2}{4+}\ \frac{2}{4+} = \frac{1}{2+}\ \frac{1}{4+}\ \frac{1}{2+}\ \frac{1}{4}$.

Problem (9.2): Without calculating convergents, prove that

$$F = \frac{2}{2+}\ \frac{3}{3+}\ \frac{4}{4+}\ \frac{5}{5} = \frac{1}{1+}\ \frac{1}{2+}\ \frac{2}{3+}\ \frac{3}{4}.$$

Solution: $b_2 = 3$, $a_2 = 3$, $b_3 = 4$.

Let them be multiplied by 2.

Then $F = \frac{2}{2+}\ \frac{6}{6+}\ \frac{8}{4+}\ \frac{5}{5}$.

Now $b_3 = 8$, $a_3 = 4$, $b_4 = 5$ be multiplied by 3,

then $F = \frac{2}{2+}\ \frac{6}{6+}\ \frac{24}{12+}\ \frac{15}{5} = \frac{1}{1+}\ \frac{3}{6+}\ \frac{24}{12+}\ \frac{15}{5}$ [cancelling 2 from b_1, a_1, b_2]

$= \frac{1}{1+}\ \frac{1}{2+}\ \frac{8}{12+}\ \frac{15}{5}$ [cancelling 3 from b_2, a_2, b_3]

$= \frac{1}{1+}\ \frac{1}{2+}\ \frac{8}{12+}\ \frac{3 \times 4}{4}$ $\left[\frac{15}{5} = 3 = \frac{3 \times 4}{4}\right]$

$= 1 + \frac{1}{2+}\ \frac{2}{3+}\ \frac{3}{4}$ [cancelling 4 from b_3, a_3, b_4].

Simple continued fraction (S.C.F.) and their properties

Definition (9.2): *A continued fraction of the form* $a_1 + \frac{1}{a_2+} \frac{1}{a_3+} \cdots$ *is called simple if* a_i *are all (+)ve integers except that* a_1 *may be zero and it is denoted by* $\langle a_1, a_2, \ldots, a_n, \ldots \rangle$.

If it is terminated at the nth stage, then it is defined to be a **finite simple continued** fraction.

That is, $\langle 1, 2, 3, 4 \rangle$ is a finite simple continued fraction and it is of the form $1 + \frac{1}{2+} \frac{1}{3+} \frac{1}{4}$.

$\underset{n\to\infty}{\text{Lt}} \langle a_1, a_2, \ldots, a_n \rangle = \langle a_1, a_2, \ldots, a_{n-1} \ldots \infty \rangle$ be called an **infinite simple continued** fraction.

Ex:- $\langle 0, 1, 2, 5, 7, \ldots \rangle$, $\langle 1, 2, 3, 7, 9, \ldots \rangle$ are the examples of infinite S.C.F.

Definition (9.3): Partial Quotient, Complete Quotient and Convergents.

In a S.C.F., $a_1 + \dfrac{1}{a_2+} \dfrac{1}{a_3+} \cdots + \dfrac{1}{a_n+} \dfrac{1}{a_{n+1}+} \cdots$

$= \langle a_1, a_2, \ldots a_n, a_{n+1}, \ldots \rangle$.

(i) a_i are called partial quotients or simply quotients, for $i = 1, 2, \ldots, n, \ldots$

That is, $a_1 =$ 1st quotient, $a_n = n$th quotient.

(ii) 1st complete quotient in $\langle a_1, a_2, \ldots, a_n \ldots \rangle$ is $a_1 + \frac{1}{a_2+} \frac{1}{a_3+} \cdots$, started from the 1st quotient and ended to the last quotient, if any.

2nd complete quotient $= a_2 + \dfrac{1}{a_3+} \cdots$, started from the 2nd quotient.

nth complete quotient $= a_n + \dfrac{1}{a_{n+1}+} \dfrac{1}{a_{n+2}+} \cdots$, started from the nth quotient.

Thus, in a S.C.F. $\langle a_1, a_2, \ldots, a_n, a_{n+1} \ldots \rangle$.

the kth complete quotient $k \geq 1$ is defined by $\langle a_k, a_{k+1} \cdots \rangle$

If S.C.F. is finite, i.e., $\langle a_1, a_2, a_3, \ldots, a_n \rangle$, then kth complete quotient $= \langle a_k, a_{k+1}, \ldots, a_n \rangle$ for $1 \leq k \leq a_n$.

So nth complete quotient $= a_n =$ partial quotient in a finite S.C.F. $\langle a_1, a_2, \ldots a_n \rangle$.

(iii) **Convergents**

In a S.C.F. $\langle a_1, a_2, \ldots, a_k, \ldots, a_n \rangle$, $1 \leq k \leq n$. the kth convergent is defined by $\langle a_1, a_2, \ldots, a_k \rangle$ which is terminated at the kth quotient and is denoted by $\frac{p_k}{q_k}$.

Thus, 1st convergent $= \dfrac{p_1}{q_1} = \langle a_1 \rangle$

2nd convergent $= \dfrac{p_2}{q_2} = \langle a_1, a_2 \rangle = a_1 + \dfrac{1}{a_2}$

kth convergent $= \dfrac{p_k}{a_k} = \langle a_1, a_2, \ldots, a_k \rangle = a_1 + \dfrac{1}{a_2+} \dfrac{1}{a_3+} \cdots + \dfrac{1}{a_k}$.

Example: In a S.C.F. $\langle 0, 1, 2, 5, 7 \rangle$,

$$\frac{p_1}{q_1} = \langle 0 \rangle = 0$$

$$\frac{p_2}{q_2} = \langle 0; 1 \rangle = 0 + \frac{1}{1} = \frac{1}{1}$$

$$\frac{p_3}{q_3} = 0 + \frac{1}{1+}\frac{1}{2} = \frac{1}{1+\frac{1}{2}}$$

$$\frac{p_4}{q_4} = 0 + \frac{1}{1+}\frac{1}{2+}\frac{1}{5} = \frac{1}{1+\frac{1}{2+\frac{1}{5}}}$$

$$\frac{p_5}{q_5} = 0 + \frac{1}{1+}\frac{1}{2+}\frac{1}{5+}\frac{1}{7} = \frac{1}{1+}\frac{1}{2+}\frac{1}{5+}\frac{1}{7}.$$

Theorem (9.1): *Every rational number can be expressed as a finite simple continued fraction $\langle a_1, a_2, \ldots, a_n \rangle$. It can be further arranged so that $\langle a_1, a_2, \ldots, a_n \rangle$ has either an odd or even number of convergents.*

Proof: Let $\frac{A}{B}$ be a rational number such that $A > B > 0$.

Applying divisional algorithm to the pair (A, B), we have

$A = a_1 B + r_1,\ 0 < r_1 < B$

$$\Rightarrow \quad \frac{A}{B} = a_1 + \frac{r_1}{B} = a_1 + \frac{1}{\frac{B}{r_1}}. \tag{1}$$

$B > r_1$, so applying divisional algorithm to the pair (B, r_1),

$B = a_2 r_1 + r_2,\ 0 < r_2 < r_1$

$$\Rightarrow \quad \frac{B}{r_1} = a_2 + \frac{1}{\frac{r_1}{r_2}}. \tag{2}$$

Again, $r_1 > r_2$. Applying divisional algorithm to the pair (r_1, r_2), we have

$r_1 = a_3 r_2 + r_3,\ 0 < r_3 < r_2$

$$\Rightarrow \quad \frac{r_1}{r_2} = a_3 + \frac{1}{\frac{r_2}{r_3}}. \tag{3}$$

Let the above process be continued and let at the nth stage the remainder be zero, then $r_{n-2} = a_n r_{n-1}$. (n)

Thus we have, by using the results (1), (2), (3) and (n), simultaneously,

$$\frac{A}{B} = a_1 + \frac{1}{a_2+}\frac{1}{a_3+} \cdots \frac{1}{a_{n-1}+}\frac{1}{a_n} = \langle a_1, a_2, \ldots, a_n \rangle.$$

Thus, $\frac{A}{B}$ can be expressed as s finite S.C.F.

Case(2): Let $\frac{A}{B}$ be a rational number such that $A < B,\ A, B > 0$.

Then we write, $\frac{A}{B} = \frac{1}{\frac{B}{A}},\ B > A$.

We now apply the same process as in Case (1), and we have

$\frac{B}{A} = \frac{1}{b_1+}\frac{1}{b_2+} \cdots + \frac{1}{b_m} = \langle 0, b_1, b_2, \ldots, b_m \rangle$, a finite simple continued fraction.

Now if the last quotient a_n in $\frac{A}{B} = \langle a_1, a_2, \ldots, a_n \rangle$ be greater than 1,

i.e., if $a_n > 1$, we write $\frac{1}{a_n} = \frac{1}{a_{n-1}+}\frac{1}{1}$ and if $a_n = 1$, we write $\frac{1}{a_{n-1}+}\frac{1}{a_n} = \frac{1}{a_{n-1}+1}$.

Then $\frac{A}{B}$ has an either odd or even number of quotients.

Problem (9.3): Express $\frac{147}{38}$ as an S.C.F. Also express it to an odd number of quotients.

Solution:

1	38	147	3
	33	114	
1	5	33	6
	3	30	
2	2	3	1
	2	2	
	0	1	

$$\frac{147}{38} = 3 + \frac{1}{1+}\ \frac{1}{6+}\ \frac{1}{1+}\frac{1}{1+}\frac{1}{2}$$

having even number of quotients.

Here $\quad a_n = a_6 = 2 > 1.$

So we write, $\dfrac{1}{a_6} = \dfrac{1}{a_{6-1}+}\ \dfrac{1}{1} = \dfrac{1}{2-1+}\ \dfrac{1}{1} = \dfrac{1}{1+}\ \dfrac{1}{1}.$

Then $\quad \dfrac{147}{38} = 3 + \dfrac{1}{1+}\ \dfrac{1}{6+}\ \dfrac{1}{1+}\ \dfrac{1}{1+}\ \dfrac{1}{1+}\ \dfrac{1}{1},$

which is converted to an odd number of quotients.

Problem (9.4): Express $\frac{172}{51}$ into a S.C.F. having even number of quotients.

Solution:

2	51	172	3
	38	153	
2	13	19	1
	12	13	
	1	6	6
		6	
		0	

$$\frac{172}{51} = 3 + \frac{1}{2+}\ \frac{1}{1+}\ \frac{1}{2+}\ \frac{1}{6}$$

having odd number of quotients.

Now $\quad a_5 = 6 > 1.$

We write, $\dfrac{1}{a_5} = \dfrac{1}{a_5 - 1+}\ \dfrac{1}{1} = \dfrac{1}{6-1+}\ \dfrac{1}{1} = \dfrac{1}{5+}\ \dfrac{1}{1}.$

So we express $\dfrac{172}{51} = 3 + \dfrac{1}{2+}\ \dfrac{1}{1+}\ \dfrac{1}{2+}\ \dfrac{1}{5+}\ \dfrac{1}{1},$

having even number of quotients.

Problem (9.5): Express $\langle 2, 3, 1, 2, 1\rangle$ as a S.C.F. having even number of quotients.

Solution: For $\langle 2, 3, 1, 2, 1\rangle$, S.C.F. has an expression of the form : $2 + \dfrac{1}{3+}\ \dfrac{1}{1+}\ \dfrac{1}{2+}\ \dfrac{1}{1}.$

Here $a_n = 1$. Then we write, $\dfrac{1}{a_{n-1}+}\ \dfrac{1}{a_n} = \dfrac{1}{a_4+}\ \dfrac{1}{a_5} = \dfrac{1}{a_4+1} = \dfrac{1}{2+1} = \dfrac{1}{3}.$

We replace $\frac{1}{2+}\frac{1}{1}=\frac{1}{3}$ in the given expression.

Thus, we have $\langle 2,3,1,2,1\rangle$

$$=2+\frac{1}{3+}\frac{1}{1+}\frac{1}{3}, \text{ having even number of quotients.}$$

Problem (9.6): Express $(1-x)^{-2}$ as a continued fraction by using the method of H.C.F.

Solution: $(1-x)^{-1}=\dfrac{1}{(1-x)^2}=\dfrac{1}{1-2x+x^2}$

$\frac{1}{2x}$	$1-2x+x^2$	1	1
	$1-\frac{x}{2}$	$1-2x+x^2$	
$-\frac{9}{2x}$	$-\frac{3x}{2}+x^2$	$2x-x^2$	$-\frac{4}{3}$
	$-\frac{3x}{2}$	$2x-\frac{4}{3}x^2$	
	x^2	$\frac{1}{3}x^2$	$\frac{1}{3}$
		$\frac{1}{3}x^2$	

So $$\frac{1}{(1-x)^2}=1+\frac{1}{\frac{1}{2x}+}\frac{1}{-\frac{4}{3}+}\frac{1}{-\frac{9}{2x}+}\frac{1}{\frac{1}{3}}.$$

Theorem (9.2): *If* $\dfrac{p_n}{q_n}$ *be the nth convergent of a simple continued fraction* $\langle a_1,a_2,\ldots,a_n,a_{n+1},\ldots\rangle$, *then*

(i) $\left.\begin{aligned}p_n&=a_np_{n-1}+p_{n-2}\\ q_n&=a_nq_{n-1}+q_{n-2}\end{aligned}\right\}\ \forall\, n\geq 2.$

(ii) $\dfrac{p_n}{p_{n-1}}=a_n+\dfrac{1}{a_{n-1}+}\dfrac{1}{a_{n-2}+}\cdots+\dfrac{1}{a_2+}\dfrac{1}{a_1}$

$\dfrac{q_n}{q_{n-1}}=a_n+\dfrac{1}{a_{n-1}+}\dfrac{1}{a_{n-2}+}\cdots+\dfrac{1}{a_2}$

(iii) $p_nq_{n-1}-p_{n-1}q_n=(-1)^n$

(iv) $\dfrac{p_n}{q_n}-\dfrac{p_{n-1}}{q_{n-1}}=\dfrac{(-1)^n}{q_n\cdot q_{n-1}}$

(v) Every convergent $\dfrac{p_n}{q_n}$ is in its lowest term and also p_n is prime to p_{n-1}, q_n is prime to q_{n-1}.

(vi) If $\alpha=\langle a_1,a_2,\ldots,a_n,a_{n+1},\ldots\rangle$, then $\alpha=\dfrac{\alpha_{n+1}p_n+p_{n-1}}{\alpha_{n+1}q_n+q_{n-1}}$.

(vii) Every convergent is a closure approximation to the value of the continued fraction than its preceding convergent.

(viii) If $\alpha = \langle a_1, a_2, \ldots, a_n, a_{n+1}\rangle$ and $u_n = \frac{p_n}{q_n}$, then $\{u_{2n-1}\}$ is an increasing sequence of odd convergents and $\{u_{2n}\}$ is a decreasing sequence of even convergents.

Also (i) $u_{2n-1} < \alpha < u_{2n}$ (ii) Every infinite S.C.F. is convergent.

(ix) Let $\frac{p}{q}$ be any convergent of $\alpha = \langle a_1, a_2, \ldots, a_n, \ldots\rangle$ and $\frac{r}{s}$ be any rational number where $s < q$. Then $\frac{p}{q}$ is a closer approximation to α than $\frac{r}{s}$.

(x) If ϵ be the error in taking $\frac{p_n}{q_n}$ as the value of α, then $\frac{a_{n+2}}{q_n q_{n+2}} < \epsilon < \frac{1}{q_n q_{n+1}}$.

Also the limits of error (ϵ) is given by $\frac{1}{2q_{n+1}^2} < \epsilon < \frac{1}{q_n^2}$.

Proof: Let u_n denote the nth convergent of $\langle a_1, a_2, \ldots, a_n, a_{n+1}, \ldots\rangle$.

Then $\quad u_1 = \frac{a_1}{1} = \frac{p_1}{q_1}$, where $p_1 = a_1$ and $q_1 = 1$

$$u_2 = a_1 + \frac{1}{a_2} = \frac{a_2 a_1 + 1}{a_2} = \frac{p_2}{q_2},$$

where $\quad p_2 = a_2 a_1 + 1 = a_2 p_1 + p_0$

$q_2 = a_2 \cdot 1 = a_2 q_1 + q_0$, assuming $p_0 = 1$ and $q_0 = 0$.

Then the results holds for $n = 2$.

Let it be true for $n = k$, $(2 \leq n)$.

Therefore,
$$\left.\begin{aligned} p_k &= a_k p_{k-1} + p_{k-2} \\ q_k &= a_k q_{k-1} + q_{k-2} \end{aligned}\right\} \tag{1}$$

So
$$u_k = \frac{p_k}{q_k} = \frac{a_k p_{k-1} + p_{k-2}}{a_k q_{k-1} + q_{k-2}}. \tag{2}$$

Now $(k+1)$th convergent, u_{k+1}, is obtained by replacing the kth quotient a_k by $a_k + \frac{1}{a_{k+1}}$ in (2).

Hence,
$$u_{k+1} = \frac{p_{k+1}}{q_{k+1}} = \frac{\left(a_k + \frac{1}{a_{k+1}}\right) p_{k-1} + p_{k-2}}{\left(a_k + \frac{1}{a_{k+1}}\right) q_{k-1} + q_{k-2}}$$

$$= \frac{(a_k a_{k+1} + 1)p_{k-1} + a_{k+1}p_{k-2}}{(a_k a_{k+1} + 1)q_{k-1} + a_{k+1}q_{k-2}}$$

$$= \frac{a_{k+1}(a_k p_{k-1} + p_{k-2}) + p_{k-1}}{a_{k+1}(a_k q_{k-1} + q_{k-2}) + q_{k-1}}$$

$$= \frac{a_{k+1}p_k + p_{k-1}}{a_{k+1}q_k + q_{k-1}}, \text{ by (1)},$$

where $\quad p_{k+1} = a_{k+1}p_k + p_{k-1}$ and $q_{k+1} = a_{k+1}q_k + q_{k-1}$.

Therefore, it is true for $n = k+1$. So it is true $\forall n$.

Hence, $\quad p_n = a_n p_{n-1} + p_{n-2}$ and $q_n = a_n q_{n-1} + q_{n-2}$.

(ii) $\dfrac{p_n}{p_{n-1}} = \dfrac{a_n p_{n-1} + p_{n-2}}{p_{n-1}} = a_n + \dfrac{p_{n-2}}{p_{n-1}} = a_n + \dfrac{1}{\frac{p_{n-1}}{p_{n-2}}}$

$$= a_n + \frac{1}{\frac{a_{n-1}p_{n-2}+p_{n-3}}{p_{n-2}}} = a_n + \frac{1}{a_{n-1} + \frac{1}{\frac{p_{n-2}}{p_{n-3}}}}$$

$$= a_n + \frac{1}{a_{n-1}+}\,\frac{1}{\frac{a_{n-2}p_{n-3}+p_{n-4}}{p_{n-3}}} = a_n + \frac{1}{a_{n-1}+}\,\frac{1}{a_{n-2}+\cdots+\frac{1}{\frac{p_2}{p_1}}}$$

$$= a_n + \frac{1}{a_{n-1}+}\,\frac{1}{a_{n-2}+}\cdots+\frac{1}{\frac{a_2a_1+1}{a_1}} \quad [\because\ p_2 = a_2a_1 + 1,\ p_1 = a_1]$$

$$= a_n + \frac{1}{a_{n-1}+}\,\frac{1}{a_{n-2}+\cdots}\,\frac{1}{+a_2+}\,\frac{1}{a_1}.$$

$$\frac{q_n}{q_{n-1}} = \frac{a_n q_{n-1} + q_{n-2}}{q_{n-1}} = a_n + \frac{1}{\frac{q_{n-1}}{q_{n-2}}} = a_n + \frac{1}{a_{n-1}+\cdots} + \frac{1}{\frac{q_2}{q_1}}$$

$$= a_n + \frac{1}{a_{n-1}+}\,\frac{1}{a_{n-2}+\cdots+}\,\frac{1}{\frac{a_2}{1}} \quad [q_2 = a_2,\ q_1 = 1]$$

$$= a_n + \frac{1}{a_{n-1}+\cdots}\,\frac{1}{a_2}.$$

(iii) We have $p_n q_{n-1} - p_{n-1} q_n = \begin{vmatrix} p_n & q_n \\ p_{n-1} & q_{n-1} \end{vmatrix}$

$$= \begin{vmatrix} a_n p_{n-1} + p_{n-2} & a_n q_{n-1} + q_{n-2} \\ p_{n-1} & q_{n-1} \end{vmatrix} = a_n \begin{vmatrix} p_{n-1} & q_{n-1} \\ p_{n-1} & q_{n-1} \end{vmatrix} + \begin{vmatrix} p_{n-2} & q_{n-2} \\ p_{n-1} & q_{n-1} \end{vmatrix}$$

$$= 0 + (-1)\begin{vmatrix} p_{n-1} & q_{n-1} \\ p_{n-2} & q_{n-2} \end{vmatrix} = (-1)\begin{vmatrix} a_{n-1}p_{n-2} + p_{n-3} & a_{n-1}q_{n-2} + q_{n-3} \\ p_{n-2} & q_{n-2} \end{vmatrix}$$

$$= (-1)\left[a_{n-1}\begin{vmatrix} p_{n-2} & q_{n-2} \\ p_{n-2} & q_{n-2} \end{vmatrix} + (-1)\begin{vmatrix} p_{n-2} & p_{n-2} \\ p_{n-3} & p_{n-3} \end{vmatrix}\right].$$

$$= (-1)\left[0 + (-1)\begin{vmatrix} p_{n-2} & q_{n-2} \\ p_{n-3} & q_{n-3} \end{vmatrix}\right]$$

$$= (-1)^2 \begin{vmatrix} p_{n-2} & q_{n-2} \\ p_{n-3} & q_{n-3} \end{vmatrix}.$$

Continuing this process $(n-2)$ times, we get

$$p_n q_{n-1} - p_{n-1} q_n = (-1)^{n-2}\begin{vmatrix} p_2 & q_2 \\ p_1 & q_1 \end{vmatrix} = (-1)^{n-2}\begin{vmatrix} a_2a_1 + 1 & a_2 \\ a_1 & 1 \end{vmatrix}$$

$$= (-1)^{n-2}\{a_2a_1 + 1 - a_2a_1\} = (-1)^{n-2}(-1)^2 = (-1)^n.$$

(iv) $\dfrac{p_n}{q_n} - \dfrac{p_{n-1}}{q_{n-1}} = \dfrac{p_n q_{n-1} - p_{n-1} q_n}{q_n q_{n-1}} = \dfrac{(-1)^n}{q_n q_{n-1}}$ [by (iii)].

(v) If possible, let $(p_n, q_n), (p_n, p_{n-1}), (q_n, q_{n-1}) = d$. Then d divides $(-1)^n$, since $p_n q_{n-1} - p_{n-1} q_n = (-1)^n = \pm 1$. So d divides ± 1, which is possible only when $d = 1$ and hence $(p_n, q_n), (p_n, p_{n-1}), (q_n, q_{n-1}) = 1$.

Then $\dfrac{p_n}{q_n}$ is its lowest term.

Also (p_n, p_{n-1}) and (q_n, q_{n-1}) are prime to each other.

(vi) $\alpha = \langle a_1, a_2, \ldots, a_n, a_{n+1}, \ldots \rangle$

Then, $\alpha_{n+1} = (n+1)$th complete quotient

$= \langle a_{n+1}, a_{n+2}, \ldots \rangle.$

Now $$u_n = \frac{p_n}{q_n} = \frac{a_n p_{n-1} + p_{n-2}}{a_n q_{n-1} + q_{n-2}}, \tag{1}$$

where $p_n = a_n p_{n-1} + p_{n-2}$ and $q_n = a_n q_{n-1} + q_{n-2}$.

$u_n = \langle a_1, a_2, \ldots, a_n \rangle.$

So α is obtained from u_n if a_n is replaced by $\langle a_n, \alpha_{n+1} \rangle$, since $\alpha_{n+1} = \langle a_{n+1}, a_{n+2}, \ldots \rangle$.

Hence, from (1) we have

$$\alpha = \frac{\left(a_n + \frac{1}{\alpha_{n+1}}\right) p_{n-1} + p_{n-2}}{\left(a_n + \frac{1}{\alpha_{n+1}}\right) q_{n-1} + q_{n-2}} = \frac{\alpha_{n+1}(a_n p_{n-1} + p_{n-2}) + p_{n-1}}{\alpha_{n+1}(a_n q_{n-1} + q_{n-2}) + q_{n-1}}$$

$$= \frac{\alpha_{n+1} p_n + p_{n-1}}{\alpha_{n+1} q_n + q_{n-1}}.$$

(vii) If $\frac{p_n}{q_n}$ is not the last convergent, then $\alpha = \frac{\alpha_{n+1} p_n + p_{n-1}}{\alpha_{n+1} q_n + q_{n-1}}$

$\Rightarrow \quad \alpha \cdot \alpha_{n+1} q_n + \alpha q_{n-1} = \alpha_{n+1} p_n + p_{n-1}$

$\Rightarrow \quad \alpha_{n+1}(p_n - \alpha q_n) = \alpha q_{n-1} - p_{n-1}$

$\Rightarrow \quad \alpha_{n+1} q_n \left(\frac{p_n}{q_n} - \alpha\right) = q_{n-1}\left(\alpha - \frac{p_{n-1}}{q_{n-1}}\right)$

$\Rightarrow \quad \frac{p_n}{q_n} - \alpha = \frac{q_{n-1}}{\alpha_{n+1} q_n}\left(\alpha - \frac{p_{n-1}}{q_{n-1}}\right)$

$$\Rightarrow \quad \left|\frac{p_n}{q_n} - \alpha\right| = \frac{q_{n-1}}{\alpha_{n+1} q_n}\left|\alpha - \frac{p_{n-1}}{q_{n-1}}\right|. \tag{1}$$

$\alpha_{n+1} \geq 1$ as a_i are (+)*ve* integers, where a_1 may be zero.

$\Rightarrow \quad \frac{1}{\alpha_{n+1}} \leq 1$

and $q_n > q_{n-1} \Rightarrow \frac{q_{n-1}}{q_n} < 1.$

So $\frac{q_{n-1}}{\alpha_{n+1} q_n} < 1.$

Therefore, from (1),

$$\left|\frac{p_n}{q_n} - \alpha\right| < \left|\alpha - \frac{p_{n-1}}{q_{n-1}}\right|. \tag{2}$$

Again, if $\frac{p_n}{q_n}$ be the last convergent of α,

then $$\left|\frac{p_n}{q_n} - \alpha\right| = 0 < \left|\alpha - \frac{p_{n-1}}{q_{n-1}}\right|. \tag{3}$$

Hence, the result follows from (2) and (3).

(viii) Let $U_n = \dfrac{p_n}{q_n}$. Then $U_{n-1} = \dfrac{p_{n-1}}{q_{n-1}}$.

Now $\quad U_n - U_{n-1} = \dfrac{p_n}{q_n} - \dfrac{p_{n-1}}{q_{n-1}} = \dfrac{p_n q_{n-1} - q_n p_{n-1}}{q_n \cdot q_{n-1}} = \dfrac{(-1)^n}{q_n \cdot q_{n-1}}.$ (1)

Also $\quad U_n - U_{n-2} = \dfrac{p_n}{q_n} - \dfrac{p_{n-2}}{q_{n-2}} = \dfrac{p_n q_{n-2} - p_{n-2} q_n}{q_n \cdot q_{n-2}}$

$$= \frac{(a_n p_{n-1} + p_{n-2})q_{n-2} - (a_n q_{n-1} + q_{n-2})p_{n-2}}{q_n \cdot q_{n-2}}$$

$$= \frac{a_n(p_{n-1}q_{n-2} - q_{n-1}p_{n-2})}{q_n \cdot q_{n-2}}$$

$$= (-1)^{n-1}\frac{a_n}{q_n \cdot q_{n-2}} \quad (a_n > 0). \tag{2}$$

From (1) and (2) it follows that $U_n - U_{n-1}$ and $U_n - U_{n-2}$ are of opposite signs, since $a_n, q_n, q_{n-1}, q_{n-2}$ are all (+)ve and consequently, U_n lies between U_{n-1} and U_{n-2}. It immediately follows that **every convergent** lies between its two preceding convergents.

When n is odd, $(-1)^{n-1} = 1$.

So from (2), $U_n > U_{n-2}$

$n = 3 \Rightarrow U_3 > U_1$

$n = 5 \Rightarrow U_5 > U_3$

... ...

$n = 2m+1 \Rightarrow U_{2m+1} > u_{2m-1}$

Thus, we have $U_1 < U_3 < U_5 \cdots < U_{2m-1} < U_{2m+1} < \cdots$ (3)

Again, when n is even, $(-1)^{n-1} = -1$.

So $\quad U_n < U_{n-2}$.

$n = 4 \Rightarrow U_4 < U_2$

$n = 6 \Rightarrow U_6 < U_4$

... ...

$n = 2m \Rightarrow U_{2m} < U_{2m-2}$

Thus, we have $\cdots U_{2m+2} < U_{2m} < \cdots < U_6 < U_4 < U_2$ (4)

Also $\quad n = 2m+2 \Rightarrow U_{2m+2} - U_{2m+1} > 0$ [by (1)]

$\Rightarrow \quad U_{2m+1} < U_{2m+2} < U_{2m}$. (5)

Combining (3), (4) and (5), we have

$$U_1 < U_3 < U_5 \cdots < U_{2m+1} \cdots < U_{2m+2} < U_{2m} \cdots < U_6 < U_4 < U_2. \tag{6}$$

It follows from (6) that $\{U_{2n-1}\}$, a sequence of odd convergents, is an increasing sequence while $\{U_{2n}\}$, a sequence of even convergents, is a decreasing sequence. It also follows that $\{U_{2n-1}\}$ and $\{U_{2n}\}$ are both convergent.

Therefore, $\operatorname*{Lt}_{n\to\infty} U_{2n-1}$ and $\operatorname*{Lt}_{n\to\infty} U_{2n}$ both exist finitely.

Let $\quad \operatorname*{Lt}_{n\to\alpha} U_{2n-1} = \alpha_1$ and $\operatorname*{Lt}_{n\to\alpha} U_{2n} = \alpha_2$, say.

Again, $U_{2n} - U_{2n-1} = \dfrac{1}{q_{2n}q_{2n-1}} \to 0$, because the integers q_i are increasing.

as $\quad n \to \infty. \quad \therefore \quad \underset{n\to\alpha}{\text{Lt}}\ U_{2n} = \underset{n\to\alpha}{\text{Lt}}\ U_{2n-1}.$

So $\quad \alpha_2 = \alpha_1 = \alpha$ (say).

Hence, $U_n \longrightarrow \alpha$ such that $\forall n$, $U_{2n-1} < \alpha < U_{2n}$.

Therefore, we conclude that

(i) Every infinite S.C.F. is convergent.

(ii) It converges to α, the value of S.C.F. such that $\alpha > U_{2n-1}$ and $\alpha < U_{2n}$, $\forall\ n$.

(ix) Let $\dfrac{p'}{q'}$ be the convergent immediately preceding $\dfrac{p}{q}$.

If possible, let $\left|\alpha - \dfrac{r}{s}\right| < \left|\alpha - \dfrac{p}{q}\right|$, when $s < q$. (1)

Now α lies between $\dfrac{p}{q}$ and $\dfrac{p'}{q'}$ [by viii]

$\Rightarrow \dfrac{r}{s}$, (any rational number) with $s < q$ lies between $\dfrac{p}{q}$ and $\dfrac{p'}{q'}$.

So $\quad \left|\dfrac{r}{s} - \dfrac{p'}{q'}\right| < \left|\dfrac{p}{q} - \dfrac{p'}{q'}\right| = \dfrac{|pq' - p'q|}{q \cdot q'} = \dfrac{1}{q \cdot q'} \Rightarrow \dfrac{q'|rq' - sp'|}{sq'} < \dfrac{1}{q}$

$\Rightarrow \quad \dfrac{1}{s} < \dfrac{1}{q} \Rightarrow q < s$, which contradicts that $s < q$.

$\therefore \quad \left|\alpha - \dfrac{p}{q}\right| < \left|\alpha - \dfrac{r}{s}\right|$, when $s < q$.

(x) Let ϵ be the error in taking $\dfrac{p_n}{q_n}$ as the value of $\alpha = \langle a_1, a_2, \ldots, a_n, a_{n+1}, \ldots\rangle$.

Then $\quad \epsilon = \left|\alpha - \dfrac{p_n}{q_n}\right| < \left|\dfrac{p_n}{q_n} - \dfrac{p_{n+1}}{q_{n+1}}\right|$, since $u_n < \alpha < u_{n+1}$, i.e., $\dfrac{p_n}{q_n} < \alpha < \dfrac{p_{n+1}}{q_{n+1}}$

$$< \frac{|p_nq_{n+1} - q_np_{n+1}|}{q_n \cdot q_{n+1}} = \left|\frac{(-1)^{n+1}}{q_n \cdot q_{n+1}}\right| = \frac{1}{q_nq_{n+1}}. \quad (1)$$

Again, $u_n > u_{n+2} > \alpha$, when n is even

and $\quad u_n < u_{n+2} < \alpha$, when n is odd.

So in either case we have $|\alpha - u_n| > |u_{n+2} - u_n|$

$$\Rightarrow \left|\alpha - \frac{p_n}{q_n}\right| > \left|\frac{p_{n+2}}{p_{n+2}} - \frac{p_n}{q_n}\right| = \left|\frac{p_{n+2}q_n - p_nq_{n+2}}{q_{n+2}q_n}\right|$$

$$= \frac{|(a_{n+2}p_{n+1} + p_n)q_n - p_n(a_{n+2}q_{n+1} + q_n)|}{q_{n+2} \cdot q_n}$$

$$= \frac{|a_{n+2}(p_{n+1}q_n - p_nq_{n+1})|}{q_{n+2} \cdot q_n}$$

$$= \frac{a_{n+2} \cdot |(-1)^{n+1}|}{q_{n+2} \cdot q_n} = \frac{a_{n+2}}{q_{n+2}q_n} \quad [\because a_{n+2} > 0]. \quad (2)$$

So $\quad \dfrac{a_{n+2}}{q_{n+2}q_n} < \epsilon < \dfrac{1}{q_nq_{n+1}}$ [by (1) and (2)]

Again, from (1), $\epsilon < \dfrac{1}{q_n q_{n+1}}$

$$\Rightarrow \qquad \epsilon < \frac{1}{q_n^2}, \quad \text{since } q_{n+1} > q_n. \tag{3}$$

Also from (2), $\dfrac{a_{n+2}}{q_n(a_{n+2}q_{n+1}+q_n)} < \epsilon,$

i.e., $\qquad \epsilon > \dfrac{1}{q_n\left(q_{n+1}+\frac{q_n}{a_{n+2}}\right)}.$

Now $\qquad \dfrac{q_n}{a_{n+2}} < q_n \Rightarrow q_{n+1}+\dfrac{q_n}{a_{n+2}} < q_{n+1}+q_n$

$$\Rightarrow \qquad \frac{1}{q_{n+1}+\frac{q_n}{a_{n+2}}} > \frac{1}{q_{n+1}+q_n} > \frac{1}{2q_{n+1}}.$$

$$\therefore \qquad \epsilon > \frac{1}{q_n 2q_{n+1}} > \frac{1}{2q_{n+1}^2}\left[q_n < q_{n+1} \Rightarrow \frac{1}{q_n} > \frac{1}{q_{n+1}}\right]. \tag{4}$$

So combining (3) and (4) we obtain the limit of ϵ, i.e., $\dfrac{1}{2q_{n+1}^2} < \epsilon < \dfrac{1}{q_n^2}.$

Corollary [9.1]: Every convergent $\dfrac{p_n}{q_n}$ to α has the property $|\alpha - \dfrac{p_n}{q_n}| < \dfrac{1}{q_n^2}.$

[**Hints:** From (x) of Th. [9.2] it follows that $\left|\alpha - \dfrac{p_0}{q_n}\right| < \dfrac{1}{q_n^2}.$]

Problem (9.7): Find the error in taking the convergent $\frac{355}{113}$ as an approximation to the value of π,

where $\quad \pi = 3 + \dfrac{1}{7+}\,\dfrac{1}{15+}\,\dfrac{1}{1+}\,\dfrac{1}{292+}\cdots$ is less than 0.000000027.

Solution: $\dfrac{p_1}{q_1} = 3,\ \dfrac{p_2}{q_2} = \dfrac{22}{7},\ \dfrac{p_3}{q_3} = 3 + \dfrac{15}{106} = \dfrac{333}{106}$

$$\frac{p_4}{q_4} = \frac{a_4p_3+p_2}{a_4q_3+q_2} = \frac{1\cdot 333+22}{1\cdot 106+7} = \frac{355}{113}.$$

Also $\quad q_5 = a_5q_4 + q_3 = 292 \cdot 113 + 106 = 33102.$

$$\begin{aligned}\epsilon < \frac{1}{q_n \cdot q_{n+1}} &= \frac{1}{q_4 \cdot q_5} = \frac{1}{113 \times 33102}\\ &< \frac{1}{113 \times 33100} = \frac{1}{3740300}\\ &< \frac{1}{374} \times 10^{-4} = 0.0027 \times 10^{-4}\\ &= 0.00000027.\end{aligned}$$

Theorem (9.3): ***Every irrational number can be represented by an infinite S.C.F. and the representation is unique.***

Proof: Let α be any irrational number.

Let $a_1 = [\alpha]$ = the greatest integer not exceeding α.

Then $\quad \alpha = [\alpha] + \alpha - [\alpha] = a_1 + \dfrac{1}{\frac{1}{\alpha-[\alpha]}}.$

Let $\alpha_1 = \dfrac{1}{\alpha - [\alpha]} = \dfrac{1}{\alpha - a_1}$, where $\alpha - a_1 = \alpha - [\alpha]$

is a (+)ve irrational number < 1.

So $\alpha = a_1 + \dfrac{1}{\alpha_1}$, α_1 is an irrational number > 1. (1)

Let $a_2 = [\alpha_1]$. Then $\alpha_1 = [\alpha_1] + \alpha_1 - [\alpha_1] = a_2 + \dfrac{1}{\frac{1}{\alpha_1 - [\alpha_1]}}$.

From (1), we have $\alpha = a_1 + \dfrac{1}{a_2 + \frac{1}{\frac{1}{\alpha_1 - [\alpha_1]}}}$. (2)

Let $\alpha_2 = \dfrac{1}{\alpha_1 - [\alpha_1]} = \dfrac{1}{\alpha_1 - a_2}$ such that α_2 is an irrational number > 1. Then $\alpha = a_1 + \dfrac{1}{a_2 + \frac{1}{\alpha_2}}$.

Let $a_3 = [\alpha_2]$. Then $\alpha_2 = [\alpha_2] + \alpha_2 - [\alpha_2] = a_3 + \dfrac{1}{\frac{1}{\alpha_2 - [\alpha_2]}}$.

Hence, $\alpha = a_1 + \dfrac{1}{a_2+} \dfrac{1}{a_3+} \dfrac{1}{\alpha - [\alpha_2]}$ (3)

We write, $\alpha_3 = \dfrac{1}{\alpha_2 - [\alpha_2]}$.

Then (3) reduces to the form, $\alpha = a_1 + \dfrac{1}{a_2 + \frac{1}{a_3 + \frac{1}{\alpha_3}}}$.

Thus, we get $\alpha = a_1 + \dfrac{1}{a_2+} \dfrac{1}{a_3+} \cdots \dfrac{1}{+\alpha_{k+1}}$, where α_{k+1} is an irrational number > 1

and $\alpha_{k+1} = \dfrac{1}{\alpha_k - [\alpha_k]} = \dfrac{1}{\alpha_k - a_k}$, where $a_k = [\alpha_k]$.

So the process never ends after finite number terms. Therefore, the quotients $a_1, a_2, \ldots, a_k$ are infinite in numbers.

We have $\alpha = \langle a_1, a_2, \ldots, a_k, \alpha_{k+1} \rangle = \dfrac{\alpha_{k+1} p_k + p_{k-1}}{\alpha_{k+1} q_k + q_{k-1}}$

So
$$\alpha - \frac{p_k}{q_k} = \frac{\alpha_{k+1} p_k + p_{k-1}}{\alpha_{k+1} q_k + q_{k-1}} - \frac{p_k}{q_k} = \frac{p_{k-1} q_k - p_k q_{k-1}}{q_k(\alpha_{k+1} q_k + q_{k-1})}$$
$$= \frac{(-1)^{k-1}}{q_k(\alpha_{k+1} q_k + q_{k-1})} \to 0 \text{ as } k \to \infty$$

$$\Rightarrow \quad \alpha = \operatorname*{Lt}_{k \to \alpha} \frac{p_k}{q_k} = \operatorname*{Lt}_{k \to \infty} \langle a_1, a_2, \ldots, a_k \rangle$$

That is $\alpha = \langle a_1, a_2, \ldots a_k \ldots \rangle$.

If possible, let α cannot be represented uniquely.

So let $\alpha = \langle a_1, a_2, \ldots \rangle = \langle b_1, b_2, \ldots \rangle$

$$a_1 + \frac{1}{a_2 + \frac{1}{a_3+}} = b_1 + \frac{1}{b_2 + \frac{1}{b_3+}}.$$

Integral parts of these continued fractions are equal, therefore, $a_1 = b_1$.

Also
$$\frac{1}{a_2 + \frac{1}{a_3+}} = \frac{1}{b_2 + \frac{1}{b_3+}} \Rightarrow a_2 + \frac{1}{a_3+} = b_2 + \frac{1}{b_3+}$$

$\Rightarrow a_2 = b_2$, on the same ground. Thus we have $a_k = b_k \ \forall\ k$.

Hence α can be represented as a S.C.F. uniquely.

Theorem (9.4): *An infinite continued fraction represents an irrational number.*

[Go through the following examples.]

(i) Let us consider an infinite S.C.F., $1 + \frac{1}{2+}\frac{1}{3+}\frac{1}{2+}\frac{1}{3+}\frac{1}{2+}\frac{1}{3+\cdots}$

Let $\alpha = 1 + \frac{1}{2+}\frac{1}{3+}\frac{1}{2+}\frac{1}{3+\cdots} = 1 + \frac{1}{2+\frac{1}{3+(\alpha-1)}} = 1 + \frac{1}{2+\frac{1}{\alpha+2}}$

$= 1 + \frac{\alpha+2}{2\alpha+5} = \frac{3\alpha+7}{2\alpha+5}$

or, $2\alpha^2 + 5\alpha = 3\alpha + 7$ or, $2\alpha^2 + 2\alpha - 7 = 0$

or, $\alpha = \frac{-2 \pm \sqrt{4+56}}{2} = -1 \pm \sqrt{\frac{60}{4}} = -1 \pm \sqrt{15}.$

Here $\alpha = \sqrt{15} - 1$ represents an irrational number. [since each term is (+)ve]

(ii) Let us consider the infinite S.C.F., $1 + \frac{1}{2+}\frac{1}{2+}\frac{1}{2+\cdots}$

Let $\alpha = 1 + \frac{1}{2+}\frac{1}{2+}\frac{1}{2+\cdots} = 1 + \frac{1}{2+\alpha-1} = 1 + \frac{1}{\alpha+1} = \frac{\alpha+2}{\alpha+1}$

$\Rightarrow \alpha^2 + \alpha = \alpha + 2 \Rightarrow \alpha^2 = 2$

$\Rightarrow \alpha = \sqrt{2} =$ an irrational number.

***Proof*:** If possible, let the theorem be not true. Then there exists a rational number α such that $\alpha = \langle a_1, a_2, \ldots, a_n, \ldots \rangle$. But we have proved that a rational number can be expressed as a finite S.C.F., say, $\langle b_1, b_2, \ldots, b_m \rangle$.

Equating the integral parts, we have $a_1 = b_1$, $a_2 = b_2$, $\ldots a_{n-1} = b_{m-1}$ and lastly $a_n + \frac{1}{a_{n+1}+}\frac{1}{a_{n+2}+} = b_m$, which is not possible as b_m is an integer. Hence the theorem.

Definition (9.3): Recurring continued Fraction : *A C.F. is said to be recurring if after certain stages, the elements of the C.F., i.e., the quotients of the C.F. recur in the same order.*

The recurring elements form a cyclic, while the non-recurring elements, if exist, form acyclic.

Example: $5 + \frac{1}{2+}\frac{1}{3+}\frac{1}{4+}\frac{1}{2+}\frac{1}{3+}\frac{1}{4+\cdots}$

is a recurring continued fraction, where the quotients 2, 3, 4 are recurring, i.e., they form a cyclic while the 1st quotient 5 is acyclic.

We denote this C.F. by $5 + \frac{1}{2+}\frac{1}{3+}\frac{1}{4+}$ or $\langle 5, \overline{2, 3, 4} \rangle$.

Problem (9.8): Express the following irrational numbers as a S.C.F.

(i) $\sqrt{14}$ (ii) $\frac{5+\sqrt{37}}{4}$ (iii) $\pi = 3.141592653\ldots$

Solution: (i) $\alpha = \sqrt{14}$, $[\alpha] = 3$.

$\alpha = 3 + (\sqrt{14} - 3)$

$$\alpha_1 = \frac{1}{\sqrt{14}-3} = \frac{\sqrt{14}+3}{5} = 1 + \frac{\sqrt{14}-2}{5}, \ [\alpha_1] = 1$$

$$\alpha_2 = \frac{1}{\frac{\sqrt{14}-2}{5}} = \frac{5\cdot(\sqrt{14}+2)}{10} = \frac{\sqrt{14}+2}{2} = 2 + \frac{\sqrt{14}-2}{2}, \ [\alpha_2] = 2$$

$$\alpha_3 = \frac{1}{\frac{\sqrt{14}-2}{2}} = \frac{2(\sqrt{14}+2)}{10} = \frac{\sqrt{14}+2}{5} = 1 + \frac{\sqrt{14}-3}{5}, \ [\alpha_3] = 1$$

$$\alpha_4 = \frac{1}{\frac{\sqrt{14}-3}{5}} = \frac{5}{\sqrt{14}-3} = \frac{5(\sqrt{14}+3)}{5} = \sqrt{14}+3 = 6 + (\sqrt{14}-3), \ [\alpha_4] = 6$$

$$\alpha_5 = \frac{1}{\sqrt{14}-3} = \alpha_1.$$

So the next quotients will recur.

Hence, $\sqrt{14} = \langle 3, \overline{1, 2, 1, 6} \rangle$.

(ii) $\alpha = \dfrac{5+\sqrt{37}}{4} = 2 + \dfrac{5+\sqrt{37}}{4} - 2 = 2 + \dfrac{\sqrt{37}-3}{4}, \ [\alpha] = 2$

$$\alpha_1 = \frac{1}{\frac{\sqrt{37}-3}{4}} = \frac{4(\sqrt{37}+3)}{28} = \frac{\sqrt{37}+3}{7} = 1 + \frac{\sqrt{37}-4}{7}, \ [\alpha_1] = 1$$

$$\alpha_2 = \frac{1}{\frac{\sqrt{37}-4}{7}} = \frac{7(\sqrt{37}+4)}{21} = \frac{\sqrt{37}+4}{3} = 3 + \frac{\sqrt{37}-5}{3}, \ [\alpha_2] = 3$$

$$\alpha_3 = \frac{1}{\frac{\sqrt{37}-5}{3}} = \frac{3(\sqrt{37}+5)}{12} = \frac{\sqrt{37}+5}{4} = 2 + \frac{\sqrt{37}-3}{4}, \ [\alpha_3] = 2$$

$$\alpha_4 = \frac{1}{\frac{\sqrt{37}-3}{4}} = \alpha_1.$$

So the next quotients will recur.

Hence, $\dfrac{5+\sqrt{37}}{4} = \langle 2, \overline{1, 3, 2} \rangle$

(iii) $\pi = 3.14\,15\,92\,653$

$$\alpha = \pi = 3 + \pi - 3, \quad [\pi] = 3$$

$$\alpha_1 = \frac{1}{\pi - 3} = \frac{1}{0.141592653} = 7.06251330, \ [\alpha_1] = 7$$

$$\alpha_2 = \frac{1}{0.06251330} = 15.9965944, \ [\alpha_2] = 15$$

$$\alpha_3 = \frac{1}{0.9965944} = 1.00341723, \ [\alpha_3] = 1$$

$$\alpha_4 = \frac{1}{0.00341723} = 292.63467, \ [\alpha_4] = 292$$

Thus, $\pi = \langle 3, 7, 15, 1, 292, \ldots \rangle$

Definition (9.4): Symmetric continued fraction : *If in a continued fraction the quotients of equidistant terms from the beginning and end are equal, then the C.F. is called symmetric.*

Example: $\frac{77}{34} = 2 + \frac{1}{3+}\,\frac{1}{1+}\,\frac{1}{3+}\,\frac{1}{2}$ having odd number of quotients.

Here $a_1 = a_5$

$a_2 = a_4$

$a_3 =$ mid-quotient,

$\frac{425}{132} = 3 + \frac{1}{4+}\,\frac{1}{1+}\,\frac{1}{1+}\,\frac{1}{4+}\,\frac{1}{3}$ having even number of quotients.

$a_1 = a_6$

$a_2 = a_5$

$a_3 = a_4$.

Pell's Equation

The equation $x^2 - ry^2 = 1$, where $r \in z^+$ and $r \neq a^2$, $a \in z^+$ is known as *Pell's equation.*

Though the name of John Pell is attached to the equation, the problem was initiated by Fermat. So it will be more justified if it is called Fermat's equation. But unfortunately Fermat's solution of this problem was not published. However, the complete solution of the problem was given by Lagrange, by using continued fraction. Lagrange has large contribution in Number theory. Here we shall consider the (+)ve integral solution of it.

Theorem (9.5): *If (p, q) be a solution of $x^2 - ry^2 = a$ $(0 < a < \sqrt{r})$, then $\frac{p}{q}$ is a convergent to $\sqrt{r}$.*

Proof: (p, q) is a solution of $x^2 - ry^2 = a$

$\Rightarrow \quad p^2 - rq^2 = a \Rightarrow (p - \sqrt{r}q)(p + \sqrt{r}q) = a$

$\Rightarrow \quad p - \sqrt{r}q = \dfrac{a}{p + \sqrt{r}q}$

$\Rightarrow \quad \dfrac{p}{q} - \sqrt{r} = \dfrac{a}{q(p + \sqrt{r}q)} = \dfrac{a}{q^2\left(\frac{p}{q} + \sqrt{r}\right)} > 0, \; a, p, q, > 0$

$\Rightarrow \quad \dfrac{p}{q} - \sqrt{r} > 0 \Rightarrow \dfrac{p}{q} > \sqrt{r}.$

Therefore, $\left|\dfrac{p}{q} - \sqrt{r}\right| = \dfrac{a}{q^2\left(\frac{p}{q} + \sqrt{r}\right)} < \dfrac{a}{q^2 2\sqrt{r}}$, since $\dfrac{p}{q} > \sqrt{r}$

$\Rightarrow \quad \dfrac{p}{q} + \sqrt{r} > 2\sqrt{r} \Rightarrow \dfrac{1}{\left(\frac{p}{q} + \sqrt{r}\right)} < \dfrac{1}{2\sqrt{r}}.$

Also $\quad a < \sqrt{r} \Rightarrow \left|\dfrac{p}{q} - \sqrt{r}\right| < \dfrac{\sqrt{r}}{q^2 2\sqrt{r}} = \dfrac{1}{2q^2} < \dfrac{1}{q^2}.$

$\Rightarrow \dfrac{p}{q}$ is a convergent to $\sqrt{r}$. [by corollary [9.1]]

Theorem (9.6): *If $\frac{p}{q}$ be a convergent to the continued fraction of $\sqrt{r}$, then (p, q) is a solution of one of the equations, $x^2 - ry^2 = a$, when $|a| < 1 + 2\sqrt{r}$.*

Proof: $\frac{p}{q}$ is a convergent to the continued fraction of $\sqrt{r}$, then $\left|\frac{p}{q}-\sqrt{r}\right|<\frac{1}{q^2}$, [by (x) of Th. [9.2]]

$$\Rightarrow \qquad |p-q\sqrt{r}|<\frac{1}{q} \tag{1}$$

$$\begin{aligned}\text{Again,}\quad |p+q\sqrt{r}| &= |p-q\sqrt{r}+2q\sqrt{r}| \\ &\le |p-q\sqrt{r}|+|2q\sqrt{r}| \\ &< \frac{1}{q}+2q\sqrt{r} \quad \text{[by(1)]} \\ &= q\left(\frac{1}{q^2}+2\sqrt{r}\right) \\ &< q(1+2\sqrt{r}) \text{ as } q>1. \end{aligned} \tag{2}$$

Therefore, from (1) and (2), we get

$$|p^2-rq^2|<\frac{1}{q}\cdot q(1+2\sqrt{r})=(1+2\sqrt{r})$$

$\Rightarrow$ (p,q) is a solution of $x^2-ry^2=a$, when $a<(1+2\sqrt{r})$.

Example: (1) We have $\sqrt{14}=\langle 3,\overline{1,2,1,6}\rangle$, by problem (9.8).

We find its some convergents,

$$\frac{p_1}{q_1}=\frac{3}{1},\ \frac{p_2}{q_2}=3+\frac{1}{1}=\frac{4}{1},\ \frac{p_3}{q_3}=3+\frac{1}{1+\frac{1}{2}}=\frac{11}{3}.$$

$$\frac{p_4}{q_4}=3+\frac{1}{1+\frac{1}{2+\frac{1}{1}}}=\frac{15}{4}.$$

Now $15^2-14\cdot 4^2=1$.

So (15, 4) is a solution of $x^2-ry^2=1$, $r=14$,

which is one of the equations, $x^2-ry^2=a$, where $a=1<1+2\sqrt{r}$.

(2) $\sqrt{11}=\langle 3,\overline{3,6}\rangle$

$$\alpha=\sqrt{11}=3+\sqrt{11}-3,\ [\alpha]=3$$

$$\alpha_1=\frac{1}{\sqrt{11}-3}=\frac{\sqrt{11}+3}{2}=3+\frac{\sqrt{11}-3}{2},\ [\alpha_1]=3$$

$$\alpha_2=\frac{1}{\frac{\sqrt{11}-3}{2}}=\frac{2(\sqrt{11}+3)}{2}=\sqrt{11}+3=6+(\sqrt{11}-3),\ [\alpha_2]=6$$

$$\alpha_3=\frac{1}{\sqrt{11}-3}=\alpha_1.$$

So $\sqrt{11}=\langle 3,\overline{3,6}\rangle$

$$\frac{p_1}{q_1}=\frac{3}{1},\ \frac{p_2}{q_2}=3+\frac{1}{3}=\frac{10}{3}$$

$$10^2-11\cdot 3^2=1$$

$\Rightarrow$ (10, 3) is a solution of $x^2-ry^2=1$, $r=11$.

Theorem (9.7): *If (x_1, y_1) be a fundamental solution of $x^2 - ry^2 = 1$, then every pair of integers (x_n, y_n) satisfying the condition,*

$$x_n + y_n\sqrt{r} = (x_1 + y_1\sqrt{r})^n, \; n = 1, 2, 3, \ldots$$

is also a solution of $x^2 - ry^2 = 1$.

Proof: We first show that, $(x_1 - y_1\sqrt{r})^n = x_n - y_n\sqrt{r}$.

We prove it by the method of induction.

$$(x_1 + y_1\sqrt{r})^2 = (x_1^2 + ry_1^2) + 2x_1y_1\sqrt{r}$$
$$= x_2 + y_2\sqrt{r} \text{ [by the given condition].}$$

$$\Rightarrow \quad x_2 = x_1^2 + ry_1^2 \quad \text{and} \quad y_2 = 2x_1y_1. \tag{1}$$

Now $\quad (x_1 - y_1\sqrt{r})^2 = (x_1^2 + ry_1^2) - 2\sqrt{r}x_1y_1 = x_2 - \sqrt{r}y_2$ [by (1)].

The result holds for $n = 2$.

Let it be true for $n = k$. That is, the result holds up to $n = k$.

Thus, $$(x_1 - y_1\sqrt{r})^k = x_k - y_k\sqrt{r}. \tag{2}$$

Now $$(x_1 - y_1\sqrt{r})^{k+1} = (x_1 - \sqrt{r}y_1)^k(x_1 - \sqrt{r}y_1) = (x_k - \sqrt{r}y_k)(x_1 - \sqrt{r}y_1)$$
$$= (x_kx_1 + ry_ky_1) - \sqrt{r}(x_ky_1 + y_kx_1). \tag{3}$$

We have by the given condition that

$$x_{k+1} + \sqrt{r}\,y_{k+1} = (x_1 + y_1\sqrt{r})^{k+1}$$
$$= (x_1 + y_1\sqrt{r})^k \cdot (x_1 + y_1\sqrt{r})$$
$$= (x_k + y_k\sqrt{r})(x_1 + y_1\sqrt{r})$$
$$= (x_kx_1 + ry_ky_1) + \sqrt{r}(x_ky_1 + y_kx_1).$$

Comparing: $$\left.\begin{aligned} x_{k+1} &= x_kx_1 + ry_ky_1 \\ \text{and} \quad y_{k+1} &= x_ky_1 + y_kx_1 \end{aligned}\right\} \tag{4}$$

So from (3) by (4), it follows that $(x_1 - \sqrt{r}y_1)^{k+1} = x_{k+1} - \sqrt{r}y_{k+1}$.

So the result holds for $n = k + 1$. Hence, it is true for all $n = 1, 2, 3, \ldots$

Therefore, $\quad (x_1 - \sqrt{r}y_1)^n = x_n - \sqrt{r}y_n$.

Let (x_1, y_1) be a fundamental solution of $x^2 - \sqrt{r}y^2 = 1$.

Then $$x_1^2 - ry_1^2 = 1. \tag{5}$$

Now $$x_n^2 - ry_n^2 = (x_n - \sqrt{r}y_n)(x_n + \sqrt{r}y_n)$$
$$= (x_1 - \sqrt{r}y_1)^n(x_1 + \sqrt{r}y_1)^n$$
$$= (x_1^2 - ry_1^2)^n = 1 \quad \text{[by (5)],}$$

which shows that (x_n, y_n) is also a solution of $x^2 - ry^2 = 1$, for $n = 1, 2, \ldots$.

Example: (1) The equation $x^2 - 63y^2 = 1$ (1)

has a fundamental solution, $(8, 1)$, since $8^2 - 63.1 = 1$.

$$x_2 + y_2\sqrt{63} = (x_1 + y_1\sqrt{63})^2 = (8 + 1\sqrt{63})^2$$
$$= (64 + 63) + 16\sqrt{63} = 127 + 16\sqrt{63}$$

$$\Rightarrow \quad x_2 = 127, \; y_2 = 16.$$

Then (x_2, y_2) is a solution of (1),

since $(127)^2 - 63(16)^2 = 16129 - 16128 = 1.$

(2) The equation $x^2 - 15y^2 = 1$ (1)

has fundamental solution $(4, 1)$, since $4^2 - 15 \cdot 1^2 = 1$

$$x_1 + y_2\sqrt{15} = (x_1 + y_1\sqrt{15})^2 = (4 + \sqrt{15})^2 = (16 + 15) + 8\sqrt{15} = 31 + 8\sqrt{15}$$

$$\Rightarrow \quad x_2 = 31, \; y_2 = 8.$$

$(x_2, y_2) = (31, 8)$ is a solution of (1), since $31^2 - 15 \cdot 64 = 1.$

Again,
$$\begin{aligned} x_3 + y_3\sqrt{15} &= (x_1 + y_1\sqrt{15})^3 = (x_1 + y_1\sqrt{15})^2(x_1 + y_1\sqrt{15}) \\ &= (x_2 + y_2\sqrt{15})(x_1 + y_1\sqrt{15}) \\ &= (31 + 8\sqrt{15})(4 + \sqrt{15}) = (124 + 120) + (32 + 31)\sqrt{15} \\ &= 244 + 63\sqrt{15}. \end{aligned}$$

$$\left.\begin{aligned} x_3 &= 244 \\ y_3 &= 63 \end{aligned}\right\}$$

Then (x_3, y_3) is also a solution of (1),

since $x_3^2 - 15y_3^2 = 244^2 - 15 \times 63^2 = 59536 - 59535 = 1.$

Problem (9.9): Find the fundamental solution of $x^2 - 29y^2 = 1.$

Solution: We find the continued fraction of $\sqrt{29}$.

$$\alpha = \sqrt{29} = 5 + (\sqrt{29} - 5), \; [\alpha] = 5$$

$$\alpha_1 = \frac{1}{\sqrt{29} - 5} = \frac{\sqrt{29} + 5}{4} = 2 + \frac{\sqrt{29} - 3}{4}, \; [\alpha_1] = 2$$

$$\alpha_2 = \frac{1}{\frac{\sqrt{29}-3}{4}} = \frac{4(\sqrt{29} + 3)}{20} = \frac{\sqrt{29} + 3}{5} = 1 + \frac{\sqrt{29} - 2}{5}, \; [\alpha_2] = 1$$

$$\alpha_3 = \frac{1}{\frac{\sqrt{29}-2}{5}} = \frac{5(\sqrt{29} + 2)}{25} = \frac{\sqrt{29} + 2}{5} = 1 + \frac{\sqrt{29} - 3}{5}, \; [\alpha_3] = 1$$

$$\alpha_4 = \frac{1}{\frac{\sqrt{29}-3}{5}} = \frac{5(\sqrt{29} + 3)}{20} = \frac{\sqrt{29} + 3}{4} = 2 + \frac{\sqrt{29} - 5}{4}, \; [\alpha_4] = 2$$

$$\alpha_5 = \frac{1}{\frac{\sqrt{29}-5}{4}} = \frac{4(\sqrt{29} + 5)}{4} = 10 + (\sqrt{29} - 5), \; [\alpha_5] = 10$$

$$\alpha_6 = \frac{1}{\sqrt{29} - 5} = \alpha_1.$$

So $\sqrt{29} = \langle 5, \overline{2, 1, 1, 2, 10} \rangle$

$$\frac{p_1}{q_1} = 5, \; \frac{p_2}{q_2} = \frac{11}{2}, \; \frac{p_3}{q_3} = \frac{16}{3}$$

$$\frac{p_4}{q_4} = \frac{1 \cdot p_3 + p_2}{1 \cdot q_3 + q_2} = \frac{16 + 11}{3 + 2} = \frac{27}{5}$$

$$\frac{p_5}{q_5} = \frac{2 \cdot p_4 + p_3}{2 \cdot q_4 + q_3} = \frac{2 \cdot 27 + 16}{2 \cdot 5 + 3} = \frac{70}{13}$$

$$\frac{p_6}{q_6} = \frac{10 \cdot 70 + 27}{10 \cdot 13 + 5} = \frac{727}{135}$$

$$\frac{p_7}{q_7} = \frac{2 \cdot 727 + 70}{2 \cdot 135 + 13} = \frac{1524}{283}$$

$$\frac{p_8}{q_8} = \frac{1 \cdot 1524 + 727}{1 \cdot 283 + 135} = \frac{2251}{418}$$

$$\frac{p_9}{q_9} = \frac{1 \cdot 2251 + 1524}{1 \cdot 418 + 283} = \frac{3775}{701}$$

$$\frac{p_{10}}{q_{10}} = \frac{2 \cdot 3775 + 2251}{2 \cdot 7013 + 418} = \frac{9801}{1820}.$$

Now $(9801)^2 - 29(1820)^2 = 1.$

So $(9801, 1820)$ is a fundamental solution of $x^2 - 29y^2 = 1$. It is to be noted that no other convergents from $\frac{p_1}{q_1}$ to $\frac{p_9}{q_9}$ satisfies it.

Problem (9.10): Let (x_1, y_1) be a fundamental solution of $x^2 - ry^2 = 1$ and $x_n + y_n\sqrt{r} = (x_1 + y_1\sqrt{r})^n$, $n = 1, 2, \ldots$, then the pair of integers x_n, y_n can be obtained from the formulas,

$$x_n = \frac{1}{2}[(x_1 + y_1\sqrt{r})^n + (x_1 - y_1\sqrt{r})^n]$$

$$y_n = \frac{1}{2\sqrt{r}}[(x_1 + y_1\sqrt{r})^n - (x_1 - y_1\sqrt{r})^n],$$

and hence deduce that

$$x_{n+1} = x_1x_n + ry_1y_n$$

$$y_{n+1} = x_1y_n + x_ny_1.$$

Also show that $x_{n+1} = 2x_1x_n - x_{n-1}$ and $y_{n+1} = 2x_1y_n - y_{n-1}$.

Solution: Since (x_1, y_1) is a fundamental solution of $x^2 - ry^2 = 1$, it follows that (x_n, y_n) is also a solution of $x^2 - ry^2 = 1$ [by Th. (9.7)]

and so $x_n^2 - ry_n^2 = 1$,

which implies that $(x_n + \sqrt{r}y_n)(x_n - \sqrt{r}y_n) = 1$.

So $$x_n - \sqrt{r}y_n = \frac{1}{x_n + \sqrt{r}y_n} = \frac{(x_1 - \sqrt{r}y_1)^n}{(x_1 + \sqrt{r}y_1)^n(x_1 - \sqrt{r}y_1)^n}$$

$$= \frac{(x_1 - \sqrt{r}y_1)^n}{(x_1^2 - ry_1^2)^n} = (x_1 - \sqrt{r}y_1)^n \quad [\text{since } x_1^2 - ry_1^2 = 1].$$

Therefore, we have

$$x_n + \sqrt{r}y_n = (x_1 + \sqrt{r}y_1)^n \quad [\text{given condition}] \tag{1}$$

and $$x_n - \sqrt{r}y_n = (x_1 - \sqrt{r}y_1)^n. \tag{2}$$

(1) + (2) gives $$x_n = \frac{1}{2}[(x_1 + \sqrt{r}y_1)^n + (x_1 - \sqrt{r}y_1)^n] \tag{3}$$

(1) – (2) gives $$y_n = \frac{1}{2\sqrt{r}}[(x_1 + \sqrt{r}y_1)^n - (x_1 - \sqrt{r}y_1)^n. \tag{4}$$

Also $x_n + \sqrt{r}y_n = (x_1 + y_1\sqrt{r})^n$, $\forall\, n = 1, 2, \ldots$

$$\Rightarrow \quad x_{n+1} + \sqrt{r}y_{n+1} = (x_1 + y_1\sqrt{r})^{n+1} = (x_1 + y_1\sqrt{r})^n(x_1 + y_1\sqrt{r})$$
$$= (x_n + \sqrt{r}y_n)(x_1 + \sqrt{r}y_1)$$
$$= (x_nx_1 + ry_ny_1) + \sqrt{r}(x_ny_1 + y_nx_1)$$

$$\Rightarrow \quad x_{n+1} = x_nx_1 + ry_ny_1 \tag{5}$$

and $y_{n+1} = x_ny_1 + y_nx_1.$ (6)

From (3), $x_{n-1} = \frac{1}{2}[(x_1 + \sqrt{r}y_1)^{n-1} + (x_1 - \sqrt{r}y_1)^{n-1}]$

$$= \frac{1}{2}\left[\frac{(x_1 + \sqrt{r}y_1)^n}{x_1 + \sqrt{r}y_1} + \frac{(x_1 - \sqrt{r}y_1)^n}{x_1 - \sqrt{r}y_1}\right]$$

or, $x_{n-1} = \frac{1}{2}[(x_n + y_n\sqrt{r})(x_1 - y_1\sqrt{r}) + (x_n - y_n\sqrt{r})(x_1 + y_1\sqrt{r})]$ $[\because\ x_1^2 - ry_1^2 = 1]$

$$= \frac{1}{2}[2(x_1x_n - ry_1y_n) + 0] = x_1x_n - ry_1y_n. \tag{7}$$

(5) + (7) gives $x_{n+1} + x_{n-1} = 2x_1x_n$ or, $x_{n+1} = 2x_1x_n - x_{n-1}$.

Similarly, we can obtain $y_{n+1} = 2x_1y_n - y_{n-1}$.

Theorem (9.8): *Let* $\alpha = \frac{\sqrt{r}}{q} = \langle a_1, \overline{a_2, a_3, \ldots, a_{k+1}}\rangle$, *then* $p_k^2 - \frac{r}{q^2}q_k^2 = (-1)^k$, *where* $\frac{p_k}{q_k}$ *is the kth convergent of* $\langle a_1, \overline{a_2, a_3, \ldots, a_{k+1}}\rangle$.

Proof: $\alpha = \langle a_1, \overline{a_2, \ldots, a_{k+1}}\rangle$

$$\Rightarrow \quad \alpha = a_1 + \frac{1}{\langle a_2, \ldots, a_{k+1}\rangle} \Rightarrow \alpha - a_1 = \frac{1}{\langle a_2, a_3, \ldots, a_{k+1}\rangle}$$

$$\Rightarrow \quad \langle a_2, a_3, \ldots, a_{k+1}\rangle = \frac{1}{\alpha - a_1}.$$

Now $\alpha = \langle a_1, a_2, \ldots, a_{k+1}, a_2, a_3, \ldots, a_{k+1}, \ldots\rangle$

$$= \langle a_1, a_2, \ldots, a_{k+1}, \frac{1}{\alpha - a_1}\rangle.$$

Therefore, $\alpha = \dfrac{\frac{1}{\alpha - a_1}p_{k+1} + p_k}{\frac{1}{\alpha - a_1}q_{k+1} + q_k} = \dfrac{p_{k+1} + (\alpha - a_1)p_k}{q_{k+1} + (\alpha - a_1)q_k}$

$$\Rightarrow \quad \alpha\{q_{k+1} + (\alpha - a_1)q_k\} = p_{k+1} + (\alpha - a_1)p_k$$

$$\Rightarrow \quad \alpha^2q_k + \{q_{k+1} - q_ka_1 - p_k\}\alpha - (p_{k+1} - a_1p_k) = 0$$

$$\Rightarrow \quad \alpha^2q_k - \{q_ka_1 + p_k - q_{k+1}\}\alpha - (p_{k+1} - a_1p_k) = 0.$$

Since α is an irrational number, it contains no rational part.

So $p_k + a_1q_k - q_{k+1} = 0$ (1)

and hence, $q_k\alpha^2 = p_{k+1} - a_1p_k$

$$\Rightarrow \quad q_k^2\alpha^2 = q_kp_{k+1} - a_1q_kp_k = q_kp_{k+1} - (q_{k+1} - p_k)p_k \text{ [by (1)]}$$
$$= q_kp_{k+1} - q_{k+1}p_k + p_k^2$$

$$\Rightarrow \quad q_k^2\alpha^2 - p_k^2 = (-1)^{k+1} = -(-1)^k \Rightarrow p_k^2 - q_k^2\alpha^2 = (-1)^k$$

$$\Rightarrow \quad p_k^2 - q_k^2 \cdot \frac{r}{q^2} = (-1)^k \quad \left[\text{since } \alpha = \frac{\sqrt{r}}{q}\right].$$

Note: If $q = 1$, then $p_k^2 - rq_k^2 = (-1)^k$, (2)

where $\alpha = \sqrt{r}$.

Then $p_{2k}^2 - rq_{2k}^2 = (-1)^{2k} = 1$ and $p_{2k-1}^2 - rq_{2k-1}^2 = (-1)^{2k-1} = -1$.

So we can write (2) by

$$p_{kn}^2 - rq_{kn}^2 = (-1)^{kn}, \text{ where } k = 1, 2, 3, \ldots$$

and n is the length of the expansion.

Theorem (9.9): *If* $\sqrt{r} = \langle a_1, \overline{a_2, a_3, \ldots, a_{k+1}} \rangle$, *where k is odd and let* $n = sk$, *where* $s = 1, 3, 5, \ldots$, *then* (p_n, q_n) *is solution of* $x^2 - ry^2 = -1$.

Proof: We have $p_n^2 - rq_n^2 = (-1)^n$,

Then, $n = sk$, $p_{sk}^2 - rq_{sk}^2 = (-1)^{sk} = -1$, where n is odd as k is odd and $s = 1, 3, 5, \ldots$

$\Rightarrow$ $p_n^2 - rq_n^2 = (-1)$, for $sk = n$.

$\Rightarrow$ (p_n, q_n) is a solution of $x^2 - ry^2 = -1$.

Thus, an odd convergent becomes a solution of $x^2 - ry^2 = -1$.

Problem (9.11): Solve $x^2 - 13y^2 = -1$.

Solution: $\sqrt{13} = \langle 3, \overline{1, 1, 1, 1, 6} \rangle$ [From table]

$$\frac{p_1}{q_1} = \frac{3}{1}, \; \frac{p_2}{q_2} = 3 + 1 = \frac{4}{1}$$

$$\frac{p_3}{q_3} = \frac{1 \cdot p_2 + p_1}{1 \cdot q_2 + q_1} = \frac{4+3}{1+1} = \frac{7}{2}$$

$$\frac{p_4}{q_4} = \frac{1 \cdot p_3 + p_2}{1 \cdot q_3 + q_2} = \frac{1 \cdot 7 + 4}{1 \cdot 2 + 1} = \frac{11}{3}$$

$$\frac{p_5}{q_5} = \frac{1 \cdot 11 + 7}{1 \cdot 3 + 2} = \frac{18}{5} = \text{odd convergent,}$$

such that $p_5^2 - 13q_5^2 = -1$,

i.e., $18^2 - 13 \cdot 5^2 = 324 - 325 = -1$.

So $(p_5, q_5) = (18, 5)$ is a solution of it.

Also $n = sk$, $s = 1, 3, 5, \ldots, k$ is any odd $= 5$. (say)

Then, (p_{15}, q_{15}), (p_{25}, q_{25}), (p_{35}, q_{35}) are its other solutions.

Problem (9.12): Show that $\dfrac{449}{120}$ differs from $\sqrt{14}$ by less than $\dfrac{1}{90\,000}$.

Solution: $\sqrt{14} = \langle 3, \overline{1, 2, 1, 6} \rangle$ [from Table]

$$\frac{p_1}{q_1} = \frac{3}{1}, \; \frac{p_2}{q_2} = 3 + 1 = \frac{4}{1}, \; \frac{p_3}{q_3} = \frac{2 \cdot 4 + 3}{2 \cdot 1 + 1} = \frac{11}{3}$$

$$\frac{p_4}{q_4} = \frac{1 \cdot 11 + 4}{1 \cdot 3 + 1} = \frac{15}{4}, \; \frac{p_5}{q_5} = \frac{6 \cdot 15 + 11}{6 \cdot 4 + 3} = \frac{101}{27}$$

$$\frac{p_6}{q_6} = \frac{1 \cdot 101 + 15}{1 \cdot 27 + 4} = \frac{116}{31}, \; \frac{p_7}{q_7} = \frac{2 \cdot 116 + 101}{2 \cdot 31 + 27} = \frac{333}{89}$$

$$\frac{p_8}{q_8} = \frac{1 \cdot 333 + 116}{1 \cdot 89 + 31} = \frac{449}{120}, \; \frac{p_9}{q_9} = \frac{6 \cdot 449 + 333}{6 \cdot 120 + 89} = \frac{3027}{809}.$$

Let ϵ be the error in taking $\dfrac{449}{120}$ as an approximation to $\sqrt{14}$.

Then $\quad \epsilon < \dfrac{1}{q_n \cdot q_{n+1}} = \dfrac{1}{q_8 \cdot q_9} = \dfrac{1}{120 \times 809} = \dfrac{1}{97\,080} < \dfrac{1}{90\,000}.$

Problem (9.13): Given that 1 metre = 3.2809 ft. Show that 8 km is approximately equal to 5 miles. Also find the error in taking 8 km = 5 miles.

Solution: 1 m = 3.2809 ft

$$= \frac{3.2809}{3 \times 1760} \text{ miles.}$$

$$\Rightarrow \quad 1 \text{ km} = \frac{3280.9}{5\,280} \text{ miles}$$

$$= \frac{32\,809}{52\,800}.$$

1	52800	32809	0
	32809	19991	1
1	19991	12818	1
	12818	7173	
1	7173	5645	3
	5645	4584	
1	1528	1061	2
	1061	934	
	467	127	

$$\frac{32809}{52800} = 0 + \frac{1}{1+}\ \frac{1}{1+}\ \frac{1}{1+}\ \frac{1}{1+}\ \frac{1}{1+}\ \frac{1}{3+}\ \frac{1}{1+}\ \frac{1}{2+}\ .$$

$$\frac{p_1}{q_1} = \frac{1}{1}, \quad \frac{p_2}{q_2} = \frac{1}{1+\frac{1}{1}} = \frac{1}{2}$$

$$\frac{p_3}{q_3} = \frac{1 \cdot p_2 + p_1}{1 \cdot q_2 + q_1} = \frac{1+1}{2+1} = \frac{2}{3}$$

$$\frac{p_4}{q_4} = \frac{1 \cdot p_3 + p_2}{1 \cdot q_3 + q_2} = \frac{2+1}{3+2} = \frac{3}{5}$$

$$\frac{p_5}{q_5} = \frac{1 \cdot 3 + 2}{1 \cdot 5 + 3} = \frac{5}{8}.$$

Therefore, 1 km $= \dfrac{5}{8}$ miles (approximately)

$\Rightarrow \quad$ 8 km = 5 miles.

Let ϵ be the error in taking 8 km = 5 miles.

Then $\quad \epsilon < \dfrac{1}{q_n^2} = \dfrac{1}{q_5^2} = \dfrac{1}{64} = 0.015.$

$\therefore \quad \epsilon < 0.015.$

Problem (9.14): Solve the Diophantine equation, $18x + 5y = 24$, by means of continued fraction.

Solution: $\dfrac{18}{5} = 3 + \dfrac{1}{1+}\,\dfrac{1}{1+}\,\dfrac{1}{2}$

$$\frac{p_1}{q_1} = 3,\ \frac{p_2}{q_2} = 3 + \frac{1}{1} = \frac{4}{1},\ \frac{p_3}{q_3} = \frac{7}{2},\ \frac{p_4}{q_4} = \frac{18}{5}.$$

Now $\quad p_4 q_3 - p_3 q_4 = (-1)^4 \Rightarrow 18 \cdot 2 - 7 \cdot 5 = 1$

$\Rightarrow \quad 18 \cdot 2 + 5 \cdot (-7) = 1 \Rightarrow 18 \cdot (48) + 5(-168) = 24$

$\Rightarrow \quad x = 48,\ y = -168,$

as the given equation is $18x + 5y = 24$.

Now the general solution is given by $x = 48 + 5t,\ y = -168 - 18t,\ t = 0, \pm 1, \pm 2, \cdots$

TABLE:

A list of continued fraction of $\sqrt{r}$ where r is not the square of any integer.

$\sqrt{2} = \langle 1; \overline{2} \rangle$

$\sqrt{3} = \langle 1; \overline{1, 2} \rangle$

$\sqrt{5} = \langle 2; \overline{4} \rangle$

$\sqrt{6} = \langle 2; \overline{2, 4} \rangle$

$\sqrt{7} = \langle 2; \overline{1, 1, 1, 4} \rangle$

$\sqrt{8} = \langle 2; \overline{1, 4} \rangle$

$\sqrt{10} = \langle 3; \overline{6} \rangle$

$\sqrt{11} = \langle 3; \overline{3, 6} \rangle$

$\sqrt{12} = \langle 3; \overline{2, 6} \rangle$

$\sqrt{13} = \langle 3; \overline{1, 1, 1, 1, 6} \rangle$

$\sqrt{14} = \langle 3; \overline{1, 2, 1, 6} \rangle$

$\sqrt{15} = \langle 3; \overline{1, 6} \rangle$

$\sqrt{17} = \langle 4; \overline{8} \rangle$

$\sqrt{18} = \langle 4; \overline{4, 8} \rangle$

$\sqrt{19} = \langle 4; \overline{2, 1, 3, 1, 2, 8} \rangle$

$\sqrt{20} = \langle 4; \overline{2, 8} \rangle$

$\sqrt{21} = \langle 4; \overline{1, 3, 1, 8} \rangle$

$\sqrt{22} = \langle 4; \overline{1, 2, 4, 2, 1, 8} \rangle$

$\sqrt{23} = \langle 4; \overline{1, 3, 1, 8} \rangle$

$\sqrt{24} = \langle 4; \overline{1, 8} \rangle$

$\sqrt{26} = \langle 5; \overline{10} \rangle$

$\sqrt{27} = \langle 5; \overline{5, 10} \rangle$

$\sqrt{28} = \langle 5; \overline{3, 2, 3, 10} \rangle$

$\sqrt{29} = \langle 5; \overline{2, 1, 1, 2, 10} \rangle$

$\sqrt{30} = \langle 5; \overline{2, 10} \rangle$

$\sqrt{31} = \langle 5; \overline{1, 1, 3, 5, 3, 1, 1, 10} \rangle$

$\sqrt{32} = \langle 5; \overline{1, 1, 1, 10} \rangle$

$\sqrt{33} = \langle 5; \overline{1, 2, 1, 10} \rangle$

$\sqrt{34} = \langle 5; \overline{1, 4, 1, 10} \rangle$

$\sqrt{35} = \langle 5; \overline{1, 10} \rangle$

$\sqrt{37} = \langle 6; \overline{12} \rangle$

$\sqrt{38} = \langle 6; \overline{6, 12} \rangle$

$\sqrt{39} = \langle 6; \overline{4, 12} \rangle$

$\sqrt{40} = \langle 6; \overline{3, 12} \rangle$

$\sqrt{41} = \langle 6; \overline{2, 2, 12} \rangle$

$\sqrt{42} = \langle 6; \overline{2, 12} \rangle$

$\sqrt{43} = \langle 6; \overline{1, 1, 3, 1, 5, 1, 3, 1, 1, 12} \rangle$

$\sqrt{44} = \langle 6; \overline{1, 1, 2, 1, 1, 1, 12} \rangle$

$\sqrt{45} = \langle 6; \overline{1, 2, 2, 2, 1, 12} \rangle$

$\sqrt{46} = \langle 6; \overline{1, 3, 1, 1, 2, 6, 2, 1, 1, 3, 1, 12} \rangle$

$\sqrt{47} = \langle 6; \overline{4, 5, 1, 12} \rangle$

$\sqrt{48} = \langle 6; \overline{1, 12} \rangle$

$\sqrt{50} = \langle 7; \overline{14} \rangle$

Exercise 9

1. Express $\sqrt{\frac{17}{11}}$ and $\sqrt{61}$ as simple continued fractions.
2. For the C.F., $\frac{1}{a+}\,\frac{1}{a+}\,\frac{1}{a+}\cdots$, prove that $p_n = q_{n-1}$.
3. Show that $\sqrt{a^2+1} = a + \frac{1}{2a+}\,\frac{1}{2a+}\,\frac{1}{2a+}\cdots$
4. Find the value of $1 + \frac{1}{1+}\,\frac{1}{1+}\,\frac{1}{3+}\,\frac{1}{2}$.
5. If 1 metre = 3.2809 ft, then find the upper limit of the error in taking an approximation of 103 km = 64 miles.
6. Find the general solutions of (i) $13x - 17y = 5$ (ii) $13x - 17y = 906$ by using the method of S.C.F.
7. Find all (+)ve integral solutions of $x^2 - 2y^2 = 1$, when $y < 250$.
8. If r is divisible by a prime m, where $m \equiv 3 \pmod 4$, then show that the equation $x^2 - ry^2 = 1$ has no solution.
9. If $1 \le b \le q$, then show that the rational number $\frac{a}{b}$ satisfies $\left|x - \frac{p_n}{q_n}\right| \le \left|x - \frac{a}{b}\right|$, where $\frac{p_n}{q_n}$ be the nth convergent of the C.F. of the irrational number x.
10. Prove that at least one of any two consecutive convergents of the irrational number x, the rational number $\frac{a}{b}$ satisfies the inequality $\left|x - \frac{a}{b}\right| < \frac{1}{2b^2}$.
11. Determine the unique irrational number represented by the infinite continued fraction $1 + \frac{1}{2^*} + \frac{1}{3^*}$.
12. If $\frac{p_r}{q_r}$ and $\frac{p'_r}{q'_r}$ be the rth convergents of $a_1 + \frac{1}{a_2+}\,\frac{1}{a_3+}\cdots\frac{1}{+a_n}$ and $\frac{1}{a_n+}\,\frac{1}{a_{n-1}+}\cdots\frac{1}{+a_2+}\,\frac{1}{a_1}$ respectively, then show that, $p'n = p_n$, $q'_n = p_{n-1}$, $p'_{n-1} = q_n$ and $q'_{n-1} = q_{n-1}$.
13. If x_0, y_0 is a (+)ve solution of the equation $x^2 - dy^2 = 1$, prove that $x_0 > y_0$.
14. Find a solution of $x^2 - 41y^2 = -1$.
15. Let $\langle a_1, a_2, \ldots, a_n,\ a_n, a_{n-1}, \ldots, a_1\rangle$ be a symmetric continued fraction with even number of partial quotients, then show that

    (i) $p_{2n} = p_n^2 + p_{n-1}^2$ (ii) $q_{2n-1} = q_n^2 + q_{n-1}^2$.

Chapter 10

Perfect Number and Fibonacci Sequence

Proper divisors of an integer $n(> 1)$.

Each of the divisors of n excluding n itself is called a *proper divisor of n.*

$d_i|n,\ d_i \neq n$ are the proper divisors of n.

Let $n = 12$. Then proper divisors of 12 are 1, 2, 3, 4, 6.

$\sigma_0(n)$ denotes the sum of the proper divisors of n. Then $\sigma_0(n) = \Sigma d_i,\ d_i \neq n$.

For $n = 12,\ \sigma_0(n) = 1 + 2 + 3 + 4 + 6 = 16$.

Perfect Number

Definition (10.1): *An integer n is called a perfect number if* $\sigma_0(n) = n$.

$$n = 6,\ \sigma_0(6) = 1 + 2 + 3 = 6.$$

So 6 is a perfect number.

$$n = 28,\ \sigma_0(28) = 1 + 2 + 4 + 7 + 14 = 28.$$

28 is a perfect number.

Thus, we have $\sigma_0(n) + n = \sigma(n)$.

When n is perfect, $\sigma_0(n) = n$.

Hence, $n + n = \sigma(n)$, $\quad\therefore\quad 2n = \sigma(n)$.

So n is perfect if $\sigma(n) = 2n$.

Theorem (10.1): $2^{k-1}(2^k - 1)$ *is a perfect number if* $2^k - 1$ *is prime* $(k > 1)$ *and conversely every even perfect number is of the form* $2^{k-1}(2^k - 1)$. *[Euclid form]*

Proof: Let $n = 2^{k-1}(2^k - 1)$ and let $2^k - 1$ is prime $(k > 1)$.

Now $\quad (2^{k-1}, 2^k - 1) = 1$.

So $\quad \sigma(n) = \sigma(2^{k-1})\sigma(2^k - 1)$, since $\sigma(n)$ is multiplicative

$$= \frac{2^{k-1+1} - 1}{2 - 1} \cdot 2^k \quad [\because\ 2^k - 1 = p \text{ (prime)} \Rightarrow \sigma(2^k - 1) = \sigma(p) = p + 1 = 2^k - 1 + 1 = 2^k]$$

$$= (2^k - 1)2^k = 2[2^{k-1}(2^k - 1)] = 2n.$$

$\therefore \quad \sigma(n) = 2n \Rightarrow n$ is perfect.

Conversely, let n be an even perfect number.

Then $\quad n = 2^{k-1} \cdot m$ where m is an odd integer.

Also $\quad (2^{k-1}, m) = 1$.

Then $\quad \sigma(n) = \sigma(2^{k-1})\sigma(m) = (2^k - 1)\sigma(m)$.

Since n is a perfect number, $\sigma(n) = 2n$.

So $(2^k - 1)\sigma(m) = 2n = 2 \cdot 2^{k-1}m = 2^k m$ (1)

$\Rightarrow$ $(2^k - 1)|2^k \cdot m$ $[\because \sigma(m) \in Z^+]$

$\Rightarrow$ $2^k - 1|m$ as $(2^k - 1, 2^k) = 1$

$\Rightarrow$ $m = (2^k - 1)t$, for same $t \in Z^+$. (2)

Putting m in (1),

$(2^k - 1)\sigma(m) = 2^k(2^k - 1)t$ (3)

$\Rightarrow$ $\sigma(m) = 2^k \cdot t;\ t \in Z^+$.

From (2), it follows that t is a divisor of m.

Thus, t and m are two (+)ve divisor of m.

Also $\sigma(m) = 2^k \cdot t \Rightarrow \sigma(m) \geq m + t = (2^k - 1)t + t = 2^k \cdot t$,

$\Rightarrow$ $2^k \cdot t \geq m + t = 2^k t$

$\Rightarrow$ $\sigma(m) = m + t \Rightarrow m$ has only two divisors m and t and this is true when m is prime and $t = 1$.

$\therefore$ from (1), $n = 2^{k-1} \cdot m = 2^{k-1}(2^k - 1)$ [by (2) for $t = 1$]

Hence the result.

Problem (10.1): If n be a perfect number, then show that $\sum_{d|n} \frac{1}{d} = 2$.

Solution: Let $n = 2^{k-1}(2^k - 1)$ $(k > 1)$.

Then $\sigma_0(n) = 1 + 2^{k-1} + (2^k - 1) = n$. (1)

Again, $$\sum_{d|n} \frac{1}{d} = 1 + \frac{1}{2^{k-1}} + \frac{1}{(2^k - 1)} + \frac{1}{2^{k-1}(2^k - 1)} = 1 + \frac{(2^k - 1) + 2^{k-1} + 1}{2^{k-1}(2^k - 1)}$$

$$= 1 + \frac{n}{n} = \text{[by (1)]}.$$

$$= 1 + 1 = 2.$$

Theorem (10.2): *For any integer $a > 1$, $a^k - 1$ is prime if $a = 2$ and k is prime.*

Proof: $a^k - 1 = (a - 1)(a^{k-1} + a^{k-2} + \cdots + a + 1) \geq a + 1 > 1$.

Also $a^k - 1$ is prime $\Rightarrow$ $a - 1 = 1$ $\Rightarrow$ $a = 2$.

If possible, let k be not prime. Then k is a composite number. Say, $k = mn$, $1 < m < k$, $1 < n < k$.

Now $a^{mn} - 1 = (a^m)^n - 1 = (a^m - 1)((a^m)^{n-1} + (a^m)^{n-2} + \cdots + a^m + 1)$

$\Rightarrow$ $a^{mn} - 1$ is composite $\forall$ $a > 1$.

i.e., $a^k - 1$ is composite, a contradiction.

Hence, k is prime.

Thus, we have **$2^k - 1$ is prime when k is prime.**

Prime number of the form $2^k - 1$, when k is prime, is known as **Mersenne prime**, in the name of the Mathematician Mersenne.

Theorem (10.3): *The last digit of every even perfect number is 6 or 8.*

Proof: Let n be any even perfect number.

Then $n = 2^{k-1}(2^k - 1)$, when k is prime.

Since k is prime, it is of the form: $4a+1$ $(a>0)$ or $4a+3$.

Now for $k=4a+1$,

$$n = 2^{4a}(2^{4a}\cdot 2-1) = (16)^a(2\times 16^a-1). \qquad (1)$$

For any (+)ve integer a, last digit of 16^a is 6,

while the last digit in $(2\cdot 16^a-1)$ is 1.

Therefore, the last digit in (1) is 6.

Again, for $k=4a+3$,

$$n = (2^4)^a\cdot 2^2(2^{4a}\cdot 2^3-1) = (16)^a\cdot 4(16^a\cdot 8-1).$$

For any integer a $(a\geq 0)$, the last digit in $(16)^a\cdot 4$ is 4 while the last digit in $16^a\cdot 8-1$ is 7. So the last digit in the product $16^a\cdot 4(16^a\cdot 8-1)$ is 8.

Hence the result.

Problem (10.2): Show that the sum of the first $2^{\frac{k-1}{2}}$ odd cubes is an even perfect number.

Solution: Let us consider the sum of the first m odd cubes, where $m=2^{\frac{k-1}{2}}$.

Then $1^3+3^3+\cdots+(2m-1)^3 = \sum_{r=1}^{m}(2r-1)^3 = \sum_{r=1}^{m}(8r^3-12r^2+6r-1)$

$$= 8(1^3+2^3+\cdots+m^3)-12(1^2+2^2+\cdots+m^2)+6(1+2+\cdots+m)-m$$

$$= 8\left(\frac{m(m+1)}{2}\right)^2 - 12\frac{m(m+1)(2m+1)}{6}+\frac{6m(m+1)}{2}-m$$

$$= 2m^2(m+1)^2-2m(m+1)(2m+1)+3m(m+1)-m$$

$$= m(m+1)\{2m(m+1)-2(2m+1)+3]-m$$

$$= m(m+1)\{2m^2+2m-4m+1\}-m$$

$$= m(m+1)2m(m-1)+m(m+1)-m$$

$$= 2m^2\{m^2-1\}+m^2$$

$$= m^2\{2m^2-2+1\} = m^2(2m^2-1)$$

$$= (2^{\frac{k-1}{2}})^2\{(2^{\frac{k-1}{2}})^2\cdot 2-1\}$$

$$= 2^{k-1}\{2\cdot 2^{k-1}-1\} = 2^{k-1}(2^k-1) = n,$$

an even perfect number.

k-Perfect number

Definition (10.2): *A positive integer n is called a k-perfect number, $(k\geq 3)$, if $\sigma(n)=kn$.*

Ex: Let $n=2^9\cdot 3\cdot 11\cdot 31$.

$$\sigma(n) = \frac{2^{9+1}-1}{2-1}\cdot\frac{3^2-1}{3-1}\cdot\frac{11^2-1}{11-1}\cdot\frac{31^2-1}{31-1}$$

$$= (2^{10}-1)(3+1)(11+1)(31+1) = 31\cdot 33\cdot 4\cdot 12\cdot 32$$

$$= 31\cdot 3\cdot 11\cdot 2^2\cdot 2^2\cdot 3\cdot 2^5 = 3.2^9.3.11.31 = 3\cdot n,\ k=3$$

$\therefore$ n is a 3-Perfect number.

Problem (10.3): Show that 30240 is a 4-Perfect number.

Solution: $n = 30240 = 3024 \times 10.$

Now $3024 = 2^4 \times 3^3 \cdot 7$ since

2	3024
2	1512
2	756
2	378
3	189
3	63
3	21
	7

$\therefore \quad 30240 = 2^4 \cdot 3^3 \cdot 7 \cdot 2 \cdot 5 = 2^5 \cdot 3^3 \cdot 5 \cdot 7$

$$\sigma(n) = (2^6 - 1)\frac{(3^4 - 1)}{2} \cdot \frac{5^2 - 1}{5 - 1} \cdot \frac{7^2 - 1}{7 - 1} = 63 \cdot 40 \cdot 6 \cdot 8 = 3^2 \cdot 7 \cdot 2^3 \cdot 5 \cdot 2 \cdot 3 \cdot 2^3$$
$$= 2^7 \cdot 3^3 \cdot 5 \cdot 7 = 2^2[2^5 \cdot 3^3 \cdot 5 \cdot 7] = 4n.$$

Therefore, 30240 is a 4-Perfect number.

Definition (10.3): *An integer $n \geq 1$ is called multiplicative perfect if n is equal to the product of all divisors excluding n itself.*

That is, if $n = \prod_{\substack{d|n, \\ d \neq n}} d.$

In other words, $n^2 = \prod_{d|n} d.$

Example: $n = 6, \quad d|6 \Rightarrow d = 1, 2, 3, 6$

$$\prod_{\substack{d|6, \\ d \neq 6}} d = 1 \cdot 2 \cdot 3 = 6 = n,$$

otherwise, $\prod_{d|6} d = 1 \cdot 2 \cdot 3 \cdot 6 = 36 = 6^2 = n^2.$

$\therefore$ 6 is multiplicative perfect.

Problem (10.4): The harmonic mean $H(n)$ of the divisors of a (+)ve integer n is defined by the formula $\dfrac{1}{H(n)} = \dfrac{1}{\tau(n)} \sum_{d|n} \dfrac{1}{d}.$

Hence, show that $\dfrac{1}{H(n)} = \dfrac{\sigma(n)}{n\tau(n)}$ and deduce that $H(n)$ is an integer if n is perfect.

Solution: Let $n = p_1^{a_1} \cdot p_2^{a_2} \cdots p_k^{a_k},\ a_i \geq 0,\ p_i$ are prime.

$d|n \Rightarrow d = 1,\ p_i^{x_i}, p_i^{x_i}p_j^{x_j}, p_i^{x_i}p_j^{x_j}p_k^{x_k}, \ldots, p_1^{a_1}p_2^{a_2} \ldots p_k^{a_k}.$

$1 \leq x_1 \leq a_1,\ 1 \leq x_2 \leq a_2, \ldots, 1 \leq x_k \leq a_k$

$h \neq i \neq j, \quad h, i, j : 1 \to k,$

where $\sigma(n) =$ sum of the divisors $= 1 + \sum p_i^{x_i} + \sum p_i^{x_i}p_j^{x_j} + \cdots + p_1^{a_1}p_2^{a_2} \cdots p_k^{a_k}.$

Now the A.M. of the reciprocals of the (+)ve divisors of n is defined by

$$\text{A.M.} = \frac{1+\sum \frac{1}{p_i^{x_i}} + \sum \frac{1}{p_i^{x_i} p_j^{x_j}} + \cdots + \frac{1}{p_1^{a_1} p_2^{a_2} \cdots p_k^{a_k}}}{\text{number of divisors of } n}.$$

$$\therefore \quad \frac{1}{H(n)} = \frac{1+\sum \frac{1}{p_i^{x_i}} + \sum \frac{1}{p_i^{x_i} p_j^{x_j}} + \cdots + \frac{1}{p_1^{a_1} p_2^{a_2} \cdots p_k^{a_k}}}{\tau(n)} \tag{1}$$

$$\text{or,} \quad \frac{1}{H(n)} = \frac{\sum\limits_{d|n} \frac{1}{d}}{\tau(n)} = \frac{1}{\tau(n)} \sum_{d|n} \frac{1}{d}. \tag{2}$$

$$\text{Also from (1),} \quad \frac{1}{H(n)} = \frac{1+\sum p_i^{x_i} + \sum p_i^{x_i} p_j^{x_j} + \cdots + p_1^{a_1} p_2^{a_2} \cdots p_k^{a_k}}{p_1^{a_1} p_2^{a_2} \cdots p_k^{a_k} \cdot \tau(n)}$$

$$= \frac{\sigma(n)}{n \cdot \tau(n)}. \tag{3}$$

If n is a perfect number, then $\sigma(n) = 2n$.

$$\therefore \quad \frac{1}{H(n)} = \frac{2n}{n \cdot \tau(n)}$$

$$\therefore \quad H(n) = \frac{\tau(n)}{2} = \text{an integer, } \forall \text{ perfect number } n. \tag{4}$$

The results followed from (2), (3) & (4).

Fibonacci sequence

Definition (10.4): *The sequence $f : N \to N$ defined by $f(1) = 1$, $f(2) = 1$ and $f(n) = f(n-1) + f(n-2)$, $\forall\, n \geq 3$, i.e., the sum of any two consecutive terms of the sequence $\{f(n)\}$ follows the next term of $\{f(n)\}$ is called a Fibonacci sequence.*

The elements of $\{f(n)\}$ are called the Fibonacci numbers.

Thus, $\{f(n)\} = \{1, 1, 2, 3, 5, 8, 13, 21, 34, 55, 89, \ldots\}$.

Some observations

(i) When n is prime, $n > 2 \not\Rightarrow f(n)$ is prime $\forall\, n$.

$f(1) = 1,\ f(2) = 1,\ f(3) = 2,\ f(4) = 3$

$f(5) = 5,\ f(6) = 8,\ f(7) = 13,\ f(8) = 21$

$f(9) = 34,\ f(10) = 55,\ f(11) = 89,\ f(12) = 144$

$f(13) = 233,\ f(14) = 377,\ f(15) = 610,\ f(16) = 987$

$f(17) = 1597,\ f(18) = 2584,\ f(19) = 4191 (= 37 \times 113)$.

Here, $f(3) = 2,\ f(5) = 5,\ f(7) = 13,\ f(11) = 89$

$f(13) = 233,\ f(17) = 1597$.

When n is prime, $f(n)$ is prime for $3, 5, 7, 11, 13, 17, \ldots$

But $f(19) = 4181 = 37 \times 113$, which is not prime though $n = 19$ is prime.

So for all prime numbers $n (> 2)$, $f(n)$ is not prime.

(ii) G.C.D. $(f(n), f(n+1)) = 1\ \forall\, n$.

But G.C.D. $(f(n), f(n+k)) \neq 1;\ \forall\, k (> 1)$

$$(f(2), f(3)) = (1, 2) = 1$$

$$(f(4), f(5)) = (2, 3) = 1, \ (f(10), f(11)) = (55, 89) = 1.$$

But $(f(3), f(6)) = (2, 8) = 2 \neq 1.$

Therefore, G.C.D. of any two consecutive terms of $\{f(n)\}$ is 1, but G.C.D. of any two terms of $\{f(n)\}$ is not necessarily 1.

(iii) It follows that G.C.D. of any two Fibonacci numbers is again a Fibonacci number.

$$(f(5), f(10)) = (5, 55) = 5 = f(5)$$

$$(f(8), f(12)) = (21, 144) = 3 = f(4).$$

(iv) $1|3 \Leftrightarrow f(1)|f(3)$

$4|8 \Leftrightarrow f(4)|f(8) \quad [\because f(4) = 3, f(8) = 21]$

So in general, $m|n \Leftrightarrow f(m)|f(n)$.

Theorem (10.4): *Two consecutive Fibonacci numbers are prime to each other* $\forall n \geq 1$.

That is, $(f(n), \ f(n+1)) = 1 \forall n \geq 1$.

Proof: From Euclidian Algorithm for obtaining $(f(n+2), \ f(n+1)) \forall n \geq 1$, we have

$$f(n+2) = 1 \cdot f(n+1) + f(n)$$

$$f(n+1) = 1 \cdot f(n) + f(n-1)$$

$$f(n) = 1 \cdot f(n-1) + f(n-2)$$

$$\ldots \quad \ldots \quad \ldots \quad \ldots$$

$$f(4) = 1 \cdot f(3) + f(2)$$

$$f(3) = 2 \cdot f(2) + 0 \ [\because f(3) = 2, \ f(2) = 1 \rightarrow f(3) = 2 \cdot f(2) + 0]$$

which determines $(f(n+2), f(n+1)) = f(2) = 1$.

Hence the result, i.e., $(f(n), f(n+1)) = 1$.

Go through the Example: $(f(10), f(9)) = (55, 34)$

Applying Euclidian divisional algorithm, we have

$$55 = 1 \cdot 34 + 21$$

$$34 = 1 \cdot 21 + 13$$

$$21 = 1 \cdot 13 + 8$$

$$13 = 1 \cdot 8 + 5$$

$$8 = 1 \cdot 5 + 3$$

$$5 = 1 \cdot 3 + 2$$

$$3 = 1 \cdot 2 + 1$$

$$2 = 2 \cdot 1 + 0$$

$\Rightarrow \quad (55, 34) = 1$

i.e., $(f(10), f(9)) = 1$.

Remarks: $(f(n), \ f(n+k)) \neq 1 \ \forall k > 1$.

Go through the example

Let us consider $(f(9),\ f(12)) = (34, 144)$.

By divisional algorithm, we have

$$144 = 4 \cdot 34 + 8$$

$$34 = 4 \cdot 8 + 2$$

$$8 = 4 \cdot 2 + 0$$

$\Rightarrow \quad (34, 144) = 2 \neq 1$ i.e., $(f(9), f(12)) \neq 1$.

Theorem (10.5): *For a Fibonacci sequence there exists the identity* $f(m+n) = f(m-1)f(n) + f(m)f(n+1) \forall m \geq 2$ *and* $\forall n \geq 1$.

***Proof*:** We shall prove it by method of induction. When $n = 1$ and m is fixed.

$$f(m+1) = f(m-1)f(1) + f(m) \cdot f(2) \ \forall m \geq 2$$

$$[\because f(m-1) + f(m) = f(m+1) \Rightarrow f(m-1) \cdot f(1) + f(m) \cdot 2 = f(m+1) \text{ as } f(1) = f(2) = 1]$$

$\Rightarrow$ the result is true for $n = 1$ and fixed $m \geq 2$.

Let it be true for $n = 1,\ 2,\ k-1,\ k$ where m is fixed.

Then for $n = k,\ k-1,\ \ldots,\ 3,\ 2,\ 1$

$$f(m+k) = f(m-1)f(k) + f(m)f(k+1) \tag{1}$$

$$f(m+\overline{k-1}) = f(m-1)f(k-1) + f(m) \cdot f(k). \tag{2}$$

$(1) + (2)$ gives

$$f(m+k) + f(m+\overline{k-1}) = f(m-1)\{f(k) + f(k-1)\} + f(m)\{f(k+1) + f(k)\}.$$

$$\Rightarrow \quad f(m+\overline{k+1}) = f(m-1)f(k+1) + f(m)f(k+2)$$

$\Rightarrow$ the result is true for $n = k+1$.

Hence, it is true $\forall n \geq 1$ and $\forall m \geq 2$.

Lemma (10.1): If $m = nq + r$, then $(f(nq+r),\ f(n)) = (f(r),\ f(n))$.

***Proof*:** $(f(m), f(n)) = (f(nq+r), f(n)) = (f(nq-1)f(r) + f(nq)f(r+1), f(n))$

[by identity in Theorem (10.5) in choosing $m = nq$ and $n = r$]

Now $\quad \forall m \geq 1,\ n \geq 1,\ f(mn)$ is divisible by $f(m)$. (1)

since for $n = 1,\ f(mn) = f(m)$ and it is divisible by $f(m)$, which shows the result is true for $n = 1$.

Let it be true for $n = 1,\ 2,\ \ldots,\ k$.

Now $\quad f(m(k+1)) = f(mk+m)$

$$= f(mk-1)f(m) + f(mk)f(m+1) \text{ [by Th. (10.5)]} \tag{2}$$

By assumption $f(m)|f(mk)$ and from (1) it follows that $f(m)|f(mk+m)$.

Hence the result holds. That is $f(m)|f(mn) \forall m \geq 1,\ n \geq 1$.

Also $\quad (x+y, z) = (x, z), \quad$ if $z|y$. (3)

Therefore, applying (1) and (2) successively, we get

$$(f(nq-1)f(r) + f(nq)f(r+1), f(n)) = (f(nq-1)f(r),\ f(n)) \ [\because f(n)|f(nq)]$$

If possible, let $(f(nq-1), f(n)) = d$.

Then $d|f(n)$. Also $f(n)|f(nq)$. Hence, $d|f(nq)$.

Again, $d|f(nq-1)$. But $(f(nq-1),\ f(nq)) = 1$,

since the two successive Fibonacci numbers are prime to other. So $d = 1$

Therefore, $(f(m), f(n)) = (f(nq-1)f(r), f(n))$ [by (1)]

$$= (f(r), f(n))$$

because of the face that $(a, b) = 1 \Rightarrow (ac, b) = (c, b)$ for any integer c by Th. [1.15].
Here, $(f(nq-1), f(n)) = 1$.

Theorem (10.6): *G.C.D. of two Fibonacci numbers is again a Fibonacci number. That is, $(f(m),\ f(n)) = f(d)$, where $d = (m, n)$.*

Proof: Let $m \geq n$. Then by applying Euclidian algorithm to m and n, we get the following system of equations:

$$m = q_1 n + r_1,\ 0 < r_1 < n$$
$$n = q_2 r_1 + r_2,\ 0 < r_2 < r_1$$
$$r_1 = q_3 r_2 + r_3,\ 0 < r_3 < r_2$$
$$\cdots\cdots\cdots\cdots\cdots\cdots\cdots\cdots\cdots$$
$$r_{n-2} = q_n r_{n-1} + r_n,\ 0 < r_n < r_{n-1}$$
$$r_{n-1} = q_{n-1} r_n + 0$$

Therefore, $(f(m),\ f(n)) = (f(r_1),\ f(n)) = (f(r_1),\ f(r_2)) \cdots = (f(r_{n-1}), f(r_n))$

[by Lemma (10.1)]

Since $r_n | r_{n-1}$ and r_n is a non-zero remainder in the above divisional algorithm of m and n, it follows that $(m, n) = r_n.$ $\cdots$ (i)

Also $\quad r_n | r_{n-1} \Rightarrow f(r_n)|f(r_{n-1}) \Rightarrow (f(r_n), f(r_{n-1})) = f(r_n).$

Thus $\quad (f(m),\ f(n)) = f(r_n) = f(m, n).$ [by (i)]

Theorem (10.7): *In a Fibonacci sequence there exists the identity that the sum of the 1st n-Fibonacci numbers is equal to $f(n+2) - 1\ \forall n > 1$.*

That is, $\displaystyle\sum_{r=1}^{n} f(r) = f(n+2) - 1.$

[*Go through the Example*] : *we consider the* 1*st* 10 *Fibonacci numbers*

1, 1, 2, 3, 5, 8, 13, 21, 34, 55.

The sum of these 1*st 10 Fibonacci numbers*

$$= 1 + 1 + 2 + 3 + 5 + 8 + 13 + 21 + 34 + 55 = 143 = 144 - 1 = f(12) - 1.$$

[$\because f(12) = 89 + 55 = 144$].

Thus, $\displaystyle\sum_{r=1}^{10} f(r) = f(10+2) - 1.$

Proof: For $n > 1$, taking the sum of 1st two terms,

$$f(1) + f(2) = 1 + 1 = 2 = 3 - 1 = f(4) - 1 = f(2+2) - 1.$$

Result holds for $n = 2$. Let it hold for $n = k$.

Then, $\sum_{r=1}^{k} f(r) = f(k+2) - 1.$

Now, $\sum_{r=1}^{k+1} f(r) = \sum_{r=1}^{k} f(r) + f(k+1) = f(k+2) - 1 + f(k+1)$

$$= f(k+1) + f(k+2) - 1$$
$$= f(k+3) - 1 = f(\overline{k+1}+2) - 1$$

[by def. $f(k+1) + f(k+2) = f(k+3)$]

Therefore, the identity is true for $n = k+1$.

Hence, it is true for all $n > 1$.

That is, $\sum_{r=1}^{n} f(r) = f(n+2) - 1,\ n > 1.$

Alternative proof: We have the following results from definition of Fibonacci numbers.

$$f(1) + f(2) = f(3)$$
$$\Rightarrow \quad f(1) = f(3) - f(2)$$

Similarly, $f(2) = f(4) - f(3)$

$$f(3) = f(5) - f(4)$$
$$\cdots\cdots\cdots\cdots\cdots\cdots$$
$$f(n-1) = f(n+1) - f(n)$$
$$f(n) = f(n+2) - f(n+1)$$

Adding the above results in both sides, we get

$$\sum_{r=1}^{n} f(r) = f(n+2) - f(2) = f(n+2) - 1 \quad [\because f(2) = 1]$$

Theorem (10.8): *Establish the following identities.*

(i) $f^2(n) = f(n+1) \cdot f(n-1) + (-1)^{n-1}$

(ii) $f^2(1) + f^2(2) + \cdots + f^2(n) = f(n) \cdot f(n+1)$

(iii) $f(2) + f(4) + f(b) + \cdots + f(2n) = f(2n+1) - 1$

(iv) $f(1) + f(3) + f(5) + \cdots + f(2n-1) = f(2n) - 1$

(v) $f(1) - f(2) + f(3) - f(4) + \cdots + (-1)^{k+1} f(k) = 1 + (-1)^{k+1} f(k-1)$

Proof: (i) $f^2(n) - f(n+1)f(n-1) = f(n) \cdot f(n) - f(n+1)f(n-1)$

$$= f(n)\{f(n-1) + f(n-2)\} - f(n+1)f(n-1)$$
$$= [f(n) - f(n+1)]f(n-1) + f(n) \cdot f(n-2).$$
$$= -f(n-1)f(n-1) + f(n)f(n-2) \quad [\because f(n-2) + f(n-1) = f(n)]$$
$$= -f^2(n-1) + f(n)f(n-2)$$
$$= (-1)[f^2(n-1) - f(n)f(n-2)]. \tag{1}$$

It follows immediately from (1) for $n = n - 1$,

$$f^2(n-1) - f(n)f(n-2) = (-1)[f^2(n-2) - f(n-1)(n-3)]. \qquad (2)$$

Thus, from (1) and (2) it follows that

$$f^2(n) - f(n+1)\cdot f(n-1) = (-1)^2[f^2(n-2) - f(n-1)f(n-3)].$$

Continuing the above process, after $(n-2)$th steps we get

$$f^2(n) - f(n-1)f(n-1) = (-1)^{n-2}[f^2(2) - f(3)f(1)] = (-1)^{n-2}[1^2 - 2\cdot 1]$$
$$= (-1)^{n-2}(-1) = (-1)^{n-1}.$$

Hence, $f^2(n) = (-1)^{n-1} + f(n+1)f(n-1)$.

Example: For $n = 6$.

$$f^2(6) = 8^2 = 64 = 65 - 1 = 13\cdot 5 - 1 = f(7)\cdot f(5) + (-1)^{6-1}.$$

For $n = 9$, $f^2(9) = 34^2 = 1156 = 55\cdot 21 + 1 = f(10)\cdot f(8) + (-1)^{9-1}$.

Proof: (ii) $f^2(1) + f^2(2) = 1^2 + 1^2 = 2 = f(2)\cdot f(3)$, that is, true for $n = 2$.

Let it hold for $n = k$.

Then $\quad f^2(1) + f^2(2) + \cdots + f^2(k) = f(k)\cdot f(k+1)$

Therefore, $f^2(1) + f^2(2) + \cdots + f^2(k) + f^2(k+1) = f(k)f(k+1) + f^2(k+1)$

$$= f(k+1)[f(k) + f(k+1)] = f(k+1)\cdot f(k+2).$$

So, the result is true for $n = k + 1$. Hence, it is true $\forall n \geq 2$.

That is, $\quad f^2(1) + f^2(2) + \cdots + f^2(n) = f(n)\cdot f(n+1)$.

Example: $f^2(1) + f^2(2) + f^2(3) + f^2(4) + f^2(5) + f^2(6)$

$$= 1^2 + 1^2 + 2^2 + 3^2 + 5^2 + 8^2$$
$$= 1 + 1 + 4 + 9 + 25 + 64 = 104 = 8\cdot 13 = f(6)\cdot f(7) \qquad [\because f(7) = 13]$$

Proof: (iii) Sum of 1st two even terms,

$$= f(2) + f(4) = 1 + 3 = 4 = 5 - 1 = f(5) - 1 = f(2\cdot 2 + 1) - 1.$$

Sum of 1st three even terms $= f(2) + f(4) + f(6)$

$$= 1 + 3 + 8 = 12 = 13 - 1 = f(7) - 1 = f(2\cdot 3 + 1) - 1.$$

So the result is true for $n = 2$ and 3.

Let the result be true for the sum of first k even terms.

Then, $f(2) + f(4) + \cdots + f(2k) = f(2k+1) - 1$.

Therefore, the sum of 1st $(k+1)$ even terms

$$= f(2) + f(4) + \cdots + f(2k) + f(2k+2)$$
$$= f(2k+1) + f(2k+2) - 1 = f(2k+3) - 1 = f(2(k+1)+1) - 1.$$

So, it is true for $n = k + 1$.

Hence, it is true $\forall n$ i.e., $f(2) + f(4) + \cdots + f(2n) = f(2n+1) - 1$

Proof: (iv) Follow (iii).

Proof: (v) For the first two terms

L.H.S $= f(1) - f(2) = 1 - 1 = 0$

$\text{R.H.S} = 1 + (-1)^{2+1} f(1) = 1 - 1 = 0$

$\Rightarrow$ L.H.S = R.H.S $\Rightarrow$ identity holds for $n = 2$.

For the 1st 3-terms,

$\text{L.H.S} = f(1) - f(2) + f(3) = 1 - 1 + 2 = 2.$

$\text{R.H.S} = 1 + (-1)^{3+1} f(2) = 1 + 1 = 2.$

L.H.S = R.H.S $\Rightarrow$ identity holds for $n = 3$.

So the identity holds for 2 & 3.

Let the identity hold for $n = k$.

Then $f(1) - f(2) + f(3) - f(4) + \cdots + (-1)^{k+1} f(k) = 1 + (-1)^{k+1} f(k-1)$

Therefore, $f(1) - f(2) + \cdots + (-1)^{k+1} f(k) + (-1)^{k+2} f(k+1)$

$$= 1 + (-1)^{k+1} f(k-1) + (-1)^{k+2} f(k+1)$$
$$= 1 + (-1)^{k+1} [f(k-1) - f(k+1)]$$
$$= 1 + (-1)^{k+1} [-f(k)] \quad [\because\ f(k-1) + f(k) = f(k+1)]$$
$$= 1 + (-1)^{k+2} [f(k)$$
$$= 1 + (-1)^{(k+1)+1} \cdot f(\overline{k+1} - 1)$$

$\Rightarrow$ identity is true for $n = k + 1$.

Hence, it is true $\forall\ n$.

Theorem (10.9): *Every (+)ve integer can be written as a sum of distinct Fibonacci numbers.*

Proof: Fibonacci sequence is given by $\{1, 1, 2, 3, 5, 8, 13, 21, 34, 55, 89, \ldots\}$

i.e., $\{f(1), f(2), f(3), f(4), \ldots, f(11), \ldots\}$

Then $1 = f(1) = f(2)$

$2 = f(3) = f(1) + f(2)$ [Not distinct as $f(2)$ and $f(1)$ have the same value]

$3 = f(4) = f(3) + f(2)$

$4 = f(4) + f(1) = f(3) + f(2) + f(1)$

$5 = f(5) = f(4) + f(3) = f(4) + f(2) + f(1)$

$6 = 5 + 1 = f(5) + f(1) = f(4) + f(3) + f(1)$

....

Result is true for $n = 3, 4, \ldots$ (each $n > 2$)

It is now sufficient to show by method of induction on $n > 2$ that each of the integers 1, 2, 3, 4, ..., $f(n) - 1$, can be expressed as a sum of members from the set $\{f(1), f(2), f(3), \ldots, f(n-2)\}$.

For $n = 9$ (> 2), each of the integers 1, 2, 3, ..., $f(9) - 1$, i.e., 1, 2, 3, ..., $34 - 1 = 1, 2, 3, \ldots, 33$ can be expressed as a sum of Fibonacci numbers from the set

$$\{f(1), f(2), \ldots f(9-2)\} = \{f(1), f(2), \ldots f(7)\}$$

Let $n = 33$

Then, $33 = 13 + 8 + 5 + 3 + 2 + 1 + 1$

$$= f(7) + f(6) + f(5) + f(4) + f(3) + f(2) + f(1)$$

Let $n = 20$, $f(8) = 1 = 21 - 1 = 20$.

Then, $20 = 8 + 5 + 3 + 2 + 1 + 1$

$= f(6) + f(5) + f(4) + f(3) + f(2) + f(1)$

[From the set $\{f(1), f(2), \ldots f(8-2)\}$ i.e. $\{f(1), \ldots f(6)\}$]

Let the result be true for $n = k(> 2)$.

We choose N such that $f(k) - 1 < N < f(k+1))$

Then, $N - f(k-1) < f(k+1) - f(k-1) = f(k)$

$\Rightarrow$ $N - f(k-1)$ is expressible as a sum of distinct Fibonacci numbers from the set $\{f(1), f(2) \ldots f(k-2)\}$

$\Rightarrow$ N is expressible as a sum of distinct Fibonacci numbers from the set $\{f(1), f(2), \ldots f(k-2), f(k-1)\}$.

$\Rightarrow$ the result holds for $n = k + 1$. Hence the theorem.

Note: The smallest Fibonacci number in this representation is either $f(2)$ or $f(3)$ if the possibility of $f(1)$ is ignored, as $f(2)$ has also the value 1.

This representation is known as *Zeckendorf* representation.

Theorem (10.10): *Every (+)ve integer n can be expressed as a sum of distinct Fibonacci numbers, no two of which are consecutive.*

That is, $n = f(r_1) + f(r_2) + \cdots + f(r_k)$.

Hence, r_i *are distinct and* r_i, r_j *are not consecutive.*

Proof: Let n be a (+)ve integer such that $f(r) < n < f(n+1)$. (1)

We have established the identities in (iii) and (ii) of Th. [10.8]

that $f(2) + f(4) + \cdots + f(2k) = f(2k+1) - 1 = f(r) - 1$ (2)

and $f(1) + f(3) + f(5) + \cdots + f(2k-1) = f(2k) - 1 = f(r) - 1.$ (3)

where r is odd and even respectively.

We have also shown in Zeckendorf representation of any (+)ve integer n as a sum of distinct Fibonacci numbers.

From (1) it follows that $f(r) - 1 < f(r) < n < f(r+1)$.

Now n is expressible as a sum of distinct non-consecutive Fibonacci numbers if $f(r)$ must appear in the sum. For if the representation does not contain $f(r)$, then the sum of admissible Fibonacci numbers would not attain n as the resulting sum is always less than n, follows from (2) and (3). The same is true for Zeckendorf representation of an integer n as discussed in above Theorem.

Let $n = 136$. Then $f(11) < 136 < f(12)$.

Here $136 = 13 + 34 + 89 = f(7) + f(9) + f(11)$

which contains $f(11) = f(r)$. Here $r = 11$.

But if we consider the sum $f(7) + f(9) + f(10)$ which does not contain $f(11)$, then their sum $= 13 + 34 + 55 = 102 < 136$. Hence the sum of expression does not attain 136.

Theorem (10.11): *If* $f(n)$ *be the nth term of the Fibonacci sequence defined by* $f(1) = 1$, $f(2) = 1$ *and* $f(n) = f(n-2) + f(n-1)$.

Let $\alpha = \dfrac{1+\sqrt{5}}{2}$. *Then* $f(n) > \alpha^{n-2} \forall n > 2$.

Proof: $\alpha = \dfrac{1+\sqrt{5}}{2}$

$$\Rightarrow \quad \alpha^2 = \frac{6+2\sqrt{5}}{4} = \frac{3}{2} + \frac{\sqrt{5}}{2} = \left(\frac{1+\sqrt{5}}{2} + 1\right) = \alpha + 1 \qquad (1)$$

Then, $\quad \alpha^{k-1} = \alpha^{k-3} \cdot \alpha^2 = \alpha^{k-3}(\alpha+1) = \alpha^{k-2} + \alpha^{k-3}$ (2)

We have $f(3) = 2$ and $\alpha = \dfrac{1+\sqrt{5}}{2} < 2 \Rightarrow f(3) > \alpha$.

Again $\quad f(4) = f(3) + f(2)$

$\Rightarrow \quad f(4) > \alpha + 1 = \alpha^2$ [by(1)]

Thus the result is true for $n = 3$ and 4.

Let the result be true for $n = k\ (> 2)$.

Then, $\quad f(k) > \alpha^{k-2}$.

Then, $f(k+1) = f(k) + f(k-1) > \alpha^{k-2} + \alpha^{k-3} = \alpha^{k-1}$ by (2).

Therefore, it holds for $n = k+1$.

Hence, it is true $\forall n > 2$.

That is, $f(n) > \alpha^{n-2}$.

Theorem (10.12): [*Jacques-Phillipe-Marrie-Binet formula*]. *The nth term of the Fibonacci sequence can be expressed as*

$$f(n) = \frac{1}{\sqrt{5}}\left[\left(\frac{1+\sqrt{5}}{2}\right)^n - \left(\frac{1-\sqrt{5}}{2}\right)^n\right]$$

for any (+) ve integers n. Hence deduce that $f^2(n+2) - f^2(n) = f(n+2)$.

Proof: Let α, β be the roots of $x^2 - x - 1 = 0$.

Then, we can choose $\alpha = \dfrac{1+\sqrt{5}}{2}$ and $\beta = \dfrac{1-\sqrt{5}}{2}$.

Subsequently we have, $\alpha^2 = \alpha + 1$, $\beta^2 = \beta + 1$.

Then, $\quad \alpha^{n+2} = \alpha^{n+1} + \alpha^n$ and $\beta^{n+2} = \beta^{n+1} + \beta^n$ (i)

Therefore, $\alpha^{n+2} - \beta^{n+2} = (\alpha^{n+1} + \alpha^n) - (\beta^{n+1} + \beta^n)$

Dividing both sides by $(\alpha - \beta)$ $(\alpha > \beta)$

$$\frac{\alpha^{n+2} - \beta^{n+2}}{\alpha - \beta} = \frac{(\alpha^{n+1} - \beta^{n+1})}{\alpha - \beta} + \frac{\alpha^n - \beta^n}{\alpha - \beta} \qquad \text{(ii)}$$

Let $\quad H(n) = \dfrac{\alpha^n - \beta^n}{\alpha - \beta}$.

Then (ii) $\Rightarrow$ 7 $H(n+2) = H(n+1) + H(n)\ \forall n \geq 1$ (iii)

Again, $\quad \alpha = \dfrac{1+\sqrt{5}}{2}$, $\beta = \dfrac{1-\sqrt{5}}{2}$

$\Rightarrow \quad \alpha + \beta = 1$, $\alpha - \beta = \sqrt{5}$, $\alpha\beta = -1$

Hence, $\quad H(1) = 1$

$$H(2) = \frac{\alpha^2 - \beta^2}{\alpha - \beta} = \alpha + \beta = 1$$

$$H(3) = \frac{\alpha^3 - \beta^3}{\alpha - \beta} = \alpha^2 + \alpha\beta + \beta^2 = (\alpha + \beta)^2 - \alpha\beta = 1 - (-1) = 2$$

Consequently $H(1),\ H(2),\ H(3)\cdots$ become the members of the Fibonacci sequence, which follows that $H(n) = \frac{\alpha^n - \beta^n}{\alpha - \beta}$ is the nth term of the Fibonacci sequence.

That is, $f(n) = \dfrac{\alpha^n - \beta^n}{\alpha - \beta}$.

i.e. of the form $f(n) = \dfrac{1}{\sqrt{5}}\left[\left(\dfrac{1+\sqrt{5}}{2}\right)^n - \left(\dfrac{1-\sqrt{5}}{2}\right)^n\right]$

Now $\alpha\beta = -1 \Rightarrow (\alpha\beta)^{2k} = 1\ \forall k \geq 1$ (iv)

$$f(n) = \frac{\alpha^n - \beta^n}{\alpha - \beta},\ f(n+2) = \frac{\alpha^{n+2} - \beta^{n+2}}{\alpha - \beta}$$

$$\Rightarrow \quad f^2(n+2) - f^2(n) = \frac{\alpha^{2(m+2)} + \beta^{2(n+2)} - 2(\alpha\beta)^{n+2} - \alpha^{2n} - \beta^{2n} + 2(\alpha\beta)^n}{(\alpha - \beta)^2}$$

$$= \frac{\alpha^{2(n+2)} + \beta^{2(n+2)} - 2(\alpha\beta)^n - \alpha^{2n} - \beta^{2n} + 2(\alpha\beta)^n}{(\alpha - \beta)^2}$$

[by (iv)]

$$= \frac{\alpha^{2(n+2)} + \beta^{2(n+2)} - \alpha^{2n} - \beta^{2n}}{(\alpha - \beta)^2}$$

$$= \frac{\{\alpha^{2(n+2)} - (\alpha\beta)^2 \cdot \alpha^{2n} + \beta^{2(n+2)} - (\alpha\beta)^2\beta^{2n}\}}{(\alpha - \beta)^2}$$

[$\because$ by (iv) $(\alpha\beta)^2 = 1$]

$$= \frac{\{\alpha^{2n+2}(\alpha^2 - \beta^2) + \beta^{2n+2}(\beta^2 - \alpha^2)\}}{(\alpha - \beta)^2}$$

$$= \frac{\{(\alpha^{2n+2} - \beta^{2n+2})(\alpha^2 - \beta^2)\}}{(\alpha - \beta)^2}$$

$$= \frac{(\alpha + \beta)}{(\alpha - \beta)}(\alpha^{2n+2} - \beta^{2n+2}) = \frac{\alpha^{2n+2} - \beta^{2n+2}}{\alpha - \beta}.$$

[$\because \alpha + \beta = 1$]

$$= f(n+2).$$

Hence $f^2(n+2) - f^2(n) = f(n+2).$

Exercise 10

1. Evaluate $gcd(f(9), f(12))$, $gcd(f(15), f(20))$ and $gcd(f(24), f(36))$.
2. Use the fact $f(m)|f(n)$ if and only if $m|n$ to verify
   (i) $2|f(n)$ iff $3|n$ (ii) $5|f(n)$ iff $5|n$
3. For $n \geq 1$, prove that $2^{n-1}f(n) \equiv n(\text{mod}5)$
4. Prove that there is no (+)ve integer n for which $f(1)+f(2)+f(3)+\cdots+f(3n) = 16!$.

5. Prove that, for $n = 1, 2, \ldots, 10$, $5f^2(n) + 4(-1)^n$ is always a perfect square.

6. Prove that, if $p = x + 3$ is prime, then p can not divide a Fibonacci number with odd index. That is, $p \nmid f(2n-1) \ \forall \ n \geq 1$.

7. Show that, for $n \geq 1$, the Fermat number $F_n = 2^{2^n} + 1$ is prime if and only if $3^{(F_n-1)/2} \equiv -1 \pmod{F_n}$.

8. Show that the Fermat number F_5 is divisible by 64!.

9. For any even perfect number $n = 2^{k-1}(2^k - 1)$, show that $2^k | 6(n^2) + 1$.

10. For an even perfect number show that either $n \equiv 6 \pmod{10}$ or $n \equiv 8 \pmod{10}$, i.e. n ends in the digit 6 or 8.

11. Prove that the integer $n = 2^{10}\ (2^{11} - 1)$ is not a perfect number by showing $6(n) \neq 2n$.

12. If n is a perfect number, prove that $\sum_{d|n} \frac{1}{d} = 2$.

Chapter 11

Solved Examples

Problem (11.1): Prove by the method of induction that $2^{2n} - 1$ is divisible by 3 when n is a (+)ve integer.

Sol: Let $f(n) = 2^{2n} - 1$.

Then $f(1) = 2^2 - 1 = 3$. So $3|f(1)$.

Result holds for $n = 1$.

Let $3|f(n)$. Then $f(n+1) = 2^{2(n+1)} - 1$.

$\therefore$ $f(n+1) - f(n) = 2^{2n+2} - 2^{2n} = 2^{2n}(2^2 - 1) = 3 \cdot 2^{2n}$

$\Rightarrow$ $3|f(n+1)$.

Hence, $3|f(n)\ \forall$ (+)ve integer n.

Problem (11.2): If $(m,n) = 1$, $(k,n) = 1$, then show that $(mk,n) = 1$, where m, n, k are (+)ve integers.

Sol: $(m,n) = 1 \Rightarrow \exists r, s \in z$ such that $mr + ns = 1$.

$(k,n) = 1, \Rightarrow \exists x, y \in z$ such that $kx + ny = 1$.

Now $(mr + ns)(kx + ny) = 1 \Rightarrow mkxr + mnry + nksx + n^2sy = 1$

$\Rightarrow$ $mk(xr) + n(mry + ksx + nsy) = 1 \Rightarrow mkr' + ns' = 1$,

where $r' = xr$, $s' = mry + ksx + nsy \in Z$,

$\therefore$ $(mk,n) = 1$.

Problem (11.3): Solve the following system of congruences: $x \equiv 2 \pmod 3$, $x \equiv 3 \pmod 5$, $x \equiv 1 \pmod 7$.

Sol: (We apply Chinese remainder theorem).

3,5,7 are relatively prime.

So the system is solvable.

$$M = m_1m_2m_3 = 3 \cdot 5 \cdot 7 = 105$$

Let $\frac{M}{m_1}y_1 \equiv 1 \pmod 3 \Rightarrow \frac{105}{3}y_1 \equiv 1 \pmod 3$

$\Rightarrow$ $35y_1 \equiv (1 + 69) \pmod 3 \equiv 70 \pmod 3$

$\Rightarrow$ $y_1 \equiv 2 \pmod 3$. (i)

$\frac{M}{m_2}y_2 \equiv 1 \pmod{m_2} \Rightarrow \frac{105}{5}y_2 \equiv 1 \pmod 5 \Rightarrow 21y_2 \equiv 1 \pmod 5$

$\Rightarrow$ $y_2 \equiv 1 \pmod 5$. (ii)

$\frac{M}{m_3}y_3 \equiv 1 \pmod{m_3} \Rightarrow \frac{105}{7}y_3 \equiv 1 \pmod 7$

$\Rightarrow$ $15y_3 \equiv 1 \pmod 7 \Rightarrow y_3 \equiv 1 \pmod 7$. (iii)

Therefore, the solution of the system is

$$x \equiv \frac{M}{m_1}y_1b_1 + \frac{M}{m_2}y_2b_2 + \frac{M}{m_3}y_3b_3 \pmod M$$
$$\equiv 35 \cdot 2 \cdot 2 + 21 \cdot 1 \cdot 3 + 15 \cdot 1 \cdot 1 \pmod{105}$$
$$\equiv 218 \pmod{105} \equiv 8 \pmod{105}.$$

Problem (11.4): Find the integer $k > 5$ satisfying $-11 \equiv 19 \pmod k$.

Sol: $-11 \equiv 19 \pmod k$

$\Rightarrow \quad -11 = 19 + kt, \ t \in Z$

$\Rightarrow \quad kt = -30.$

Thus, the integer $k(> 5)$ is given by $\frac{-30}{t}$, where t is a negative integer such that $t| - 30$.

Here $k = 6, 10, 15, 30$.

Problem (11.5): Express (195, 285) in the form $285m + 195n$, $m, n \in z$.

Sol: By divisional algorithm,

$$285 = 195 \cdot 1 + 90 \qquad \text{(i)}$$
$$195 = 90 \cdot 2 + 15 \qquad \text{(ii)}$$
$$90 = 15 \cdot 6 + 0. \qquad \text{(iii)}$$

Therefore, $(195, 285) = 15$

$$15 = 195 - 90 \cdot 2 \quad \text{[by (ii)]}$$
$$= 195 - 2 \cdot 90 = 195 - 2[285 - 195 \cdot 1] \quad \text{[by (i)]}$$
$$= 195 - 2 \cdot 285 + 2 \cdot 195 = -2 \cdot 285 + (1 + 2) \cdot 195$$
$$= -2 \cdot 285 + 3 \cdot 195 = m \cdot 285 + n \cdot 195,$$

where $\quad m = -2, \ n = 3 \in Z.$

Problem (11.6): If $(a, b) = 1$ and $a|c$, $b|c$, then show that $ab|c$. But the result does not hold if $(a, b) = 6 (\neq 1)$.

Sol: $a|c, \ b|c \ \Rightarrow \ c = at_1$

and $\quad c = bt_2, \quad$ where $\quad (t_1, t_2) = 1$

$\Rightarrow \quad c^2 = abt_1t_2 = abt, \ t \in z$

$\Rightarrow \quad ab|c^2 \ \Rightarrow \ ab|c.$

Next let $a = 6, \ b = 12$. Then $(a, b) = 6$. Let $c = 24$ so that $a|24, b|24$.

$ab = 72$. So $72 \not| \ 24$. That is, $ab \not| \ c$.

Problem (11.7): Give an example that $(a, b) = 1$, $(b, c) = 1$ does not imply $(a, c) = 1$.

Sol: Let $(7, 5) = 1$ and $(5, 21) = 1$.

But $\quad (7, 21) = 7 \neq 1.$

Problem (11.8): Show that $f(n) = 3^{2n-1} + 2^{n+1}$ is divisible by 7, when n is a (+)ve integer.

Sol: $f(n) = 3^{2n-1} + 2^{n+1}$ (1)

$f(1) = 3 + 2^2 = 7,$

$\therefore \quad 7 | f(1).$

So the result is true for $n = 1$.

Let it be true for n, i.e., $7|f(n)$.

Now, $f(n+1) = 3^{2n+1} + 2^{n+2}$. (2)

Then, $f(n+1) - 2f(n) = 3^{2n+1} - 2 \cdot 3^{2n-1} = 3^{2n-1}(3^2 - 2) = 7 \cdot 3^{2n-1}$

$\Rightarrow \quad 7|f(n+1)$. Hence the result.

Problem (11.9): Solve the set of congruences using Chinese remainder theorem:

(i) $x \equiv 2 \pmod 3$, $x \equiv 4 \pmod 7$, $x \equiv 6 \pmod{10}$

(ii) $x \equiv 1 \pmod 4$, $x \equiv 3 \pmod 7$, $x \equiv 5 \pmod{11}$.

Sol: (i) $M = m_1 m_2 m_3 = 3 \cdot 7 \cdot 10 = 210.$

Now $\frac{M}{m_1} y_1 \equiv 1 \pmod 3 \Rightarrow \frac{210}{3} y_1 \equiv 1 \pmod 3$

$\Rightarrow \quad 70 y_1 \equiv 1 \pmod 3 \equiv 70 \pmod 3$

$\Rightarrow \quad y_1 \equiv 1 \pmod 3.$ (i)

$\frac{M}{m_2} y_2 \equiv 1 \pmod{m_2} \Rightarrow \frac{210}{7} y_2 \equiv 1 \pmod 7$

$\Rightarrow \quad 30 y_2 \equiv 1 \pmod 7 \equiv 15 \pmod 7$

$\Rightarrow \quad 2 y_2 \equiv 1 \pmod 7 \quad [(15, 7) = 1]$

$\Rightarrow \quad 2 y_2 \equiv 8 \pmod 7$

$\Rightarrow \quad y_2 \equiv 4 \pmod 7 \quad [(2, 7) = 1].$ (ii)

$\frac{M}{m_3} y_3 \equiv 1 \pmod{10}$

$\Rightarrow \quad \frac{210}{10} y_3 \equiv 1 \pmod{10} \Rightarrow 21 y_3 \equiv 21 \pmod{10}$

$\Rightarrow \quad y_3 \equiv 1 \pmod{10}.$ (iii)

$\because$ solution is $x \equiv \frac{M}{m_1} y_1 b_1 + \frac{M}{m_2} y_2 b_2 + \frac{M}{m_3} y_3 b_3 \pmod M$

$\equiv 70 \cdot 1 \cdot 2 + 30 \cdot 4 \cdot 4 + 21 \cdot 1 \cdot 6 \pmod{210}$

$\equiv 140 + 480 + 126 \pmod{210}.$

$\therefore \quad x \equiv 746 \pmod{210} \equiv 116 \pmod{210}.$

(ii) Proceed as in Ex (i).

Problem (11.10): If $(a, b) = 1$, then show that $(a+b, a-b) = 1$ or 2.

Sol: Let $(a+b, a-b) = d$. then $d | a+b$ and $d | a-b$,

which shows that $d|2a,\ d|2b$.

$\Rightarrow \quad d \leq (2a, 2b) = 2(a, b) \Rightarrow d = 1$ or 2.

Problem (11.11): If p_n be the nth prime, then prove that $p_n \leq 2^{2^{n-1}}$.

Sol: Let $p_1 = 2,\ p_2 = 3,\ p_3 = 5, \ldots$ be the primes in ascending order and let p_n be the last prime. We now consider a (+)ve integer p such that $p = p_1 p_2 \ldots, p_n + 1$.

Then $p > 1 \Rightarrow p$ has a prime factor. If there are several such prime factors, then p_{n+1} cannot exceed the smallest of such prime factors.

So $\quad p_{n+1} \le p_1 p_2 \ldots p_n + 1$.

We prove the result by the method of induction. $n = 1, 2^{2^{n-1}} = 2^1 = 2 = p_1$.

So the result holds for $n = 1$.

We assume that $n > 1$ and let it be true for n. That is, true for all integers up to n.

Then $\quad p_{n+1} \le p_1 p_2 \ldots p_n + 1 \le 2 \cdot 2^2 \ldots 2^{2^{n-1}} + 1$

$$= 2^{1+2+\cdots+2^{n-1}} + 1 = 2^{(2^n-1)} + 1$$

$$\le 2^{2^n-1} + 2^{2^n-1} \quad [\because\ 1 \le 2^{2^n-1}, \forall\, n]$$

$\therefore \quad p_{n+1} \le 2 \cdot 2^{2^n-1} = 2^{2^n}$.

Hence the result.

Problem (11.12): Prove that if n is odd, $n^2 - 1$ is divisible by 8.

Sol: Any odd integer is of the form $4q \pm 1$.

Then $\quad n = 4q \pm 1 \Rightarrow n^2 = 16q^2 \pm 8q + 1$

$\Rightarrow \quad n^2 - 1 = 8q(2q \pm 1) \Rightarrow 8 | n^2 - 1$.

Problem (11.13): Find a (+)ve integral solution of $5x + 7y = 100$.

Sol: $5x + 7y = 100$

$\Rightarrow \quad 5x = 100 - 7y \Rightarrow 5x \equiv 100 \pmod 7$

$\Rightarrow \quad x \equiv 20 \pmod 7, \quad [(5,7) = 1]$

$\Rightarrow \quad x \equiv 6 \pmod 7 \Rightarrow x = 6 + 7t,\ t = 0, 1, 2, 3, \ldots$

So for $t = 0$, $x = 6$ is a (+)ve integral solution.

For $x = 6$, $y = 10$. So $(6, 10)$ is a (+)ve integral solution.

Problem (11.14): Solve, if possible, (i) $14x \equiv 21 \pmod{12}$ (ii) $9x \equiv 3 \pmod 6$.

Sol: (i) $14x \equiv 21 \pmod{12}$

Here $\quad (14, 12) = 2, \quad$ but $\quad 2 \not| \, 21$.

So the congruence is not solvable.

(ii) $9x \equiv 3 \pmod 6$

$\Rightarrow \quad 3x \equiv 1 \pmod 2 \Rightarrow 3x \equiv 3 \pmod 2$

$\Rightarrow \quad x \equiv 1 \pmod 2 \quad [(3, 2) = 1]$

$\Rightarrow \quad x = 1 + 2t,\ t \in Z$.

Problem (11.15): Prove that if x and y are odd, then $x^2 + y^2$ is even but not divisible by 4.

Sol: Let $x = 2m + 1,\ y = 2n + 1$.

Then $\quad x^2 + y^2 = 4m^2 + 4n^2 + 4m + 4n + 2$

$$= 4(m^2 + n^2) + 4(m + n) + 2,$$

which is even, but since $4 \not| \, 2$, $4 \not| \, (x^2 + y^2)$.

Problem (11.16): Let $a = 2^4 \cdot 3^3 \cdot 7 \cdot 11 \cdot 13$

$$b = 2^3 \cdot 3^2 \cdot 5^2 \cdot 11 \cdot 17.$$

Then find $d = (a, b)$.

Sol: $d = 2^{x_1} \cdot 3^{x_2} \cdot 5^{x_3} \cdot 7^{x_4} \cdot 11^{x_5} \cdot 13^{x_6} \cdot 17^{x_7}$,

where $x_1 = \min(3,4),\ x_2 = \min(3,2),\ x_3 = \min(0,2)$

$x_4 = \min(0,1),\ x_5 = \min(1,1),\ x_6 = \min(1,0),\ x_7 = \min(1,0)$.

Hence, $d = 2^3 3^2 5^0 7^0 11^1 13^0 17^0 = 2^3 \cdot 3^2 \cdot 11 = 792$.

Problem (11.17): Prove that $n^2 - n$ is divisible by 2 and $n^3 - n$ is divisible by 6.

Sol: (i) Let $f(n) = n^2 - n$.

$f(1) = 0 \Rightarrow 2|f(1)$.

Let $2|f(n)$. (1)

Then $f(n+1) - f(n) = 2n \Rightarrow 2|f(n+1)$ [by (1)].

Hence, $2|f(n)$.

Alternative: $f(n) = n(n-1),\ 2|f(1)$

Since for $n > 1$, one of n and $(n-1)$ is even.

Hence, $2|f(n)$.

(ii) Let $f(n) = n^3 - n$.

$f(1) = 0 \Rightarrow 6|f(1)$.

Let $6|f(n)$. (1)

Now, $f(n+1) = (n+1)^3 - (n+1) = n^3 + 3n^2 + 3n + 1 - n - 1 = n^3 + 3n^2 + 2n$.

Then, $f(n+1) - f(n) = 3n^2 + 3n = 3n(n+1)$.

Also $2|n(n+1)$.

$f(n+1) - f(n) = 6k,\ k \in z$.

$\Rightarrow$ $6|f(n+1)$ [by (1)]

Hence, $6|f(n)$.

Alternative: $f(n) = n^3 - n = n(n^2 - 1) = n(n-1)(n+1) = (n-1)n(n+1)$.

$n-1, n, n+1$ are three consecutive integers, and hence, $(n-1)n(n+1)$ is divisible by 3!.

So $6|f(n)$.

Problem (11.18): Evaluate (ab, p^4) and $(a+b, p^4)$, when $(a, p^2) = p$ and $(b, p^3) = p^2$, p is prime.

Sol: $(a, p^2) = p$ and $(b, p^3) = p^2$

$\Rightarrow$ $p|a$ and $p^2|b$. Also $p^2|b \Rightarrow p|b$.

Thus $p|a,\ p^2|b \Rightarrow a = pt, b = p^2 t'; t,\ t' \in Z$.

Then $ab = p^3 \cdot tt' \Rightarrow p^3|ab$.

Also $p^3|p^4$.

Therefore, $p^3|ab,\ p^3|p^4$. (i)

Again $p|a,\ p^2|b \Rightarrow p|a$ and $p|b$

$$\left.\begin{array}{ll}\Rightarrow & p|ab,\\ \text{Also} & p|p^4.\end{array}\right\} \quad \text{(ii)}$$

(i) and (ii) shows that $(ab, p^4) = p^3$. [since $p|p^3$].

Again, $p|a, p|b \Rightarrow p|a+b$.

Also $p|p^4$.

Hence, $(a+b, p^4) = p$ [$\because$ highest power of p is 1].

[**Go through the example:**

$a = 6,\ p = 3,\ b = 18$

such that $(a, p^2) = (6, 9) = 3 = (p)$ and $(b, p^3) = (18, 81) = 3^2 = (p^2)$.

Then $(ab, p^4) = (108, 81) = 27 = 3^3 = (p^3)$.

Also $(a+b, p^4) = (24, 81) = 3 = (p)$.

Problem (11.19): Is it true if $(a, b) = (a, c)$, then $(a^2, b^2) = (a^2, c^2)$.

Sol: Let $(a, b) = (a, c) = d$.

Then $d|a, d|b \Rightarrow d|a^2,\ d|b^2$.

Also $d|a, d|c \Rightarrow d|a^2,\ d|c^2$.

Thus, $\exists\, d \in Z$ such that $d|a^2,\ d|b^2$.

and $d|d^2$.

$\therefore$ $(a^2, b^2) = d^2$.

Similarly, we can show that $(a^2, c^2) = d^2$. Hence, $(a^2, b^2) = (a^2, c^2)$.

[**Go through the example**]

Let $a = 6,\ b = 3,\ c = 9$

$$\text{such that } \left.\begin{array}{l}(a, b) = (6, 3) = 3\\ (a, c) = (3, 9) = 3\end{array}\right\}$$

Now $(a^2, b^2) = (36, 9) = 9$

$(a^2, c^2) = (36, 81) = 9$

$\therefore$ $(a^2, b^2) = (a^2, c^2) = 3^2$.

Problem (11.20): If $(a, b) = 1$ and $a|c,\ b|c$, then $ab|c$.

Sol: $(a, b) = 1$

$\Rightarrow$ $\exists x, y, \in Z$ such that $1 = ax + by$

$\Rightarrow$ $c = acx + bcy$. (1)

$$\text{Also } \left.\begin{array}{l}a|c \Rightarrow c = at\\ b|c \Rightarrow c = bt'\end{array}\right\} \text{ for same } t, t' \in Z.$$

Then from (1), $c = abxt' + bayt = ab(xt' + yt) \Rightarrow ab|c$.

Problem (11.21): What is the remainder when $1! + 2! + 3! + \cdots + 100!$ is divided by 15.

Sol: $15 | r!\ \forall\ r,\ 5 \le r \le 100$

$\Rightarrow\quad r! \equiv 0 \pmod{15},\ 5 \le r \le 100$ (1)

$\Rightarrow\quad 1! + 2! + 3! + 4! + r! \equiv 1! + 2! + 3! + 4! \pmod{15}$

$\Rightarrow\quad 1! + 2! + \cdots + 100! \equiv 1 + 2 + 6 + 24 \pmod{15},\ 5 \le r \le 100$ by (1)

$\equiv 33 \pmod{15} \equiv 3 \pmod{15}.$

So the remainder (mod 15) = 3.

Problem (11.22): Show that $n^2 < 2^n\ \forall\ n > 5$.

Sol: $n = 6,\ n^2 = 6^2 = 36,\ 2^n = 2^6 = 64.$

$\therefore\quad 6^2 < 2^6$, result holds for 6.

Let the result be true for n. $(n > 6)$

That is, $n^2 < 2^n$.

Now $(n+1)^2 = n^2 + 2n + 1 < 2^n + 2n + 1$

$$= 2^{n+1}\left[\frac{1}{2} + \frac{n}{2^n} + \frac{1}{2^{n+1}}\right]$$

$$< 2^{n+1}\left[\frac{1}{2} + \frac{1}{n} + \frac{1}{2 \cdot 2^n}\right] \quad \left[\because\ n^2 < 2^n\ \Rightarrow\ \frac{n}{2^n} < \frac{1}{n}\right]$$

$$< 2^{n+1}\left[\frac{1}{2} + \frac{1}{n} + \frac{1}{2n^2}\right] < 2^{n+1},\ \text{since}\ \frac{1}{2} < \frac{1}{2} + \frac{1}{n} + \frac{1}{2n^2} < 1\ \forall\ n > 6.$$

Therefore, it also holds for $n = n + 1$. So it is true $\forall\ n$.

Hence the result.

Problem (11.23): Let $m = (a_k a_{k-1} \ldots a_1 a_0)_{10}$ be a decimal number with $a_k \ne 0$ and let $s = a_k + a_{k-1} + \cdots + a_1 + a_0$, then prove that $9|m$, if and only if $9|s$.

Hence, examine whether the integer 85644 is divisible by 9.

Sol: $m = (a_k a_{k-1} \ldots a_1 a_0)_{10} = a_k 10^k + a_{k-1} 10^{k-1} + \cdots + a_1 \cdot 10 + a_0.$

$S = a_k + a_{k-1} + \cdots + a_1 + a_0$

Let $f(x) = a_k x^k + a_{k-1} x^{k-1} + \cdots + a_1 x + a_0.$

Then $f(10) = m, f(1) = S.$ Now $10 \equiv 1 \pmod 9$

$\Rightarrow\quad f(10) \equiv f(1) \pmod 9 \Rightarrow m \equiv S \pmod 9$

$\Rightarrow\quad m - S \equiv 0 \pmod 9 \Rightarrow 9|m$, iff $9|S$.

Now $m = 85644 = 8 \cdot 10^4 + 5 \cdot 10^3 + 6 \cdot 10^2 + 4 \cdot 10 + 4$

$S = 8 + 5 + 6 + 4 + 4 = 27.$

Now $9|27 \Rightarrow 9|m$. Hence, 85644 is divisible by 9.

Problem (11.24): For any two integers a, b of which $b > 0$, show that $\exists$ two integers q and r such that $a = bq + r$.

Sol: Let b be any (+)ve integer. We consider the sequence

$$\ldots, -3b, -2b, -b, 0, b, 2b, 3b, \ldots \tag{11.1}$$

extended to infinity to both sides about 0.

Let a be any integer, such that $a > 0$, or; $a < 0$ or, $a = 0$. It may happen that it lies between two consecutive terms of (1) or may be equal to any member of (1).

So in both cases, we have $bq \le a < b(q+1)$, when $a > 0$.

$\Rightarrow \quad 0 \le a - bq < b.$

Let $a - bq = r$. Then $a = bq + r,\ 0 \le r < b$.

Now q, r exist uniquely. For if, $a = bq + r,\ 0 \le r < b$

and $\quad a = bq' + r',\ 0 \le r' < b.$

$\Rightarrow \quad r - r' = b(q' - q)$

$\Rightarrow \quad b|r - r', \quad \text{but} \quad b > r - r';\ r, r' > 0$

and so $b|r - r'$ is not possible.

So q, r exist uniquely.

The same can be proved if $a < 0$.

For example, if $a = -19,\ b = 3$.

Then $\quad -19 = 3 \cdot (-7) + 2,\ 0 < 2 < 3.$

$\therefore \quad q = -7,\ r = 2$, exist uniquely.

Problem (11.25): Find the least (+)ve residue of 11^{30} (mod 4).

Sol: $11 \equiv -5 \pmod 4$

$\Rightarrow \quad 11 \equiv 3 \pmod 4 \Rightarrow\ 11^2 \equiv 9 \pmod 4 \equiv 1 \pmod 4$

$\Rightarrow \quad (11^2)^{15} \equiv 1 \pmod 4 \Rightarrow\ 11^{30} \equiv 1 \pmod 4.$

1 is the least (+)ve residue mod 4.

Problem (11.26): Determine the remainder when 10^{515} is divided by 7.

Sol: By Fermat's theorem,

$10^6 \equiv 1 \pmod 7$ as $(10, 7) = 1$, 7 is prime.

$\Rightarrow \quad (10^6)^{85} \equiv 1 \pmod 7 \Rightarrow\ 10^{510} \equiv 1 \pmod 7$

$\Rightarrow \quad 10^{510} \cdot 10^2 \equiv 100 \pmod 7 \equiv 2 \pmod 7$

$\Rightarrow \quad 10^{512} \cdot 10^2 \equiv 200 \pmod 7 \equiv 4 \pmod 7$

$\Rightarrow \quad 10^{514} \cdot 10 \equiv 40 \pmod 7 \Rightarrow\ 10^{515} \equiv 5 \pmod 7.$

$\therefore$ remainder is 5 (mod 7).

Problem (11.27): Show that $\dfrac{n^3}{3} + \dfrac{n^5}{5} + \dfrac{7n}{15}$ is an integer for all (+)ve integer n.

Sol: $\dfrac{n^3}{3} + \dfrac{n^5}{5} + \dfrac{7n}{15} = \dfrac{5n^3 + 3n^5 + 7n}{15}.$

Let $\quad f(n) = 3n^5 + 5n^3 + 7n.$

$f(1) = 3 + 5 + 7 = 15.$

$15|f(1) \Rightarrow \dfrac{5n^3 + 3n^5 + 7n}{15}$ is an integer for $n = 1$.

Let the result hold for n, when n is a (+)ve integer.

That is $15|f(n)$. (1)

Now $f(n+1) = 3(n+1)^5 + 5(n+1)^3 + 7(n+1)$

$$= 3(n^5 + 5n^4 + 10n^3 + 10n^2 + 5n + 1) + 5(n^3 + 3n^2 + 3n + 1) + 7n + 7$$

$$= (3n^5 + 5n^3 + 7n) + 15(n^4 + 2n^3 + 3n^2 + 2n) + 15$$

$\Rightarrow$ $15|f(n+1)$ [by (1)]

$\Rightarrow$ $\dfrac{5(n+1)^3 + 3(n+1)^5 + 7(n+1)}{15}$ is an integer.

Hence, $\dfrac{n^3}{3} + \dfrac{n^5}{5} + \dfrac{7n}{15}$ is an integer.

Problem (11.28): Prove that the 5th power of an integer is of the form $11k$ or $11k \pm 1$.

Sol: Let $p = 11$ (prime number).

Let a be any integer such that $(a, m) = 1$.

Then by F. Theorem, $a^{11-1} \equiv 1 \pmod{11}$

$\Rightarrow$ $a^{10} \equiv 1 \pmod{11} \Rightarrow (a^5 - 1)(a^5 + 1) \equiv 0 \pmod{11}$

$\Rightarrow$ either $a^5 - 1 \equiv 0 \pmod{11}$ or, $a^5 + 1 \equiv 0 \pmod{11}$

$\Rightarrow$ $a^5 = 11k + 1$ or, $a^5 = 11k - 1$.

$\therefore$ $a^5 \equiv 11k \pm 1$.

Now if a is divisible by 11, then $a^5 = 11k$.

Hence, $a^5 = 11k$ or, $11k \pm 1$.

Problem (11.29): If p be a prime > 2, then show that $1^p + 2^p + \cdots + (p-1)^p \equiv 0 \pmod{p}$.

Sol: By F. Theorem,

$1^p \equiv 1 \pmod{p}$

$2^p \equiv 2 \pmod{p}$

....

$(p-1)^p \equiv (p-1) \pmod{p}$

$\Rightarrow$ $1^p + 2^p + \cdots + (p-1)^p \equiv \{1 + 2 + \cdots + (p-1)\} \pmod{p} \equiv \dfrac{p(p-1)}{2} \pmod{p}$.

$p - 1$ is even as p is prime.

So $(p-1)$ is divisible by 2

$\Rightarrow$ $1^p + 2^p + \cdots + (p-1)^p \equiv 0 \pmod{p}$.

Problem (11.30): Find the remainder when the following sum is divided by 4, $1^5 + 2^5 + \cdots + 100^5$.

Sol: $1^5 \equiv 1 \pmod{4}$

$2^5 \equiv 0 \pmod{4}$

$3^5 \equiv 3 \pmod{4} \equiv\, = -1 \pmod{4}$

$4^5 \equiv 0 \pmod{4}$

$5^5 \equiv 5 \pmod{4} \equiv 1 \pmod{4}$.

Thus, it follows that $k^5 \equiv 0 \pmod{4}$, when k is even,

and $k^5 \equiv \pm 1 \pmod{4}$, when k is odd and this happens alternatively for $k = 1, 3, 5, \ldots$

That is, $k = 1,\ 1^5 \equiv 1 \pmod 4$

$k = 3,\ 3^5 \equiv -1 \pmod 4$

$k = 5,\ 5^5 \equiv 1 \pmod 4$

$k = 7,\ 7^5 \equiv -1 \pmod 4$ and so on.

$$\therefore \sum_{k=1,2,3,5,7,\ldots} k^5 \equiv 0 \pmod 4.$$

Hence, $1^5 + 2^5 + \cdots + 99^5 + 100^5 = \sum_1^{50}[(2m)^5 + (2m-1)^5] \equiv 0 \pmod 4$.

$\therefore$ reminder is 0 (mod 4).

Problem (11.31): Show that for any integer a, $a^3 \equiv 0, 1$ or $6 \pmod 7$.

Sol: Let $7|a$. Then $7|a^3$ and hence, $a^3 = 7k$.

So $a^3 \equiv 0 \pmod 7$. (i)

Again, by F. Th. $a^6 \equiv 1 \pmod 7$, when $7 \nmid a$.

$\Rightarrow\ (a^3 - 1)(a^3 + 1) \equiv 0 \pmod 7$

$\Rightarrow\ a^3 \equiv 1 \pmod 7$ (ii)

or, $a^3 \equiv -1 \pmod 7 \equiv 6 \pmod 7$. (iii)

Hence, $a^3 \equiv 0$ or 1 as 6 (mod 7) [by (i), (ii), (iii)].

Problem (11.32): Show that $a^{37} \equiv 0 \pmod{1729}$ for any integer a.

Sol: If p is prime and a is any integer, then

we have $a^p \equiv a \pmod p$. (1)

Now $1729 = 7 \cdot 13 \cdot 19$.

Then by (1), $a^7 \equiv a \pmod 7$ (i)

$a^{13} \equiv a \pmod{13}$ (ii)

$a^{19} \equiv a \pmod{19}$. (iii)

Now $(a^7)^5 \equiv a^5 \pmod 7 \Rightarrow a^{35} \cdot a^2 \equiv a^7 \pmod 7$

$\Rightarrow\ a^{37} \equiv a^7 \pmod 7 \equiv a \pmod 7$ [by (i)] (2)

Again, $a^{13} \equiv a \pmod{13}$

$\Rightarrow\ a^{26} \equiv a^2 \pmod{13} \Rightarrow a^{26} \cdot a^{11} \equiv a^{13} \pmod{13}$

$\Rightarrow\ a^{37} \equiv a(\bmod 13)$ [by (ii)]. (3)

Also $a^{19} \equiv a \pmod{19}$

$\Rightarrow\ a^{37} \equiv a^{19} \pmod{19} \equiv a \pmod{19}$ [by (iii)]. (4)

Therefore, by (2), (3) and (4) it follows that

$a^{37} \equiv a \pmod{[7, 1, 19]} \equiv a \pmod{1729}$.

Problem (11.33): Find the remainder when $2! + 3! + \cdots + 50!$ is divided by 48.

Sol: $(r+1)! \equiv 0 \pmod{48}\ \forall\, r \geq 5$.

Therefore, $2! + 3! + 4! + 5! + (r+1)! \equiv 2! + 3! + 4! + 5! \pmod{48}$, $5 \leq r \leq 49$

$\Rightarrow \quad 2! + 3! + 4! + 5! + (r+1)! \equiv 2 + 6 + 24 + 120 \pmod{48} \equiv 152 \pmod{48}$

$\equiv 8 \pmod{48},\ 5 \le r \le 49.$

$\Rightarrow \quad 2! + 3! + 4! + \cdots + 50! \equiv 8 \pmod{48}.$

So the remainder is 8 (mod 48).

Problem (11.34): Without performing the long division determine whether the integer 78234561 is divisible by 9 or 11.

Sol: Let $n = a_m(1000)^m + a_{m-1}(1000)^{m-1} + \cdots + a_1(1000) + a_0,$ (1)

where $0 \le a_m \le 999$, a_m are all integers.

Let $S = a_m + a_{m-1} + \cdots + a_1 + a_0$ and $T = a_0 - a_1 + a_2 - \cdots + (-1)^m a_m.$

Hence, $N = 78234561.$

Then N can be expressed by (1) as

$N = 78(1000)^2 + 234(1000) + 561,\ m = 2.$

$S = 78 + 234 + 561.$

Now let $f(x) = 78x^2 + 234x + 561.$ (2)

$1000 \equiv 1 \pmod 9 \Rightarrow f(1000) \equiv f(1) \pmod 9$

$\Rightarrow \quad N \equiv (78 + 234 + 561) \pmod 9$

$\Rightarrow \quad N \equiv S \pmod 9 \Rightarrow N - S \equiv 0 \pmod 9$

$\Rightarrow \quad 9|N,$ iff $9|S.$

Now, $S = 78 + 234 + 561 = 873$ and $9|873.$

Hence, $9|N$. So N is divisible by 9.

Again, $1000 \equiv -1 \pmod{11} \Rightarrow f(1000) \equiv f(-1) \pmod{11}$

$\Rightarrow \quad N \equiv 78 - 234 + 561 \pmod{11}$ [by (2)]

$\Rightarrow \quad N \equiv T \pmod{11} \Rightarrow N - T \equiv 0 \pmod{11}$

$\Rightarrow \quad 11|N$ iff $11|T.$

Here $T = 78 - 234 + 561 = 639 - 234 = 405.$

But $11 \nmid 405.$ Hence, $11 \nmid N.$

So N is not divisible by 11.

Problem (11.35): (i) Calculate $\phi(1024)$

(ii) Show that $17^{28} \equiv 1 \pmod{145}.$

Sol: (i) $\phi(1024) = \phi(2^{10}) = 2^{10}\left(1 - \dfrac{1}{2}\right) = 2^9.$

(ii) $145 = 5 \times 29.$

Now $17 \equiv 2 \pmod 5 \Rightarrow 17^2 \equiv 4 \pmod 5$

$\Rightarrow \quad 17^4 \equiv 16 \pmod 5 \equiv 1 \pmod 5$

$\Rightarrow \quad 17^{28} \equiv 1 \pmod 5.$ (1)

Again, $17^2 \equiv 28 \pmod{29} \Rightarrow 17^2 \equiv (-1) \pmod{29}$

$\Rightarrow \quad 17^4 \equiv 1 \pmod{29}$

$\Rightarrow \quad 17^{28} \equiv 1 \pmod{29}.$ (2)

Therefore, from (1) and (2), we have

$$17^{28} \equiv 1 \pmod{[5, 29]} \equiv 1 \pmod{145}.$$

Problem (11.36): Show that the number of (+)ve divisors of a composite number n is odd if and only if n is a perfect square.

Sol: Let n be a perfect square.

Since n is a composite number, it can be expressed in the form,

$$n = p_1^{2k_1} p_2^{2k_2} \dots p_r^{2k_r},\ p_i \text{ are all prime.}$$

Then $\tau(n)$ = total number of divisors of n

$$= (2k_1 + 1)(2k_2 + 1) \cdots (2k_n + 1)$$

$$= \text{an odd integer,}$$

since the product of odd integers is odd.

Now, let $n = \prod_{i=1}^{k} p_i^{a_i}$. Then $\tau(n) = \prod(a_i + 1)$. $\tau(n)$ is odd $\Rightarrow$ $(a_i + 1)$ is odd $\Rightarrow$ a_i = even $\forall i = 1, 2, \dots, k$ and hence, n is a perfect square.

Problem (11.37): Find pairs of distinct (+)ve integers m and n such that $\phi(m) = \phi(n)$.

Sol: If n be an odd integer, then

$$\phi(2n) = \phi(n) \quad \text{[i.e., when } m = 2n,\ n \text{ is odd]}$$

Since $(2, n) = 1,\ \phi(2n) = \phi(2) \cdot \phi(n) = \phi(n)$ $[\because\ \phi(2) = 1]$

When $n = 1,\ \phi(2) = \phi(1),\ (m, n) = (2, 1)$

$n = 3,\ \phi(2 \cdot 3) = \phi(3),\ (m, n) = (6, 3)$

$n = 5,\ \phi(2 \cdot 5) = \phi(5),\ (m, n) = (10, 5)$

and so on.

Problem (11.38): Find the numbers of (+)ve divisors and sum of the (+)ve divisors of 180.

Sol: Here $n = 180 = 2^2 \cdot 3^2 \cdot 5$

$$\tau(n) = (2 + 1)(2 + 1)(1 + 1) = 3 \cdot 3 \cdot 2 = 18$$

$$\sigma(n) = \frac{2^{2+1} - 1}{2 - 1} \cdot \frac{3^{2+1} - 1}{3 - 1} \cdot \frac{5^{1+1} - 1}{5 - 1} = (2^3 - 1)\frac{(3^3 - 1)}{2} \cdot \frac{(5^2 - 1)}{4}$$

$$= 7 \cdot 13 \cdot 6 = 546.$$

Problem (11.39): If n is not divisible by any prime $\leq \sqrt{n}$, then n is prime.

Sol: Suppose that n is not prime. Then n is a composite number.

So $n = p_1^{a_1} p_2^{a_2} \dots p_k^{a_k}$ [by fundamental theorem of Arithmetic], p_i are primes and $a_i \geq 0$.

Then $n \geq p_1 p_2$. (1)

But $p_1 > \sqrt{n}$, $p_2 > \sqrt{n}$, because n is not divisible by any prime $\leq \sqrt{n}$.

Hence, $n > \sqrt{n} \cdot \sqrt{n}$ [from (1)]

$= n$, which is not possible.

Therefore, n is prime.

Problem (11.40): If $n > 1$ is not a prime number, then show that n has a prime factor not exceeding $\sqrt{n}$.

Sol: Follows from Problem (39).

Problem (11.41): Find whether 407 is prime.

Sol: The primes $\leq \sqrt{407}$ are 2, 3, 5, 7, 11, 13, 17, 19.
[since $\sqrt{407} = 20$ approximately]
But none of these primes divides 407.
Hence, 407 is prime [by Prob. 39].

Problem (11.42): A post office has only stamps of Rs. 6.00 and Rs. 15 in its stock. Form a linear Diophantine equation so that a person can buy y stamps of Rs. 81 from the post office. Hence, find all (+)ve integral solutions of the equation.

Sol: Let there be x number of stamps of Rs. 6 and y number of stamps of Rs. 15. Then we have $6x + 15y = 81$, which is the linear Diophantine equation.

Now $\quad 6x + 15y = 81 \quad (1)$

$\Rightarrow \quad 6x \equiv 81 \pmod{15}$.

Applying divisional algorithm (6, 15),

we get $\quad 15 = 6 \cdot 2 + 3$.

$\quad 6 = 3 \cdot 2 + 0$

$\Rightarrow \quad 3 = 15 - 6 \cdot 2 = (-2) \cdot 6 + 1 \cdot 15$

$\Rightarrow \quad 81 = (-54)6 + 27 \cdot 15$ [multiplying by 27].

Thus, $\quad 6 \cdot (-54) + 15 \cdot 27 = 81 \Rightarrow x = -54,\ y = 27$, which satisfies (1).

General solution is $x = -54 + 15t$

$$y = 27 - 6t.$$

$$t = 4,\ x = 6,\ y = 3.$$

Least (+)ve integral solution is (6, 3).

$\left.\begin{array}{l} x \text{ is (+)ve for } t \geq 4 \\ y \text{ is (+)ve for } 0 \leq t \leq 4. \end{array}\right\}$

Problem (11.43): If n is the number and P the product of the divisors of N, then prove that $P^2 = N^n$.

Sol: Let $N = p_1^{a_1} p_2^{a_2} \ldots p_k^{a_k}$, $a_i \geq 0$, p_i are distinct primes.

Then $\quad n = (a_1 + 1)(a_2 + 1) \cdots (a_k + 1)$.

Each term of the product $(1 + p_1 + p_1^2 + \cdots + p_1^{a_1})(1 + p_2 + p_2^2 + \cdots + p_2^{a_2}) \cdots (1 + p_k + p_k^2 + \cdots + p_k^{a_k})$ is a divisor of N. Then in the product P, each of $p_1, p_1^2, \ldots p_1^{a_1}$ will occur $(a_2 + 1)(a_3 + 1) \cdots (a_k + 1)$ times, each of $p_2, p_2^2, \ldots, p_2^{a_2}$ will occur $(a_1 + 1)(a_3 + 1) \cdots (a_k + 1)$ times, each of $p_k, p_k^2, \ldots, p_k^{a_k}$ will occur $(a_1 + 1)(a_2 + 1) \cdots (a_{k-1} + 1)$ times. Consequently,

$$P = p_1^{(1+2+\cdots+a_1)\{(a_2+1)(a_3+1)\cdots(a_k+1)\}} \times p_2^{(1+2+\cdots+a_2)(a_1+1)(a_3+1)\cdots(a_k+1)} \cdots \times p_k^{(1+2+\cdots+a_k)(a_1+1)(a_2+1)\cdots(a_{k-1}+1)}$$

$$= p_1^{\frac{a_1(a_1+1)}{2}\cdot(a_2+1)\cdots(a_k+1)} \times p_2^{\frac{a_2(a_2+1)}{2}(a_1+1)(a_3+1)\cdots(a_k+1)}$$

$$\times p_k^{\frac{a_k(a_k+1)}{2}(a_1+1)\cdots(a_{k-1}+1)}$$

$$P^2 = p_1^{a_1 n} p_2^{a_2 n} \dots p_k^{a_k n} = (p_1^{a_1} p_2^{a_2} \dots p_k^{a_k})^n$$

$\Rightarrow \quad P^2 = N^n.$

[(**Go through the Example.**)

Let $\quad N = 2^2 \cdot 3^2 \cdot 5$

P = product of the divisors.

$$d|N \Rightarrow d = 1, 2, 3, 5, 2^2, 3^2,$$
$$2 \cdot 3, 2 \cdot 5, 2 \cdot 3 \cdot 5, 2 \cdot 3^2, 2 \cdot 3^2 \cdot 5$$
$$2^2 \cdot 3, 2^2 \cdot 5, 2^2 \cdot 3 \cdot 5, 3 \cdot 5$$
$$2^2 \cdot 3^2, 2^2 \cdot 3^2 \cdot 5, 3^2 \cdot 5$$

$$P = 2^{18} \cdot 3^{18} \cdot 5^9 = 2^{\{(1+2)(2+1)(1+1)\}} \cdot 3^{(1+2)\{(2+1)(1+1)\}} \cdot 5^{1\cdot\{(2+1)(2+1)\}}.$$

$$n = (2+1)(2+1)(1+1) = 18$$

$$P^2 = (2^2 \cdot 3^2 \cdot 5)^{18} = N^n]$$

Problem (11.44): If $p = 2^k - 1$ is prime, then k is odd except when $k = 2$.

Sol: For $k = 2$, $p = 2^2 - 1 = 3$, an odd prime.

But for $k = 2m$, $p = 2^{2m} - 1 = (2^2)^m - 1$ has a divisor $2^2 - 1 = 3\ \forall$ (+)ve integer m.

Then p cannot be prime. (1)

Again, $p = 2^k - 1 \Rightarrow 2^k = p + 1$ = even number.

In this case k must be odd, otherwise p cannot be prime [by (1)].

Problem (11.45): If $a \equiv b \pmod n$, then $(a, n) = (b, n)$.

Sol: Let $d = (a, n)$ and $(b, n) = p$

Then $\quad d|a \cdot d|n.$

Again $\quad a \equiv b \pmod n \Rightarrow n|a - b.$ (i)

So $\quad d|n$ and $n|a - b \Rightarrow d|a - b.$

and hence, $d|a$, $d|a - b \Rightarrow d|b$.

Thus, $\quad d|b,\ d|n \Rightarrow d|p$ (ii)

as $(b, n) = p$.

Again, $\quad (b, n) = p \Rightarrow p|b, p|n \Rightarrow p|b, p|a - b$ [by (i)]

$\Rightarrow \quad p|a.$

Thus, $\quad p|a, p|n \Rightarrow p|d$ (iii)

as $(a, n) = d$.

From (ii) and (iii) it follows that $d = p$.

i.e., $\quad (a, n) = (b, n).$

Problem (11.46): Prove that $n > 2$, there exists a prime p such that $n < p < n!$.

Sol: Let us consider a number $n! - 1 < n!$. If possible, let $n! - 1$ is not a prime, then it has a prime divisor p (say) and $p \leq n$, which implies that $p|n!$, which contradicts that $n < p$.

Hence the result.

Problem (11.47): Show that, for any odd integer a, $a^2 + (a+2)^2 + (a+4)^2 + 1$ is divisible by 12.

Sol: $a^2 + (a+2)^2 + (a+4)^2 + 1 = 3a^2 + 12a + 21$

$= 3(8k+1) + 12a + 21$, as the square of an odd integer is of the from $8k + 1$

$= 12(2k + a + 2) = 12q,\ q = 2k + a + 2 \in Z.$

Hence the result.

Problem (11.48): Show that, if a, b are integers not both zero, $(2a - 3b,\ 4a - 5b)|b$ and hence show that $(2a + 3,\ 4a + 5) = 1$.

Sol: Let $d = (2a - 3b,\ 4a - 5b)$.

Then $d|2a - 3b,\ d|4a - 5b$

$\Rightarrow \quad d|2(2a - 3b) - 1(4a - 5b) \ \Rightarrow\ d|b.$

Now for $b = -1$, $(2a + 3, 4a + 5) = 1$,

since $d|b \ \Rightarrow\ d|-1 \ \Rightarrow\ d = 1\ (d > 0)$.

Problem (11.49): Verify that $2^{35} - 1$ is divisible by 31 and 127.

Sol: $p = 31$ (a prime number)

By F. Theorem,

$2^{31-1} \equiv 1 \pmod{31} \Rightarrow 2^{30} \equiv 1 \pmod{31}$

$\Rightarrow \quad 2^{30} \cdot 2^5 \equiv 2^5 \pmod{31} \equiv 1 \pmod{31}$

$\Rightarrow \quad 2^{35} - 1 \equiv 0 \pmod{31} \Rightarrow 31|2^{35} - 1.$ (1)

Also $2^7 \equiv 1 \pmod{127}$ $[\because\ 2^7 = 128]$

$\Rightarrow \quad (2^7)^5 \equiv 1 \pmod{127}$

$\Rightarrow \quad 2^{35} \equiv 1 \pmod{127}.$ (2)

From (1) and (2) it follows from congruence relation,

$2^{126} \equiv 2^{91} \pmod{127} \Rightarrow 2^{35} \equiv 1 \pmod{127}$

$\Rightarrow \quad 127|2^{35} - 1.$

Problem (11.50): If $2^n + 1 = xy$, prove that $x - 1,\ y - 1$ are divisible by the same power of 2.

Sol: $2^n + 1 = xy$

$\Rightarrow$ xy is odd

$\Rightarrow \quad x$ and y are both odd

$\Rightarrow \quad x - 1$ and $y - 1$ are both even.

$\Rightarrow \quad x - 1$ and $y - 1$ are both divisible by 2.

Now $2^n = xy - 1 = x(y - 1) + 1(x - 1).$ (1)

Now let $2^p|y - 1,\ p \leq n$.

Then $y - 1 = 2^p \cdot t$ for some $t \in Z$.

From (1), $2^n = x \cdot 2^p \cdot t + (x-1)$

$\Rightarrow \quad x - 1 = 2^p[2^{n-p} - xt] = 2^p \cdot t', \; t' \in z.$

$\Rightarrow \quad 2^p|(x-1).$

Thus, $x-1$ and $y-1$ are both divisible by the same power of 2.

Problem (11.51): If $f(x)$ be a polynomial, it cannot represent prime only.

Sol: Let $u = f(x) = a_n x^n + a_{n-1}x^{n-1} + \cdots + a_1 x + a_0$, where $a_i \in Z$,

and let $v = f(x + ku)$, where $k \in Z$.

Then $\quad v = a_n(x+ku)^n + a_{n-1}(x+ku)^{n-1} + \cdots + a_1(x+ku) + a_0.$

$$\therefore \quad v = (a_n x^n + a_{n-1}x^{n-1} + \cdots + a_1 x + a_0) + a_n {}^n c_1 x^{n-1} ku$$
$$+ a_n {}^n c_2 x^{n-2}(ku)^2 + \cdots + (ku)^n + a_{n-1} {}^{n-1}c_1 x^{n-2}(ku)$$
$$+ a_{n-1} {}^{n-1}c_2 x^{n-3}(ku)^2 + \cdots + (ku)^{n-1} + \cdots + a_1 ku$$

$= f(x) +$ terms divisible by ku

$= u +$ terms divisible by $ku \Rightarrow u$ divides v

$\Rightarrow \quad u$ is a factor of v

$\Rightarrow \quad v$ is not prime $\forall\, k$.

$\therefore\; f(x)$ cannot represent prime only.

[Go through the examples]

(i) Let $f(x) = x^2 + x + 1$

$x = 1, \; f(1) = 3 > 1, \; f(1)$ is prime.

$f(1+3t) = (1+3t)^2 + (1+3t) + 1 = 9t^2 + 9t + 3$ which is divisible by 3

$\therefore\; f(x)$ is not prime $\forall\, t$.

(ii) Let $f(x) = x^2 + x + 4$

$f(0) = 4 > 1.$

$f(0 + 4t)$ is divisible by 4.

$f(4t)$ is not prime $\forall\, t$.

(iii) Let $f(x) = x^2 + x + 41.$

$f(0) = 41 > 1$, prime.

$f(0 + 41t)$ is divisible by 41.

So $f(41t)$ is not prime, $\forall t \in Z^+$.

Problem (11.52): If N is expressed as a polynomial in a prime p, with each of the co-efficients less than p and s is the sum of these coefficients, then prove that the powers of p contained in $\lfloor N$ is $\frac{N-s}{p-1}$.

Sol: Let $N = a_n p^n + a_{n-1}p^{n-1} + a_{n-2}p^{n-2} + \cdots + a_1 p + a_0$ where $a_n < p \;\forall\, n$, p is prime.

The highest power of a prime p contained in $\lfloor N$ is $I\left[\frac{N}{p}\right] + I\left[\frac{N}{p^2}\right] + \cdots + I\left[\frac{N}{p^n}\right]$.

$$I\left[\frac{N}{p}\right] = a_n p^{n-1} + a_{n-1}p^{n-2} + \cdots + a_2 p + a_1 \quad [\text{as } a_n < p]$$

$$I\left[\frac{N}{p^2}\right] = a_n p^{n-2} + a_{n-1}p^{n-3} + \cdots + a_3 p + a_2$$

$$I\left[\frac{N}{p^3}\right] = a_n p^{n-3} + a_{n-1}p^{n-4} + \cdots + a_4 p + a_3$$

$$\cdots \quad \cdots \quad \cdots \quad \cdots \quad \cdots \quad \cdots \quad \cdots$$

$$I\left[\frac{N}{p^{n-2}}\right] = a_n p^2 + a_{n-1}p + a_{n-2}$$

$$I\left[\frac{N}{p^{n-1}}\right] = a_n p + a_{n-1}$$

$$I\left[\frac{N}{p^n}\right] = a_n$$

$$\sum I\left[\frac{N}{p^n}\right] = a_n(1 + p + \cdots + p^{n-1}) + a_{n-1}(1 + p + \cdots + p^{n-2})$$

$$+ a_3(1 + p + p^2) + a_2(1 + p) + a_1$$

$$= a_n\frac{(p^n - 1)}{p-1} + a_{n-1}\left(\frac{p^{n-1} - 1}{p-1}\right) + \cdots + \frac{a_3(p^3 - 1)}{p-1} + \frac{a_2(p^2 - 1)}{p-1} + a_1$$

$$= \frac{(a_n p^n + a_{n-1}p^{n-1} + \cdots + a_1 p + a_0) - (a_n + a_{n-1} + \cdots + a_1 + a_0)}{p-1}$$

$$= \frac{N - s}{p-1}, \text{ where } s = \text{sum of the co-efficients}$$

$$= a_n + a_{n-1} + \cdots + a_1 + a_0.$$

Smallest number with n divisors

If $N = p_1^{a_1} p_2^{a_2} \ldots p_k^{a_k}$, where p_i are primes in increasing order and $a_i \geq 0$.
$n =$ number of divisors of $N = (a_1 + 1)(a_2 + 1)\cdots(a_k + 1)$.

N will be the smallest positive number with n divisors if we arrange p_i with the power a_i, where a_i are in decreasing order.

Problem (11.53): Find the smallest number with 24 divisors.

Sol: $(1+1)(1+1)(1+1)(2+1) = 24$

$(1+1)(1+1)(5+1) = 24$

$(1+1)(2+1)(3+1) = 24$

$(1+1)(11+1) = 24$

$(3+1)(5+1) = 24$

$(2+1)(7+1) = 24,$

which are the possible expression of the number of divisors 24 in the form $(a_1+1)(a_2+1)\ldots(a_k+1)$.

We take the prime 2, 3, 5, 7,... in increasing order.

The possible numbers having 24 divisors are $2^2 \cdot 3 \cdot 5 \cdot 7$, $2^5 \cdot 3^1 \cdot 5^1$, $2^3, 3^2, 5^1$, $2^{11} \cdot 3^1, 2^5 \cdot 3^3$, $2^7 \cdot 3^2$ of which $2^3 \cdot 3^2 \cdot 5^1 = 360$ is the smallest number having 24 divisors.

Problem (11.54): Find the smallest number with 120 divisors.

Sol: $n = 120$

$(1+1)(5+1)(9+1) = 120$

$$(2+1)(3+1)(9+1) = 120$$
$$(1+1)(2+1)(3+1)(4+1) = 120$$
$$(1+1)(59+1) = 120$$
$$(2+1)(39+1) = 120$$
$$(3+1)(29+1) = 120$$
$$(4+1)(23+1) = 120$$
$$(7+1)(14+1) = 120$$
$$(9+1)(11+1) = 120$$

Possible number with 120 divisors are $2^9 \cdot 3^5 \cdot 5^1$, $2^9 \cdot 3^3 \cdot 5^2$, $2^4 \cdot 3^3 \cdot 5^2 \cdot 7^1$, $2^{59} \cdot 3^1$, $2^{39} \cdot 3^2$, $2^{29} \cdot 3^3$, $2^{23} \cdot 3^4$, $2^{14} \cdot 3^7$, $2^{11} \cdot 3^9$ of which $2^4 \cdot 3^3 \cdot 5^2 \cdot 7 = 75600$ is the smallest (+)ve number with 120 divisors.

Problem (11.55): Prove that $\lfloor\underline{2n}$ is divisible by $\lfloor\underline{n} \cdot \lfloor\underline{n+1}$.

Sol: $2n+1 = n+(n+1)$

Then by Th. [2.19], $I\left[\dfrac{2n+1}{p}\right] \geq I\left[\dfrac{n}{p}\right] + I\left[\dfrac{n+1}{p}\right]$, where p is any prime.

$\Rightarrow$ highest power of a prime p in $(2n+1)! \geq$ sum of the highest powers of p in $n!$ and $(n+1)!$

$\Rightarrow \quad (2n+1)!$ is divisible by $n!(n+1)!$

$$\Rightarrow \quad \frac{(2n+1)!}{n!(n+1)!} = \text{an integer}$$

$$\Rightarrow \quad (2n+1)\frac{2n!}{n!(n+1)!} = \text{an integer}$$

$\Rightarrow \quad \dfrac{2n!}{n!(n+1)!}$ is an integer, as $(2n+1)$ is an integer.

Hence, $2n!$ is divisible by $n!(n+1)!$

Problem (11.56): If $a, b, c, \ldots$ are numbers whose sum is p, a prime number, then $\dfrac{\lfloor\underline{p}}{\lfloor\underline{a}\,\lfloor\underline{b}\,\lfloor\underline{c}\cdots}$ is an integer divisible by p.

Sol: We have $\dfrac{\lfloor\underline{a+b}}{\lfloor\underline{a}\,\lfloor\underline{b}} =$ an integer.

Replacing b by $b+c$,

$$\frac{\lfloor\underline{a+b+c}}{\lfloor\underline{a}\,\lfloor\underline{b+c}} = \text{an integer.} \qquad (1)$$

Again, $\dfrac{\lfloor\underline{b+c}}{\lfloor\underline{b}\cdot\lfloor\underline{c}} =$ an integer $= t$ (say)

$$\Rightarrow \quad \lfloor\underline{b+c} = \lfloor\underline{b}\cdot\lfloor\underline{c}\cdot t$$

From (1), $\dfrac{\lfloor\underline{a+b+c}}{\lfloor\underline{a}\cdot\lfloor\underline{b}\cdot\lfloor\underline{c}\cdot t} =$ an integer

$$\Rightarrow \quad \frac{\lfloor\underline{a+b+c}}{\lfloor\underline{a}\cdot\lfloor\underline{b}\cdot\lfloor\underline{c}} = t\cdot \text{ an integer} = \text{an integer.}$$

Continuing the above process, we have

$$\frac{\lfloor a+b+c+\cdots}{\lfloor a\cdot\lfloor b\cdot\lfloor c\cdots} = \text{an integer} = k \text{ (say)}. \qquad (2)$$

Now given that $p = a+b+c+\cdots =$ a prime number.

Therefore, from (2), $\dfrac{\lfloor p}{\lfloor a\,\lfloor b\,\lfloor c\ldots}$ is an integer.

It follows that $\dfrac{p(p-1)\cdots 2\cdot 3\cdot 1}{\lfloor a\,\lfloor b\,\lfloor c\ldots} = k,\ k\in Z.$ (3)

Now p being a prime number, $(\lfloor a\cdot\lfloor b\cdot\lfloor c\ldots, p) = 1$,

since $(\lfloor a, p) = 1\ \forall\ a.$

Hence, from (3) it follows that $\lfloor a\,\lfloor b\,\lfloor c\ldots|(p-1)!$

i.e., $\lfloor p-1 = (\lfloor a\cdot\lfloor b\cdot\lfloor c\ldots)\cdot t',\ t'\in Z.$

Therefore, from (3), $\dfrac{p(\lfloor a\,\lfloor b\,\lfloor c\ldots)t}{(\lfloor a\,\lfloor b\,\lfloor c\ldots)} = k$

$\Rightarrow \quad pt = k \Rightarrow p|k.$

i.e., $p\left|\dfrac{\lfloor p}{\lfloor a\,\lfloor b\,\lfloor c\ldots}\right.$, since from (3), $\dfrac{\lfloor p}{\lfloor a\,\lfloor b\lfloor c\cdots} = k$. Hence the result.

Problem (11.57): Show that $\dfrac{n^5}{5}+\dfrac{n^{11}}{11}+\dfrac{39n}{55}$ is an integer.

Sol: $\dfrac{n^5}{5}+\dfrac{n^{11}}{11}+\dfrac{39n}{55} = \dfrac{11n^5+5n^{11}+39n}{55}.$

Let $\quad f(n) = 11n^5+5n^{11}+39n.$

To prove the result, it is sufficient to show that $f(n) = 55t,\ t\in Z.$

i.e., $\quad 55|f(n).$

Now $\quad f(1) = 11+5+39 = 55 \Rightarrow 55|f(1).$

It is true for $n=1$.

Let it be true for n. That is, let $55|f(n)$.

Then

$$\begin{aligned} f(n+1) &= 11(n+1)^5 + 5(n+1)^{11} + 39(n+1)\\ &= (11n^5+5n^{11}+39n) + 11({}^5c_1n^4 + {}^5c_2n^3 + \cdots + {}^5c_4n + 1)\\ &\qquad + 5({}^{11}c_1n^{10} + {}^{11}c_2n^9 + \cdots + {}^{11}c_{10}n + 1) + 39\\ &= f(n) + 11 \text{ (terms divisible by 5)} + 5 \text{ (terms divisible by 11)}\\ &\qquad + 11 + 5 + 39\\ &= 55t + 55t' + 55t'' + 55;\ t, t', t''\in Z\\ &= 55(t+t'+t''+1) \end{aligned}$$

$\Rightarrow \quad 55|f(n+1),\ t+t'+t''+1\in Z.$

So the result is also true for $n+1$.

Hence it is true $\forall\ n$.

Therefore, $55|f(n)$ and consequently, $\dfrac{n^5}{5}+\dfrac{n^{11}}{11}+\dfrac{39n}{55}$ is an integer.

Problem (11.58): Let $\mu((p,d))$ denote the value of the Möbius function at the G.C.D. (p,d). Prove that for every prime p,

$$\sum_{d|n} \mu(d)\mu(p,d) = 1, \text{ if } n = 1$$
$$= 2, \text{ if } n = p^\alpha, \alpha \geq 1$$
$$= 0, \text{ otherwise.}$$

Sol: Case(1): Let $n = 1$. Then $d = 1$ and $(p,d) = 1$

$$\therefore \quad \sum_{d|n} \mu(d)\mu(p,d) = \mu(1)\mu(1) = 1.$$

Case(2): Let $n = p^\alpha$, p is prime and $\alpha \geq 1$.

Then $d = 1, p, p^2, \ldots, p^\alpha$.

$$(p,d) = 1 \Rightarrow \mu((p,d)) = \mu(1) = 1$$
$$(p,p) = p \Rightarrow \mu((p,p)) = \mu(p) = -1$$
$$(p,p^2) = p \Rightarrow \mu((p,p^2)) = \mu(p) = -1$$
$$\ldots \quad \ldots \quad \ldots \quad \ldots \quad \ldots$$
$$(p,p^\alpha) = p \Rightarrow \mu((p,p^\alpha)) = \mu(p) = -1.$$

$$\therefore \quad \sum_{d|n} \mu(d)\mu(p,d) = \mu(1)\mu(p,1) + \mu(p)\mu((p,p)) + \mu(p^2)\mu((p,p^2)) + \cdots + \mu(p^\alpha)\mu((p,p^\alpha))$$
$$= 1 \cdot 1 + (-1)(-1) + 0 + \cdots + 0$$
$$= 1 + 1 = 2 \quad [\mu(p^\alpha) = 0 \ \forall \alpha \geq 2].$$

Case(3): Let $n = p_1 p_2 \ldots p_k$, where p_i are prime and $p_i \neq p \ \forall i$.

Then $\quad d = 1, p_i, p_i p_j, p_i p_j p_k, \ldots, p_1 p_2 p_3 \ldots p_k$

$$(p,d) = 1, \ (p,p_i) = 1, \ (p, p_i p_j) = 1, \ldots, (p, p_1 p_2 \ldots p_k) = 1.$$

Therefore, $\sum_{d|n} \mu(d)\mu(p,d) = \mu(1)\mu(1) + \mu(1)\sum \mu(p_1) + \mu(1)\sum \mu(p_1p_2) + \cdots + \mu(1)\mu(p_1p_2\ldots p_k)$

$$= 1 - {}^kc_1 + {}^kc_2(-1)^2 + \cdots + (-1)^k = (1-1)^k = 0.$$

Problem (11.59): For any even positive integer $n(>1)$, $\sum_{d|n} \mu(d)\phi(d) = 0$.

Sol: Since n is an even integer, it contains a factor 2^m, $m \geq 1$.

So $\quad n = 2^m p_2^{a_2} p_3^{a_3} \ldots P_k^{a_k}$, $a_k \geq 0$

and $p_2, p_3, \ldots, p_k$ are all distinct primes in ascending order.

$\phi(n)$ is multiplicative, since $(2, p_i) = 1 \ \forall i = 2, 3, \ldots, k$.

So $\quad \sum_{d|n} \mu(d)\phi(d) = \{1 - \phi(2)\}\{1 - \phi(p_2)\} \cdots \{1 - \phi(p_k)\}$

$$= 0, \text{ since } \phi(2) = 1.$$

$$\therefore \quad \sum_{d|n} \mu(d)\phi(d) = 0.$$

Problem (11.60): Let $m = 3^k \cdot 568$, $n = 3^k \cdot 638$ $k \geq 0$. Then $\tau(m) = \tau(n)$, $\sigma(m) = \sigma(n)$ and $\phi(m) = \phi(n)$.

Sol:

2	568
2	284
2	142
	71

2	638
11	319
	29

$\therefore \quad 568 = 2^3 \cdot 71$

and $\quad 638 = 2 \cdot 11 \cdot 29.$

Thus, $m = 2^3 \cdot 3^k \cdot 71;\ n = 2 \cdot 3^k \cdot 11 \cdot 29.$

Then $\tau(m) = (3+1)(k+1)(1+1) = 8(k+1)$

$\tau(n) = (1+1)(k+1)(1+1)(1+1) = 8(k+1).$

(i) $\therefore \quad \tau(m) = \tau(n)\ \forall\ k \geq 0.$

(ii) $\sigma(m) = \dfrac{2^{3+1}-1}{2-1} \cdot \dfrac{3^{k+1}-1}{3-1} \cdot \dfrac{71^2-1}{71-1}$

$= 15 \cdot \dfrac{3^{k+1}-1}{2} \cdot 72 = 15 \cdot 36 \cdot (3^{k+1}-1).$

$\sigma(n) = \dfrac{2^2-1}{2-1} \cdot \dfrac{3^{k+1}-1}{3-1} \cdot \dfrac{11^2-1}{11-1} \cdot \dfrac{29^2-1}{29-1} = 3 \cdot \dfrac{(3^{k+1}-1)}{2} \cdot 12 \cdot 30$

$= 3 \cdot 6 \cdot 30(3^{k+1}-1) = 15 \cdot 36(3^{k+1}-1).$

$\therefore \quad \sigma(m) = \sigma(n).$

(iii) $\phi(m) = \phi(2^3 \cdot 3^k \cdot 71) = 2^3\left(1-\dfrac{1}{2}\right) \cdot 3^k\left(1-\dfrac{1}{3}\right) \cdot 71\left(1-\dfrac{1}{71}\right)$

$= 4 \cdot 2 \cdot 3^{k-1} \cdot 70 = 3^{k-1} \cdot 560.$

$\phi(n) = \phi(2 \cdot 3^k \cdot 11 \cdot 29) = 3^{k-1} \cdot 560.$

$\therefore \quad \phi(m) = \phi(n).$

Problem (11.61): $\sigma(n)$ is odd if and only if n is a perfect square or twice a perfect square.

Sol: Let n be a perfect square and let $n = p_1^{a_1} p_2^{a_2} \cdots p_k^{a_k}$, p_i are primes, $a_i \geq 0$.

Then $\quad a_i = 2m_i$, for $i = 1, 2, \ldots k$.

Therefore, $\sigma(n) = \dfrac{p_1^{2m_1+1}-1}{(p_1-1)} \cdot \dfrac{p_2^{2m_2+1}-1}{p_2-1} \cdots \dfrac{p_k^{2m_k+1}-1}{p_k-1}.$ (i)

Now $\quad \dfrac{p_i^{2m_i+1}-1}{p_i-1} = p_i^{2m_i} + p_i^{2m_i-1} + \cdots + p_i^2 + p_i + 1$

$= (p_i^{2m_i} + p_i^{2m_i-2} + \cdots + p_i^2) + (p_i^{2m_i-1} + p_i^{2m_i-3} + p_i) + 1$

$= p_i^2(p_i^{2m_1-2} + p_i^{2m_i-4} + \cdots + 1) + p_i(p_i^{2m_i-2} + p_i^{2m_i-4} + \cdots + 1) + 1$

$= (p_i^{2m_i-2} + p_i^{2m_i-4} + \cdots + 1)p_i(p_i+1) + 1.$ (ii)

Now $p_i + 1$ is even $\forall\ p_i$ and hence, $\dfrac{p_i^{2m_i+1}-1}{p_i-1}$ is odd followed by (ii) $\forall\ p_i$

Hence, $\sigma(n) =$ product of odd integers [by (i)]

$\Rightarrow$ $\sigma(n)$ is odd.

Next, let $\sigma(n)$ be odd, where $n = p_1^{a_1} \cdot p_2^{a_2} \cdots p_k^{a_k}$, p_i are primes, $a_i \geq 0$.

$$\sigma(n) = \frac{p_1^{a_1+1}-1}{p_1-1} \cdot \frac{p_2^{a_2+1}-1}{p_2-1} \cdots \frac{p_k^{a_k+1}-1}{p_k-1}$$

$\therefore$ $\sigma(n)$ is odd $\Rightarrow \dfrac{p_i^{a_i+1}-1}{p_i-1}$ is odd $\forall\, p_i$

$\Rightarrow$ $(p_i^{a_i+1}-1)$ is even as (p_i-1) is even, $p_i \neq 2$

$\Rightarrow$ $p_i^{a_i+1}$ is odd.

$\Rightarrow$ a_i+1 is odd or even as p_i are odd primes.

Let a_i+1 be odd.

Then a_i are even $\forall\, i$.

Also when $p_1 = 2$, $(2^{a_1+1}-1)$ is odd as a_1 is even or odd.

Now when a_1 is even, then a_i are even $\forall\, i$ and n becomes a perfect square.

When a_1 is odd, n is of the form $2^{a_1} \cdot p_2^{a_2} \ldots p_k^{a_k}$ and hence n becomes twice a perfect square as n is of the form $2 \cdot 2^{a_1-1} \cdot p_2^{a_2} \cdots p_k^{a_k}$, when $a_1-1, a_2, \ldots, a_k$ are all even.

Let a_{i+1} be even $\forall i$ or at least for one i.

Then $\sigma(n)$ becomes even, a contradiction arises. So we reject the case when a_{i+1} is even.

Problem (11.62): Prove that $\displaystyle\sum_{d|n} \frac{1}{d} = \frac{\sigma(n)}{n}$ for every (+)ve integer n.

Sol: Let $n = p_1^{a_1} \cdot p_2^{a_2} \cdots p_k^{a_k}$, $a_i \geq 0$, each p_i are distinct primes.

Each term of the product $(1+p_1+\cdots+p_1^{a_1})(1+p_2+\cdots+p_2^{a_2}) \times (1+p_k+\cdots+p_k^{a_k})$ is a divisor of n.

Again, each term of the product is of the form $p_1^{x_1} p_2^{x_2} \cdots p_k^{x_k}$, $0 \leq x_i \leq a_i$, $i = 1, 2, \ldots, k$ occurring in distinct form. Thus $p_1^{x_1} p_2^{x_2} \cdots p_k^{x_k} | n$.

$$\text{Therefore, } \sum_{d|n} \frac{1}{d} = \sum \frac{1}{p_1^{x_1} \cdot p_2^{x_2} \cdots p_k^{x_k}},\ 0 \leq x_i \leq a_i. \tag{1}$$

$$\text{Again, } \quad \sigma(n) = \frac{p_1^{a_1+1}-1}{p_1-1} \cdot \frac{p_2^{a_2+1}-1}{p_2-1} \cdots \frac{p_k^{a_k+1}-1}{p_k-1}$$

$$= (1+p_1+\cdots+p_1^{a_1}) \cdot (1+p_2+\cdots+p_2^{a_2}) \cdots (1+p_k+\cdots+p_k^{a_k})$$

$$= \sum p_1^{x_1} p_2^{x_2} \ldots p_k^{x_k};\ 0 \leq x_i \leq a_i.$$

$$\frac{\sigma(n)}{n} = \frac{\sum p_1^{x_1} p_2^{x_2} \ldots P_k^{x_k}}{p_1^{a_1} p_2^{a_2} \ldots p_k^{a_k}} = \sum \frac{1}{p_1^{x_1} p_2^{x_2} \ldots p_k^{x_k}}. \tag{2}$$

Therefore, from (1) and (2) it follows that

$$\sum_{d|n} \frac{1}{d} = \frac{\sigma(n)}{n}.$$

Definition (11.1): $w(n)$ *is defined by the numbers of distinct prime divisor of* $n(>1)$, *with* $w(1) = 0$. *That is, if* $n = p_1^{a_1} p_2^{a_2} \ldots p_k^{a_k}$, $a_i \geq 0$, p_i *are distinct primes. Then* $w(n) = k$, $w(180) = w(2^2.3^2.5) = 3$.

Problem (11.63): $2^{w(n)}$ is multiplicative.

Proof: Let $m = \prod_{i=1}^{k} p_i^{a_i}$ and $n = \prod_{j=1}^{l} q_j^{b_j}$,

where $p_i \neq q_j$, $\forall\ i, j, p_i, q_j$ are primes.

So $(m, n) = 1$, since $(p_i^{a_i}, p_j^{b_j}) = 1\ \forall\ i, j$.

$$2^{w(mn)} = 2^{w\left(\prod_{i=1}^{k} p_i^{a_i} \cdot \prod_{j=1}^{l} q_j^{b_j}\right)} = 2^{k+l} \quad \text{[There are } k+l \text{ distinct primes]}$$

$$= 2^k \cdot 2^l = 2^{w(m)} \cdot 2^{w(n)}.$$

[Go through the examples]

(i) Let $n = 12 = 2^2 \cdot 3$

$n^2 = 2^4 \cdot 3^2$

$\tau(n^2) = (4+1)(2+1) = 15.$ (1)

$d|n \Rightarrow d = 1, 2, 3, 4, 6, 12.$

$$\sum_{d|n} 2^{w(d)} = 2^{w(1)} + (2^{w(2)} + 2^{w(3)} + 2^{w(2^2)}) + 2^{w(2\cdot 3)} + 2^{w(2^2\cdot 3)}$$

$$= 1 + (2^1 + 2^1 + 2^1) + 2^2 + 2^2 = 1 + (2+1)\cdot 2^1 + (2\cdot 1)\cdot 2^2$$

$$= 1 + 6 + 8 = 15. \quad (2)$$

From (1) and (2), $\tau(n^2) = \sum_{d|n} 2^{w(d)}$.

(ii) Let $n = 540 = 2^2 \cdot 3^3 \cdot 5$.

$n^2 = 2^4 \cdot 3^6 \cdot 5^2$

$\tau(n^2) = (4+1)(6+1)(2+1) = 105$

$\Rightarrow d = 1, 2, 3, 5,\ 2^2, 3^3, 5,\ 2\cdot 3,\ 2\cdot 5,\ 2^2\cdot 3,\ 2^2\cdot 5,\ 2\cdot 3^2,\ 2\cdot 3^3,\ 2^2\cdot 3^2,\ 2^2\cdot 3^3$
$3\cdot 5,\ 3^2\cdot 5,\ 3^3\cdot 5,\ 2\cdot 3\cdot 5, 2^2\cdot 3\cdot 5,\ 2^2\cdot 3^2\cdot 5,\ 2^2\cdot 3^3\cdot 5,\ 2\cdot 3^2\cdot 5,\ 2\cdot 3^3\cdot 5$

$$\sum_{d|n} 2^{w(d)} = 2^{w(1)} + (2+3+1)2^1 + (2\cdot 3 + 2\cdot 1 + 3\cdot 1)2^2 + (2\cdot 3\cdot 1)2^3$$

$$= 1 + 12 + 44 + 48 = 105.$$

[Note that $w(p_i^{a_i}) = 1,\ w(p_i^{a_i} p_j^{a_j}) = 2,\ w(p_i^{a_i} p_j^{a_j} p_k^{a_k} = 3]$

Problem (11.64): Show that $\tau(n^2) = \sum_{d|n} 2^{w(d)}$ for any (+)ve integer $n(> 1)$.

Sol: Let $n = p_1^{a_1} p_2^{a_2} \ldots p_k^{a_k},\ a_i \geq 0,\ p_i$ are distinct primes.

(i) $d = 1 \Rightarrow 2^{w(d)} = 2^{w(1)} = 2^0 = 1.$

(ii) $d = p_i^{a_i}$ with single distinct primes of any power $a_i (\geq 0),\ i = 1, 2, \ldots, k$

$$\Rightarrow \sum_{i=1}^{k} 2^{w(p_i^{a_i})} = (a_1 + a_2 + \cdots + a_k)2^1.$$

(iii) $d = p_i^{a_i} p_j^{a_j}$ with two distinct primes of any powers $a_i, a_j,\ a_i \geq 0,\ a_j \geq 0$

$$\Rightarrow \sum_{ij=1}^{k} 2^{w(p_i^{a_i} p_j^{a_j})} = \left(\sum a_1 a_2\right) \cdot 2^2.$$

(iv) $d = p_i^{a_i} p_j^{a_j} p_l^{a_l}$ with 3 distinct primes of any powers a_i, a_j, a_l.

$$\Rightarrow \quad \sum_{i,j,l=1}^{k} 2^{w(p_i^{a_i} p_j^{a_j} p_l^{a_l})} = \left(\sum a_1 a_2 a_3\right) 2^3.$$

proceeding in this way we have at the last stage, where $d = p_1^{a_1} p_2^{a_2} \ldots p_k^{a_k}$, with k distinct primes.

$$\sum_{d|n} 2^{w(d)} = 2^{w(p_1^{a_1} \ldots p_k^{a_k})} = 2^k a_1 a_2 \ldots a_k.$$

Now combining all the above results,

we get $\displaystyle\sum_{d|n} 2^{w(d)} = 1 + \sum a_1 2' + \sum a_1 a_2 \cdot 2^2 + \cdots + a_1 a_2 a_k \cdot 2^k.$ (1)

$$\text{Again,} \quad \tau(n^2) = \tau(p_1^{2a_1} \cdot p_2^{2a_2} \ldots p_k^{2a_k}) = (2a_1 + 1)(2a_2 + 1) \cdots (2a_k + 1)$$

$$= 1 + \sum a_1 a_2 \cdot 2 + \sum a_1 a_2 \cdot 2^2 + \cdots + a_1 a_2 \cdots a_k \cdot 2^k. \quad (2)$$

So from (1) and (2) the result follows.

Problem (11.65): Find the four consecutive integers for which $\mu(n) = 0$.

Sol: We consider four consecutive integers $n+1$, $n+2$, $n+3$, $n+4$ such that $\mu(n+1) = \mu(n+2) = \mu(n+3) = \mu(n+4) = 0$.

Let us consider the system of linear congruences,

$$x \equiv -1 \pmod{2^2},\ x \equiv -2 \pmod{3^2},\ x \equiv -3 \pmod{5^2},\ x \equiv -4 \pmod{7^2},$$

we apply Chinese remainder theorem in finding the solution of the system of congruences.

Let the integer x be the solution of the system,

$$\text{where} \quad x = \frac{M}{m_1} y_1 b_1 + \frac{M}{m_2} y_2 b_2 + \frac{M}{m_3} y_3 b_3 + \frac{M}{m_4} y_4 b_4 \pmod{M},$$

$$\text{where} \quad M = m_1 m_2 m_3 m_4 = 2^2.3^2.5^2.7^2 = 4.9.25.49 = 44100.$$

We define y_i by $\dfrac{M}{m_i} y_i \equiv 1 \pmod{m_i}$.

Then $\dfrac{M}{m_1} y_1 \equiv 1 \pmod{2^2}$

$$\Rightarrow \quad 11025 y_1 \equiv 1 \pmod{2^2} \equiv 1 + 2756 \times 4 \pmod{2^2}$$

$$\Rightarrow \quad 11025 y_1 \equiv 11025 \pmod{2^2}$$

$$\Rightarrow \quad y_1 \equiv 1 \pmod{2^2}. \quad (1)$$

$$\frac{M}{m_2} y_2 \equiv 1 \pmod{3^2} \Rightarrow 4900 y_2 \equiv 1 \pmod 9$$

$$\Rightarrow \quad 49 y_2 \equiv 1 \pmod 9 \Rightarrow 49 y_2 \equiv -98 \pmod 9$$

$$\Rightarrow \quad y_2 \equiv -2 \pmod 9. \quad (2)$$

$$\frac{M}{m_3} y_3 \equiv 1 \pmod{5^2} \Rightarrow 1764 y_3 \equiv -49 \pmod{25}$$

$$\Rightarrow \quad 36 y_3 \equiv -1 \pmod{25} \equiv 24 \pmod{25}$$

$$\Rightarrow \quad 3 y_3 \equiv 2 \pmod{25}$$

$\Rightarrow \quad 3y_3 \equiv 27 \pmod{25} \Rightarrow y_3 \equiv 9 \pmod{25}$. (3)

$\frac{M}{m_4} y_4 \equiv 1 \pmod{7^2} \Rightarrow 900y_4 \equiv 1 \pmod{49}$

$\Rightarrow \quad 900y_4 \equiv 50 \pmod{49} \Rightarrow 18y_4 \equiv 1 \pmod{49} \equiv -48 \pmod{49}$

$\Rightarrow \quad 3y_4 \equiv -8 \pmod{49}$

$\Rightarrow \quad y_4 \equiv 30 \pmod{49}$. (4)

Therefore, $x \equiv 11025 \cdot (1) \cdot (-1) + 4900 \cdot (-2) \cdot (-2) + 1764 \cdot 9 \cdot (-3) + 900 \cdot 30(-4) \pmod{44100}$

$= -11025 + 19600 - 47628 - 10800 \pmod{44100}$.

$\therefore \quad x \equiv -147053 \pmod{44100} \equiv -147053 + 3 \times 44100 \pmod{44100}$

$\equiv -147053 + 132300 \pmod{44100} \equiv -14753 \pmod{44100}$

$\equiv -14753 + 44100 \pmod{44100} \equiv 29347 \pmod{44100}$.

$\therefore \quad x \equiv -1 \pmod{2^2}$

$\Rightarrow \quad 29348 \equiv 0 \pmod{2^2}$ (i)

$x \equiv -2 \pmod{3^2}$

$\Rightarrow \quad 29349 \equiv 0 \pmod{3^2}$ (ii)

$x \equiv -3 \pmod{5^2}$

$\Rightarrow \quad 29350 \equiv 0 \pmod{5^2}$ (iii)

$x \equiv -4 \pmod{7^2}$

$\Rightarrow \quad 29351 \equiv 0 \pmod{7^2}$. (iv)

Therefore, the four consecutive integers for which $\mu(n) = 0$ are 29348, 29349, 29350, 29351.

Twin primes

Definition (11.2): *A pair (p, q) of prime numbers is said to be twin primes if p, q are two consecutive primes.*

That is, a pair $(p, p+2)$ where $p, p+2$ are two consecutive primes and for any prime value of p, $p+2$ is prime. $(3,5)$, $(5,7)$, $(29,31)$ are all twin primes. For $p = 7$, $p+2 = 9$ is not prime. So $(7,9)$ is not a twin prime.

Problem (11.66): Show that if n is the product of twin prime, then $\phi(n)\sigma(n) = (n+1)(n-3)$.

Sol: Let $n = p(p+2)$ where $p, p+2$ are both primes.

Then $\phi(p) = p - 1$, $\phi(p+2) = (p+1)$

$n = p(p+2) \Rightarrow \phi(n) = \phi(p) \cdot \phi(p+2) = (p-1)(p+1)$.

Also $\sigma(p) = p+1$, $\sigma(p+2) = (p+2) + 1 = p+3$.

$\sigma(n) = \sigma(p)\sigma(p+2) = (p+1)(p+3)$.

Now $\phi(n)\sigma(n) = (p-1)(p+1)(p+1)(p+3) = (p+1)^2(p^2+2p-3)$

$= (n+1)(n-3)$.

as $n = p(p+2) \Rightarrow n+1 = (p+1)^2$ and $n = p^2 + 2p$.

Hence the result.

Problem (11.67): If $a_1, a_2, \ldots, a_{\phi(n)}$ is a reduced set of residues modulo n, then $a_1 + a_2 + \cdots + a_{\phi(n)} \equiv 0 \pmod{n}$ for $n > 2$.

Sol: Reduced set of residues modulo n is given by $S = \{0, 1, 2, \ldots, n-1\}$

Then $\quad a_1 \equiv p_1 \pmod{n},\ p_1 \in S$

$\quad a_2 \equiv p_2 \pmod{n},\ p_2 \in S$

$\quad \ldots \quad \ldots \quad \ldots$

$\quad a_{\phi(n)} \equiv p_n \pmod{n},\ p_n \in S,$

where $p_1, p_2, \ldots, p_n$ is a permutation of the elements of S.

Therefore, $a_1 + a_2 + \cdots + a_{\phi(n)} \equiv p_1 + p_2 + \cdots + p_n \pmod{n}$

That is, $\quad a_1 + \cdots + a_{\phi(n)} \equiv (0 + 1 + 2 + \cdots + n - 1) \pmod{n}$

$$\Rightarrow \quad a_1 + a_2 + \cdots + a_{\phi(n)} \equiv \frac{n(n-1)}{2} \pmod{n} \equiv 0 \pmod{n}$$

for $\quad n > 2,\ n(n-1) \equiv 0 \pmod{2}$.

Problem (11.68): For any prime p, establish each of the following results:
(i) $\tau(p!) = 2\tau((p-1)!)$ (ii) $\sigma(p!) = (p+1)\sigma((p-1)!)$ (iii) $\phi(p!) = (p-1)\phi((p-1)!)$.

Sol: $p! = p(p-1)(p-2)\cdots 3 \cdot 2 \cdot 1$

The terms of the product $(p-1)(p-2)\cdots 3 \cdot 2 \cdot 1$ can be expressed in the form $p_1^{a_1} \cdot p_2^{a_2} \cdots p_k^{a_k}$, where p_i are distinct primes and $a_i \geq 0$. That is, $(p-1)! = \prod_{i=1}^{k} p_i^{a_i}$.

Then $\quad (p, p_i) = 1,\ (p_i, p_j) = 1\ \forall\ i, j$.

τ is an multiplicative function. Therefore,

$$\tau(p!) = \tau(p) \cdot \tau(p_1^{a_1}, p_2^{a_2}, \ldots, p_k^{a_k}) = (1+1)(a_1+1)(a_2+1)\cdots(a_k+1)$$
$$= 2\tau((p-1)!)$$

Since $\quad \tau((p-1)!) = \tau(p_1^{a_1}, p_2^{a_2}, \ldots, p_k^{a_k}) = (a_1+1)(a_2+1)\cdots(a_k+1)$.

(ii) $\sigma(p!) = \sigma(p)\sigma(p_1^{a_1} \cdot p_2^{a_2}, \ldots, p_k^{a_k})$ [σ is also multiplicative]

$= (p+1) \cdot \sigma((p-1)!)$.

(iii) $\phi(p!) = \phi(p)\phi(p_1^{a_1} \cdot p_2^{a_2} \ldots p_k^{a_k}) = (p-1) \cdot \phi((p-1)!)$.

Problem (11.69): If $(a, n) = 1$ and $((a-1), n) = 1$, then show that $1 + a + a^2 + \cdots + a^{\phi(n)-1} \equiv 0 \pmod{n}$.

Sol: Since $(a, n) = 1$, by Euler-Fermat's theorem $a^{\phi(n)} \equiv 1 \pmod{n}$

or, $\quad a^{\phi(n)} - 1 \equiv 0 \pmod{n}$

$\Rightarrow \quad (a-1) \cdot [a^{\phi(n)-1} + a^{\phi(n)-2} + \cdots + a^2 + a + 1] \equiv 0 \pmod{n}$

i.e., $\quad 1 + a + a^2 + \cdots + a^{\phi(n)-1} \equiv 0 \pmod{n} \quad [\because\ ((a-1), n) = 1]$.

Problem (11.70): If m and n are relatively prime positive integers, then show that $m^{\phi(n)} + n^{\phi(m)} \equiv 1 \pmod{mn}$.

Sol: Since $(m, n) = 1$, by Euler-Fermat's theorem,

$$m^{\phi(n)} \equiv 1 \pmod{n} \quad \text{and} \quad n^{\phi(m)} \equiv 1 \pmod{m},$$

which are equivalent to $m^{\phi(n)} - 1 \equiv 0 \pmod{n}$

and $n^{\phi(m)} - 1 \equiv 0 \pmod{m}$.

These together imply that $\{m^{\phi(n)} - 1\}\{n^{\phi(m)} - 1\} \equiv 0 \pmod{mn}$

$$\Rightarrow \quad m^{\phi(n)} \cdot n^{\phi(m)} - m^{\phi(n)} - n^{\phi(m)} + 1 \equiv 0 \pmod{mn}$$

$$\Rightarrow \quad m^{\phi(n)} + n^{\phi(m)} - 1 \equiv m^{\phi(n)} n^{\phi(m)} \pmod{mn}$$

$$\equiv 0 \pmod{mn}$$

$$[\text{since } m^{\phi(n)} n^{\phi(m)} - (mn) \cdot m^{\phi(n)-1} \cdot n^{\phi(m)-1} = 0]$$

$$\Rightarrow \quad m^{\phi(n)} + n^{\phi(m)} \equiv 1 \pmod{mn}.$$

Problem (11.71): Show that $\sum_{d|n} (-1)^{\frac{n}{d}} \phi(d) = 0$, when n is even.
$= -n$, when n is odd.

Sol: Case(1): Let n be even.

Then $n = 2^k m$, where m is an odd integer.

$d|n \Rightarrow$ (i) $d = 2^r$, $r = 0, 1, \ldots, k$

(ii) $d = m$

(iii) $d = 2^{r-1} \cdot m$, $r > 1$ $(r = 2, \ldots, k)$

(iv) $d = 2^k m$

(i) $$\sum_{\substack{r=0 \\ 2^r|n}}^{k} (-1)^{\frac{n}{d}} \phi(d) = \phi(1) + \phi(2) + \phi(2^2) + \cdots + \phi(2^{k-1}) - \phi(2^k)$$

$$= 1 + [1 + 2 + 2^2 + \cdots + 2^{k-2}] - 2^{k-1}$$

$$= 1 + (2^{k-1} - 1) - 2^{k-1} = 0.$$

(ii) $d = m$

$$\sum_{d|n} (-1)^{\frac{n}{d}} \phi(d) = (-1)^{\frac{2^k m}{m}} \phi(m) = \phi(m).$$

(iii) $$\sum_{\substack{r=2 \\ 2^{r-1} m|n}}^{k} (-1)^{\frac{n}{d}} \phi(d) = \phi(m)[\phi(2) + \phi(2^2) + \cdots + \phi(2^{k-1})]$$

$$= \phi(m)[1 + 2 + 2^2 + \cdots + 2^{k-2}] = \phi(m)[2^{k-1} - 1].$$

(iv) $d = 2^k m$.

$$\sum_{d|n} (-1)^{\frac{n}{d}} \phi(d) = (-1)^1 \phi(2^k m) = -2^{k-1} \phi(m).$$

$$\therefore \quad \sum_{d|n} (-1)^{\frac{n}{d}} \phi(d) = \text{(i)} + \text{(ii)} + \text{(iii)} + \text{(iv)}$$

$$= 0 + \phi(m) + \phi(m)(2^{k-1} - 1) - 2^{k-1}\phi(m) = 0.$$

Case(2): Let n be odd.

Then $n = p_1^{a_1} \cdot p_2^{a_2} \cdots P_k^{a_k}$, where $a_i \geq 0$
and p_i are distinct odd primes.

Let $$F(n) = \sum_{d|n} \phi(d).$$

Since ϕ is multiplicative, F is multiplicative.

$\therefore \quad F(n) = F(p_1^{a_1})F(p_2^{a_2})\cdots F(p_k^{a_k}).$

Let $\quad d = p_1^{x_1},\ 0 \le x_1 \le a_1$

$$F(p_1^{a_1}) = \sum_{\substack{x_1=0 \\ d|p_1^{x_1}}}^{a_1} \phi(p_1^{x_1}) = [\phi(1) + \phi(p_1) + \phi(p_1^2) + \cdots + \phi(p_1^{a_1})]$$

$$= [1 + p_1 - 1 + p_1^2 - p_1 + \cdots + p_1^{a_1} - p_1^{a_1-1}] = p_1^{a_1}.$$

Similarly, $F(p_2^{a_2}) = p_2^{a_2}$

$F(p_k^{a_k}) = p_k^{a_k}.$

Now $\sum_{d|n}(-1)^{\frac{n}{d}}\phi(d)$. Let $d = d_1(=1), d_2, \ldots, d_n(=n)$.

Then
$$\sum_{d|n}(-1)^{\frac{n}{d}}\phi(d) = (-1)^n\phi(1) + (-1)^{\frac{n}{d_2}}\phi(d_2) + \cdots + (-1)^{\frac{n}{n}}\phi(n)$$

$$= -[\phi(1) + \phi(d_2) + \cdots + \phi(d_n)]$$

$$\left[\text{since here each } \frac{n}{d} \text{ is odd } \forall\, d\right]$$

$$= \sum_{d|n}\phi(d) = -F(n) = -[F(p_1^{a_1})F(p_2^{a_2})\cdots + (p_k^{a_k})]$$

$$= -[p_1^{a_1} \cdot p_2^{a_2} \ldots p_k^{a_k}] = -n.$$

Problem (11.72): Given x and y and let $m = ax + by,\ n = cx + dy$ where $ad - bc = \pm 1$. Then prove that $(m, n) = (x, y)$.

Sol: $m = ax + by,\ n = cx + dy$

$\Rightarrow \quad x = md - nb,\ y = na - mc$

Let $\quad (m, n) = p$ and $(x, y) = q$.

Now $\quad (m, n) = p \Rightarrow p|m,\ p|n \Rightarrow p|ax + by,\ p|cx + dy$

$\Rightarrow \quad p|d(ax + by) - b(cx + dy) \Rightarrow p|(ad - bc)x$

$\Rightarrow \quad p|x,$ since $ad - bc = \pm 1$.

Also $\quad p|a(cx + dy) - c(ax + by) \Rightarrow p|(ad - bc)y$

$\Rightarrow \quad p|y,$ since $ad - bc = \pm 1$.

Thus $\quad p|x, p|y \Rightarrow p|q$ as $(x, y) = q$. (1)

Again $\quad q|x, q|y \Rightarrow q|md - nb,\ q|na - mc$

$\Rightarrow q|m$ and $q|n$, proceeding as above with $ad - bc = \pm 1$.

$\Rightarrow \quad q|p$ as $(m, n) = p$. (2)

So from (1) and (2) it follows that $p = q$

i.e., $\quad (m, n) = (x, y)$.

Definition (11.2): Pell's numbers

The Pell's numbers p_n, q_n are defined by

$p_0 = 0,\ p_1 = 1$ and $p_n = 2p_{n-1} + p_{n-2}\ \forall\, n \ge 2$

$q_0 = 1,\ q_1 = 1$ and $q_n = 2q_{n-1} + q_{n-2}\ \forall\, n \ge 2$.

Problem (11.73): Show that the Pell's numbers p_n and q_n form the sequences $\{p_n\} = \{0, 1, 2, 5, 12, 29, \cdots\}$ and $\{q_n\} = \{0, 1, 3, 7, 17, 41, \ldots\}$.

Also show that the Pell's numbers p_n and q_n can be represented by $p_n = \frac{\alpha^n - \beta^n}{2\sqrt{2}}$ and $q_n = \frac{\alpha^n + \beta^n}{2\sqrt{2}}$, where $\alpha = 1 + \sqrt{2}$ and $\beta = 1 - \sqrt{2}$.

Sol: Given that $p_0 = 0,\ p_1 = 1$.

Then from $p_n = 2p_{n-1} + p_{n-2},\ \forall\ n \geq 2$

$$\Rightarrow \quad p_2 = 2p_1 + p_0 = 2$$

$$p_3 = 2p_2 + p_1 = 4 + 1 = 5$$

$$p_4 = 2p_3 + p_2 = 2 \cdot 5 + 2 = 12$$

$$p_5 = 2p_4 + p_3 = 2 \cdot 12 + 5 = 29$$

$$p_6 \doteq 2p_5 + p_4 = 2 \cdot 29 + 12 = 70.$$

Thus, $\{p_n\} = \{0, 1, 2, 5, 12, 29, 70 \cdots\}$.

Given $q_0 = 1, q_1 = 1,\ q_n = 2q_{n-1} + q_{n-2}\ \forall\ n \geq 2$

$$\Rightarrow \quad q_2 = 2q_1 + q_0 = 2 + 1 = 3$$

$$q_3 = 2q_2 + q_1 = 2 \cdot 3 + 1 = 7$$

$$q_4 = 2q_3 + q_2 = 2 \cdot 7 + 3 = 17$$

$$q_5 = 2q_4 + q_3 = 2 \cdot 17 + 7 = 41.$$

Thus, $\{q_n\} = \{1, 1, 3, 7, 17, 41, \ldots\}$

Now let α, β be the roots of $x^2 - 2x - 1 = 0$.

Then $\alpha = 1 + \sqrt{2}$ and $\beta = 1 - \sqrt{2}$.

Also $\alpha + \beta = 2,\ \alpha - \beta = 2\sqrt{2}$,

$$p_0 = 0 = \frac{\alpha^0 - \beta^0}{2\sqrt{2}}$$

$$p_1 = 1 = \frac{\alpha - \beta}{2\sqrt{2}}$$

$$p_2 = 2 = \frac{(1+\sqrt{2})^2 - (1-\sqrt{2})^2}{2\sqrt{2}} = \frac{\alpha^2 - \beta^2}{2\sqrt{2}}$$

$$p_3 = 5 = \frac{(1+\sqrt{2})^3 - (1-\sqrt{2})^3}{2\sqrt{2}} = \frac{\alpha^3 - \beta^3}{2\sqrt{2}}.$$

The result is true for $n = 0, 1, 2, 3$.

Let the result be true for n, i.e., $\forall\ k \leq n$.

$$\text{Then} \quad p_n = \frac{(1+\sqrt{2})^n - (1-\sqrt{2})^n}{2\sqrt{2}} = \frac{\alpha^n - \beta^n}{2\sqrt{2}} = 2p_{n-1} + p_{n-2}.$$

$$\text{Now} \quad p_{n+1} = 2p_n + p_{n-1} = 2\frac{(\alpha^n - \beta^n)}{2\sqrt{2}} + \frac{\alpha^{n-1} - \beta^{n-1}}{2\sqrt{2}}$$

$$= \frac{2\alpha^n + \alpha^{n-1} - (2\beta^n + \beta^{n-1})}{2\sqrt{2}} = \frac{\alpha^{n-1}(2\alpha + 1) - \beta^{n-1}(2\beta + 1)}{2\sqrt{2}}.$$

$$= \frac{\alpha^{n-1}(2+2\sqrt{2}+1) - \beta^{n-1}(2-2\sqrt{2}+1)}{2\sqrt{2}}$$

$$= \frac{\alpha^{n-1}(1+\sqrt{2})^2 - \beta^{n-1}(1-\sqrt{2})^2}{2\sqrt{2}}$$

$$= \frac{\alpha^{n-1}\cdot\alpha^2 - \beta^{n-1}\cdot\beta^2}{2\sqrt{2}} = \frac{\alpha^{n+1} - \beta^{n+1}}{2\sqrt{2}}.$$

Hence, it is true for $n+1$.

Therefore, the result holds $\forall\ n \geq 0$.

Thus $p_n = \dfrac{\alpha^n - \beta^n}{2\sqrt{2}}$.

Similarly, we can show that $q^n = \dfrac{\alpha^n + \beta^n}{2}$.

Jacobi's symbol

We have introduced Legendre's symbol. Legendre's symbol $\left(\frac{a}{p}\right)$ has s meaning if $(a,p) = 1$ and p is an odd prime.

Also $\left(\dfrac{a}{p}\right)$ is defined by

(i) $\left(\dfrac{a}{p}\right) = 1$, if aRp. (ii) $\left(\dfrac{a}{p}\right) = -1$, if $a \not R p$.

In this contest another great mathematician Jacobi in the year 1840–1851 has generalised the above concept by taking p as an odd (+)ve integer. Then p may be a prime or a composite number. So p can be expressed as $p = p_1 \cdot p_2 \cdots p_k$, where p_i are all primes but not necessarily they are all distinct.

Definition (11.3) [Jacobi's] *Let $(a,p) = 1$ and $p = p_1 \cdot p_2 \cdots p_k$, where p_i are all primes, not all distinct (or may be all distinct). Then*

$$\left(\frac{a}{p}\right) = \left(\frac{a}{p_1\cdot p_2\cdots p_k}\right) = \left(\frac{a}{p_1}\right)\left(\frac{a}{p_2}\right)\cdots\left(\frac{a}{p_k}\right),$$

where the symbols on the right hand side stand as Legendre's symbol.

Problem (11.74): Some properties.

(i) $\left(\dfrac{1}{p}\right) = 1.$

Proof: Let $p = p_1 \cdot p_2 \cdots p_k$

Then $\left(\dfrac{1}{p}\right) = \left(\dfrac{1}{p_1}\right)\left(\dfrac{1}{p_2}\right)\cdots\left(\dfrac{1}{p_k}\right) = 1\cdot 1\cdots 1 = 1.$

since $1Rp\ \forall$ primes p as $x^2 \equiv 1 \pmod{p}$ has a solution 1. $[1^2 \equiv 1 \pmod{p}]$

(ii) $\left(-\dfrac{1}{p}\right) = (-1)^{\frac{p-1}{2}}.$

Proof: Let $p = p_1 \cdot p_2 \cdots p_k$.

For any two odd integers u and v,

$$\frac{u-1}{2} + \frac{v-1}{2} \equiv \frac{uv-1}{2} \pmod{2}.$$

Extending it to $p_1, p_2, \ldots, p_k$ odd integers.

$$\frac{p_1-1}{2}+\frac{p_2-1}{2}+\cdots+\frac{p_k-1}{2} \equiv \frac{p_1\cdot p_2\cdots p_k-1}{2} \pmod 2 \qquad (1)$$

$$\equiv \frac{p-1}{2} \pmod 2.$$

Also $x \equiv y \pmod m$

$\Rightarrow \quad a^x \equiv a^y \pmod m$ $\forall$ (+)ve integer a.

Now $\left(-\frac{1}{p}\right) = \left(-\frac{1}{p_1}\right)\left(-\frac{1}{p_2}\right)\cdots\left(-\frac{1}{p_k}\right)$ [by definition]

$$\equiv (-1)^{\frac{p_1-1}{2}}\cdot(-1)^{\frac{p_2-1}{2}}\cdots(-1)^{\frac{p_k-1}{2}} \pmod 2 \quad \text{[See Th(7.5) Chap-7]}$$

$$\equiv (-1)^{\frac{p_1-1}{2}+\frac{p_2-1}{2}+\cdots+\frac{p_k-1}{2}} \pmod 2$$

$$\equiv (-1)^{\frac{p_1p_2\cdots p_k-1}{2}} \pmod 2 \quad \text{[by (1)]}$$

$\therefore \left(-\frac{1}{p}\right) \equiv (-1)^{\frac{p-1}{2}} \pmod 2, \quad \therefore \left(-\frac{1}{p}\right) = (-1)^{\frac{p-1}{2}}.$

(iii) If $p = p_1p_2\ldots p_k$ where p_i are odd primes,

then $\dfrac{p_1^2-1}{8}+\dfrac{p_2^2-1}{8}+\cdots+\dfrac{p_k^2-1}{8} \equiv \dfrac{p^2-1}{8} \pmod 8.$

Proof: Let u and v be two odd integers.

Then u^2 and v^2 are also odd integers.

So $\quad u^2-1 \equiv 0 \pmod 8$, $v^2-1 \equiv 0 \pmod 8$

$\Rightarrow \quad u^2-1 = 8t,\ v^2-1 = 8t'$, for some $t, t' \in z$.

Now $\quad (u^2-1)(v^2-1) = 64q,\ q = tt' \in z$

$\Rightarrow \quad (u^2v^2-1) - \{u^2-1+v^2-1\} = 64q$

$\Rightarrow \quad \dfrac{u^2v^2-1}{8} - \left\{\dfrac{u^2-1}{8}+\dfrac{v^2-1}{8}\right\} = 8q$

$\Rightarrow \quad \dfrac{u^2-1}{8}+\dfrac{v^2-1}{8} \equiv \dfrac{u^2v^2-1}{8} \pmod 8.$

Result holds for $k = 2$. That is, if $p = p_1p_2$,

then $\dfrac{p_1^2-1}{8}+\dfrac{p_2^2-1}{8} \equiv \dfrac{(p_1p_2)^2-1}{8} \pmod 8. \qquad (1)$

Let it be true for k.

Then $\dfrac{p_1^2-1}{8}+\dfrac{p_2^2-1}{8}+\cdots+\dfrac{p_k^2-1}{8} \equiv \dfrac{(p_1p_2\ldots p_k)^2-1}{8} \pmod 8. \qquad (2)$

Now $\dfrac{p_1^2-1}{8}+\dfrac{p_2^2-1}{8}+\cdots+\dfrac{p_k^2-1}{8}+\dfrac{p_{k+1}^2-1}{8}$

$$\equiv \frac{(p_1p_2\ldots p_k)^2-1}{8}+\frac{p_{k+1}^2-1}{8} \pmod 8$$

$$\equiv \frac{(p_1p_2\ldots p_kp_{k+1})^2-1}{8} \pmod 8 \quad \text{[by (1)]}.$$

Thus, the result is also true for $k+1$.

Hence, $\dfrac{p_1^2-1}{8}+\dfrac{p_2^2-1}{8}+\cdots+\dfrac{p_k^2-1}{8}\equiv\dfrac{(p_1p_2\cdots p_k)^2-1}{8} \pmod 8$

$$=\frac{p^2-1}{8} \pmod 8.$$

(iv) For any (+)ve odd integer p, $\left(\dfrac{2}{p}\right)=(-1)^{\frac{p^2-1}{8}}$.

Proof: Let $p=p_1\cdot p_2\cdots p_k, p_i$ are odd primes.

Then $\left(\dfrac{2}{p}\right)=\left(\dfrac{2}{p_1}\right)\left(\dfrac{2}{p_2}\right)\cdots\left(\dfrac{2}{p_k}\right)$

$$\equiv(-1)^{\frac{p_1^2-1}{8}}\cdot(-1)^{\frac{p_2^2-1}{8}}\cdots(-1)^{\frac{p_k^2-1}{8}} \pmod 8 \text{ [by Theorem (7.8)]}$$

So $\left(\dfrac{2}{p}\right)\equiv(-1)^{\frac{p_1^2-1}{8}+\frac{p_2^2-1}{8}+\cdots+\frac{p_k^2-1}{8}} \pmod 8$

$$\equiv(-1)^{\frac{p^2-1}{8}} \pmod 8. \quad \text{[by Prob. (iii) [11.74]]}$$

Therefore, $\left(\dfrac{2}{p}\right)=(-1)^{\frac{p^2-1}{8}}$.

Also $\dfrac{p^2-1}{8}\equiv\left[\dfrac{p+1}{4}\right] \pmod 2$.

$\therefore$ $\left(\dfrac{2}{p}\right)=(-1)^{\frac{p^2-1}{8}}=(-1)^{[\frac{p+1}{4}]}$.

(v) If $a=a_1a_2$ and $p=p_1p_2\ldots p_k$, then $\left(\dfrac{a}{p}\right)=\left(\dfrac{a_1}{p}\right)\left(\dfrac{a_2}{p}\right)$.

Proof: $\left(\dfrac{a}{p}\right)=\left(\dfrac{a_1a_2}{p}\right)=\left(\dfrac{a_1a_2}{p_1p_2\ldots p_k}\right)$

$$=\left(\frac{a_1a_2}{p_1}\right)\left(\frac{a_1a_2}{p_2}\right)\cdots\left(\frac{a_1a_2}{p_k}\right) \quad \text{[by Jacobi's symbol]}$$

$$=\left(\frac{a_1}{p_1}\right)\left(\frac{a_2}{p_1}\right)\left(\frac{a_1}{p_2}\right)\left(\frac{a_2}{p_2}\right)\cdots\left(\frac{a_1}{p_k}\right)\left(\frac{a_2}{p_k}\right) \quad \text{[by (iii) of Th. (7.5)]}$$

$$=\left(\frac{a_1}{p_1}\right)\left(\frac{a_1}{p_2}\right)\cdots\left(\frac{a_1}{p_k}\right)\cdot\left(\frac{a_2}{p_1}\right)\left(\frac{a_2}{p_2}\right)\cdots\left(\frac{a_2}{p_k}\right)$$

$$=\left(\frac{a_1}{p}\right)\left(\frac{a_2}{p}\right), \quad \text{[by Jacobi's symbol]}$$

In general, if $a=a_1\cdot a_2\cdots a_n$, and $p=p_1p_2\cdots p_k$, then

$$\left(\frac{a}{p}\right)=\left(\frac{a_1}{p}\right)\left(\frac{a_2}{p}\right)\cdots\left(\frac{a_k}{p}\right).$$

Problem (11.75): Evaluate the following Jacobi's symbol:

(i) $\left(\dfrac{63}{3658}\right)$ (ii) $\left(\dfrac{66}{1015}\right)$.

Sol: (i) $3658 = 2 \cdot 31 \cdot 59$

$$\left(\frac{63}{3658}\right) = \left(\frac{63}{2 \cdot 31 \cdot 59}\right) = \left(\frac{63}{2}\right)\left(\frac{63}{31}\right)\left(\frac{63}{59}\right) = \left(\frac{1}{2}\right)\left(\frac{1}{31}\right)\left(\frac{4}{59}\right)$$

$$[\because 63 \equiv 1 \pmod 2,\ 63 \equiv 1 \pmod{31},\ 63 \equiv 4 \pmod{59}]$$

$$= 1 \cdot 1 \cdot \left(\frac{2}{59}\right)\left(\frac{2}{59}\right) = (-1)^{\left[\frac{59+1}{4}\right]}(-1)^{\left[\frac{59+1}{4}\right]} = (-1)(-1) = 1.$$

(ii) $$\left(\frac{66}{1015}\right) = \left(\frac{2 \cdot 3 \cdot 11}{5 \cdot 7 \cdot 29}\right) = \left(\frac{2}{5 \cdot 7 \cdot 29}\right)\left(\frac{3}{5 \cdot 7 \cdot 29}\right)\left(\frac{11}{5 \cdot 7 \cdot 29}\right)$$

$$= \left[\left(\frac{2}{5}\right)\left(\frac{2}{7}\right)\left(\frac{2}{29}\right)\right]\left[\left(\frac{3}{5}\right)\left(\frac{3}{7}\right)\left(\frac{3}{29}\right)\right]\left[\left(\frac{11}{5}\right)\left(\frac{11}{7}\right)\left(\frac{11}{29}\right)\right]$$

$$= \left[(-1)^{\frac{25-1}{8}} \cdot (-1)^{\frac{49-1}{8}}(-1)^{\left[\frac{29+1}{4}\right]}\right]\left[-\left(\frac{5}{3}\right)\left(\frac{7}{3}\right)\left(\frac{29}{3}\right)\right] \times \left[\left(\frac{1}{5}\right)\left(\frac{4}{7}\right)\left(\frac{29}{11}\right)\right]$$

$$= [(-1)(1)(-1)]\left[-\left(\frac{2}{3}\right)\left(\frac{1}{3}\right)\left(\frac{2}{3}\right)\right]\left[\left(\frac{1}{5}\right)\left(\frac{4}{7}\right)\left(\frac{7}{11}\right)\right]$$

$$= 1 \cdot \left[-(-1)^{\frac{9-1}{8}} \cdot (-1)^{\frac{9-1}{8}}\right]\left[-1 \cdot (1) \cdot \left(\frac{11}{7}\right)\right]$$

$$\left[\text{since } \left(\frac{4}{7}\right) = \left(\frac{2}{7}\right)\left(\frac{2}{7}\right) = 1\right]$$

$$= 1 \cdot [-(-1)(-1)]\left[-1(1)\left(\frac{4}{7}\right)\right] = -1 \cdot (-1) = 1.$$

We have used the following properties of Legendre's symbol in finding the above results.

$\left(\frac{p}{q}\right) = \left(\frac{q}{p}\right)$, if one of the forms is $4n+1$, and $\left(\frac{p}{q}\right) = -\left(\frac{q}{p}\right)$ and if both of them are in the form $4n+3$.

Also $\left(\frac{p}{q}\right) = \left(\frac{r}{q}\right)$, if $p \equiv r \pmod q$.

Problem (11.76): Show that (i) the only prime $n^2 - 4$ is 5 and (ii) the only prime $n^3 - 1$ is 7.

Sol: (i) $n^2 - 4$ is prime if it has a factor 1 with some prime factor.

Now $n^2 - 4 = (n-2)(n+2)$.

Here $n - 2 = 1$, only when $n = 3$ and in that case $n + 2 = 5$, a prime factor which is same as $n^2 - 4$ for $n = 3$.

So only prime $n^2 - 4$ is 5.

(ii) $n^3 - 1 = (n-1)(n^2 + n + 1)$

Now, $n - 1 = 1$, only when $n = 2$ and in that case, $n^2 + n + 1 = 2^2 + 2 + 1 = 7$, a prime factor which is same as $n^3 - 1$ for $n = 2$ only.

So only prime $n^3 - 1$ is 7.

Problem (11.77): If $a_1, a_2, \ldots a_{\phi(n)}$ are integers less than and prime to n, then show that $a_1 + a_2 + \cdots + a_{\phi(n)} = \frac{1}{2} n\phi(n)$.

Sol: By theorem [5.9], we have

$$a_1 + a_2 + \cdots + a_{\phi(n)} = (n - a_1) + (n - a_2) + \cdots (n - a_{\phi(n)})$$
$$= n\phi(n) - (a_1 + a_2 + \cdots + a_{\phi(n)})$$

$$\Rightarrow \quad 2(a_1 + a_2 + \cdots + a_{\phi(n)}) = n\phi(n)$$

$$\Rightarrow \quad (a_1 + a_2 + \cdots + a_{\phi(n)}) = \frac{n}{2}\phi(n).$$

Remarks: $(a_1 + a_2 + \cdots + a_{\phi(n)}) \equiv 0 \pmod{n}$.

Example: Let $n = 35$.

Then $\quad \phi(n) = \phi(35) = \phi(7) \cdot \phi(5) = 6 \cdot 4 = 24.$

The integers less than and prime to 35 are

$$1, 2, 3, 4, 6, 8, 9, 11, 12, 13, 16, 17, 18, 19, 22, 23, 24, 26, 27, 29, 31, 32, 33, 34.$$

There sum $= 1 + 2 + 3 + 4 + 6 + 8 + 9 + 11 + 12 + 13 + 16 + 17 + 18 + 19 + 22 + 23 + 24 + 26 + 27 + 29 + 31 + 32 + 33 + 34 = 420$

$$= \frac{1}{2} \cdot 35 \cdot 24 = \frac{1}{2} \cdot n\phi(n).$$

Also sum of pairs in each case,

$$(1, 34),\ (2, 33),\ (3, 32), \ldots, (17, 18) \text{ is } 35.$$

It also follows that $a_1 + a_2 + \cdots + a_{\phi(n)} = 420$.

Therefore, $a_1 + a_2 + \cdots + a_{\phi(n)} \equiv 0 \pmod{35}$.

Problem (11.78): For $n > 2$, show that $\phi(n^2) + \phi(n+1)^2 \leq 2n^2$.

Solution: Let $n = 3$.

Then, $\quad \phi(n^2) + \phi(n+1)^2 = \phi(9) + \phi(16) = 6 + 8 = 14,$

while $\quad 2n^2 = 2 \cdot 3^2 = 18.$

Thus, for $n = 3(7^2),\ \phi(3^2) + \phi(3+1)^2 < 2 \cdot 3^2.$

Let $\quad n = 4.$ Then $\phi(4^2) + \phi(5^2) = 8 + 20 = 28,$

while $\quad 2n^2 = 2 \cdot 4^2 = 32.$

Thus, for $n = 4,\ \phi(4^2) + \phi(4+1)^2 < 2 \cdot 4^2.$

We can express n^2, $(n+1)^2$ as $n^2 = p_1^{a_1} p_2^{a_2} \cdots p_n^{a_n}$, p_i are primes

and $\quad (n+1)^2 = q_1^{b_1} q_2^{b_2} \cdots q_n^{b_m}$, where q_j are primes.

Then $\quad \phi(n^2) = n^2\left(1 - \frac{1}{p_1}\right)\left(1 - \frac{1}{p_2}\right) \cdots \left(1 - \frac{1}{p_n}\right) < n^2$

and $\quad \phi(n+1)^2 = (n+1)^2\left(1 - \frac{1}{q_1}\right)\left(1 - \frac{1}{q_2}\right) \cdot \left(1 - \frac{1}{q_m}\right) < (n+1)^2.$

Therefore, $\phi(n^2) + \phi(n+1)^2 < n^2 + (n+1)^2 = n^2 + n^2\left(1 + \frac{1}{n}\right)^2.$

$= 2n^2$, where n is very large.

Hence, $\quad \phi(n^2) + \phi(n+1)^2 \leq 2n^2$

Problem (11.79): Find the highest power of 15 in 500!.

Sol: $15 = 3 \times 5$. So power of 15 contains the power of 3 and power of 5.

Highest power of 3 in 500! is given by the sum of the following results.

$$I\left(\frac{500}{3}\right) = 166$$

$$I\left(\frac{500}{3^2}\right) = I\left(\frac{166}{3}\right) = 55$$

$$I\left(\frac{500}{3^3}\right) = I\left(\frac{55}{3}\right) = 18$$

$$I\left(\frac{500}{3^4}\right) = I\left(\frac{18}{3}\right) = 6$$

$$I\left(\frac{500}{3^5}\right) = I\left(\frac{6}{3}\right) = 2\ (< 3).$$

Hence, the highest power of 3 in 500! $= 166 + 55 + 18 + 6 + 2 = 247$ (i)

Now the highest power of 5 in 500! is the sum of the following result:

$$I\left(\frac{500}{5}\right) = 100$$

$$I\left(\frac{500}{5^2}\right) = I\left(\frac{100}{5}\right) = 20$$

$$I\left(\frac{500}{5^3}\right) = I\left(\frac{20}{5}\right) = 4\ (< 5).$$

Therefore, highest power of 5 in 500! $= 100 + 20 + 4 = 124$.

Here $500! = 3^{247} \times 5^{124} \times t$ for some t, $(t, 15) = 1$

$$= (3 \times 5)^{124} \times 3^{123} \times t$$

$$= (15)^{124} \times 3^{123} \times t.$$

Hence, the highest power of 15 in 500! is 124.

Problem (11.80): For any integer $n > 1$, prove that $\exists$ integers n_1, n_2 such that $\tau(n_1) + \tau(n_2) = n$.

Sol: For $n = 2\ (> 1)$, $\exists$ integers, 1, 1 such that $\tau(1) + \tau(1) = 1 + 1 = 2$.

For, $n = 3$, $\exists\ n_1 = 1$, $n_2 = 2$ such that $\tau(1) + \tau(2) = 1 + 2 = 3$.

For, $n = 4$, $\exists\ n_1 = 3$, $n_2 = 2$, such that $\tau(3) + \tau(2) = 2 + 2 = 4$.

So, let $n_1 = p^{n-3}$, where p is any prime and $n_2 = 2$ be two integers, $n > 3$.

Then $\tau(n_1) = \tau(p^{n-3}) = (n - 3) + 1 = n - 2$ [by Th. [5.2]]

$\tau(n_2) = \tau(2) = 2$.

Then for any $n (> 1) \exists\ n_1, n_2$ such that $\tau(n_1) + \tau(n_2) = n - 2 + 2 = n$.

In more general,

if $n = p_1^{a_1} p_2^{a_2} \cdots p_n^{a_n}$, where p_i are primes and $a_i > 0$ subject to $n > 1$.

$$\tau(n) = (a_1 + 1)(a_2 + 1) \cdots (a_n + 1)$$

$$= 1 + \sum a_1 + \sum a_1 a_2 + \cdots + a_1 a_2 \cdots a_n$$

$$= x, \quad \text{where } x = 1 + \sum a_1 + \sum a_1 a_2 + \cdots + a_1 a_2 \cdots a_n$$

$$\exists\, n_1 = p^{n-x-3} = p^{n-(x+3)},$$

$$\text{where } x = 1 + \sum a_1 + \sum a_1 a_2 + \cdots + a_1 a_2 \cdots a_n$$

and $\quad n_2 = 2^{x+1}$.

Then $\quad \tau(n_1) = n - (x+3) + 1 = n - (x+2)$

$\quad \tau(n_2) = (x+1) + 1 = x + 2$

so that $\quad \tau(n_1) + \tau(n_2) = n - (x+2) + (x+2) = n.$

(Go through the Example)

Let $\quad n = 2^2 \cdot 3^2 \cdot 5 \cdot 7^2 = 8820.$

We can choose two integers,

$$n_1 = 2^{8815} \quad \text{and} \quad n_2 = 6$$

such that $\tau(n_1) = (8815 + 1) = 8816$

$\tau(n_2) = \tau(6) = 4.$

Therefore, $\tau(n_1) + \tau(n_2) = 8816 + 4 = 8820 = n.$

We may also choose two integers,

$$n_1 = 2^{8817} \quad \text{and} \quad n_2 = 2$$

such that $\tau(n_1) = 8817 + 1 = 8818$ and $\tau(n_2) = 2.$

Therefore, $\tau(n_1) + \tau(n_2) = 8818 + 2 = 8820.$

Problem (11.81): If $n(>1) = p_1^{k} p_2^{k_2} \cdots p_r^{k_r}$ where p_i are primes and $k_r > 0$. Then show that

$$1 > \frac{n}{\sigma(n)} > \left(1 - \frac{1}{p_1}\right)\left(1 - \frac{1}{p_2}\right) \cdots \left(1 - \frac{1}{p_r}\right)$$

Sol: $\sigma(n) = \dfrac{p_1^{k_1+1}}{p_1 - 1} \cdot \dfrac{p_2^{k_2+1} - 1}{p_2 - 1} \cdots \dfrac{p_r^{k_r+1} - 1}{p_r - 1}$ [by Th. [5.2]]

$$= \frac{\Pi_{i=1}^{r}\left(p_i^{k_i+1} - 1\right)}{p_1 p_2 \cdots p_r \left(1 - \frac{1}{p_1}\right)\left(1 - \frac{1}{p_2}\right) \cdots \left(1 - \frac{1}{p_r}\right)}$$

$$\frac{n}{\sigma(n)} = \frac{p_1^{k_1} p_2^{k_2} \cdots p_r^{k_r} \cdot p_1 p_2 \cdots p_r \left(1 - \frac{1}{p_1}\right) \cdots \left(1 - \frac{1}{p_r}\right)}{\Pi_{i=1}^{n} p_i^{k_i+1}\left(1 - \frac{1}{p_i^{k_{i+1}}}\right)}$$

$$= \frac{p_1^{k_1+1} \cdot p_2^{k_2+1} \cdots p_r^{k_r+1}\left(1 - \frac{1}{p_1}\right)\left(1 - \frac{1}{p_2}\right) \cdots \left(1 - \frac{1}{p_r}\right)}{\Pi_{i=1}^{n} p_i^{k_i+1}\left(1 - \dfrac{1}{p_i^{k_i+1}}\right)}$$

$$= \frac{\Pi_{i=1}^{n}\left(1-\dfrac{1}{p_i}\right)}{\Pi_{i=1}^{r}\left(1-\frac{1}{p_i^{k_i+1}}\right)}. \tag{1}$$

Since $\left(1-\dfrac{1}{p_i^{k_i+1}}\right) < 1\ \forall i = 1, 2, \ldots, r,\ \Pi\left(1-\dfrac{1}{p_i^{k_i+1}}\right) < 1$

$$\Rightarrow \quad \frac{1}{\Pi\left(1-\frac{1}{p_i^{k_i+1}}\right)} > 1.$$

So from (1), $\dfrac{n}{\sigma(n)} > \Pi_{i=1}^{n}\left(1-\dfrac{1}{p_i}\right) = \left(1-\dfrac{1}{p_1}\right)\cdots\left(1-\dfrac{1}{p_r}\right)$ (2)

Also $\sigma(n) = (1 + p_1 + p_1^2 + \cdots + p_1^{k_1})(1 + p_2 + p_2^2 + \cdots + p_2^{k_2})$

$$\cdots(1 + p_r + p_r^2 + \cdots + p_r^{k_r})$$

$$> p_1^{k_1} p_2^{k_2} \cdots p_r^{k_r} = n.$$

$$\therefore \quad 1 > \frac{n}{\sigma(n)}. \tag{3}$$

So from (2) and (3) it follows that

$$1 > \frac{n}{\sigma(n)} > \left(1-\frac{1}{p_1}\right)\left(1-\frac{1}{p_2}\right)\cdots\left(1-\frac{1}{p_r}\right).$$

Problem (11.82): (i) If n be an odd integer, show that $n^4 + 4n^2 + 11$ is of the form $16k,\ k \in Z$.

(ii) Show that any integer of the form $6k + 5$ is also of the form $3r + 2$, but the converse does not hold.

(iii) Prove that no integer in the following sequence is a perfect square: 11, 111, 1111, 11111, ...

(iv) If r be the remainder when a is divided by $b(> 0)$, then show that kr is the remainder when ka is divided by kb, k is any arbitrary (+)ve integer.

(v) Prove that if a and b are integers with $b > 0$, then $\exists$ unique integers q and r such that $a = bq + r,\ 2b \le r < 3b$.

(vi) For $n \ge 1$, show that $\dfrac{n(n+1)(2n+1)}{6}$ is an integer.

Sol: (i): If n be an odd integer, then $n^2 = 8k' + 1$ by problem [1.3].

Therefore, $n^4 = 64k'^2 + 16k' + 1$

and hence, $n^4 + 4n^2 + 11 = 64k'^2 + 16k' + 1 + 32k' + 4 + 11 = 64k'^2 + 48k' + 16$

$$= 16(4k'^2 + 3k' + 1) = 16k,$$

where $\quad k = 4k'^2 + 3k' + 1 \in Z$.

(ii) Let $n \in Z$ and $n = 6k + 5$.

Then, $n = 3 \cdot 2k + 3 + 2 = 3(2k + 1) + 2$

$$= 3r + 2, \quad \text{where } r = 2k + 1,\ k \in Z$$

$3r + 2$ represents an integer $\forall\ r \in Z$.

For $r = 0$, $3r + 2 = 2$ is an integer.

But $r = 0 \Rightarrow 2k + 1 = 0 \Rightarrow k = -\frac{1}{2} \notin Z$.

So, though $3r+2$ represents an integer for $r = 0$, $6k+5$ represents the same if $k = -\frac{1}{2}$.

But $k = -\frac{1}{2} (\notin Z)$. Hence, the converse is not true.

(iii) $11 = 8 + 3 = 4 \cdot 2 + 3$

$111 = 108 + 3 = 4 \cdot 27 + 3$

$1111 = 11108 + 3 = 4 \cdot 2777 + 3$

....

$111\,111 \cdots 111 = 111 \cdots 108 + 3 = 4k + 3$ for some $k \in Z$.

So each term of the given sequence is of the form $4k+3$, which is not a perfect square. [See Th [1.6]].

(iv) We have, by divisional algorithm theorem [1.3],

$a = bq + r. \quad 0 \le r < b.$

Then $ka = (kb)q + kr$, $0 \le kr < kb$ $(k > 0)$.

$\Rightarrow$ kr is the remainder when ka is divided by kb.

(v) By theorem [1.3], we have

$a = bq' + r' \quad 0 \le r' < b,\ q', r' \in Z$ exist uniquely. (1)

Then $a + 2b = bq' + r' + 2b$, $2b \le r' + 2b < 3b$

$\Rightarrow$ $a = b(q' - 2) + (r' + 2b)$

$\Rightarrow$ $a = bq + r$, when $q = q' - 2$, $r \in Z$, exist uniquely satisfying $2b \le r < 3b$.

(vi) Any integer $n(> 1)$ is of the form $6k,\ 6k + 1,\ 6k + 2, \ldots, 6k + 5$. So for any one of the values of n (i.e., $n = 6k$ or $6k + 1$ or $6k + 2 \cdots$ or $6k + 5$), $n(n + 1)(n + 2)$ is multiple of 6.

Hence, $\frac{n(n+1)(n+2)}{6}$ is an integer.

Problem (11.83): Find all the primes that divide 50!.

Sol: After breaking up the factorial parts the first prime occurring in the series is considered the maximal range of the primes that divide 50!.

$50! = 50 \cdot 49 \cdot 48 \cdot 47 \cdot 46, \ldots.$

The first prime in the series is 47.

So all primes less than and equal to 47 divide 50! i.e., $2, 3, 5, 7, 11, 13, 17, 19, 23, 29, 31, 37, 41, 43, 47$ will divide 50!.

Example: $25! = 25 \cdot 24 \cdot 23 \cdot 22, \ldots$

The first prime in the series is 23.

So all primes ≤ 23 will divide 25!.

2, 3, 5, 7, 11, 13, 17, 19, 23 will divide 25!.

Problem (11.84): (i) Find the smallest integer $a > 2$ such that $2|a,\ 3|a + 1,\ 4|a + 2,\ 5|a + 3,\ 6|a + 4$.

(ii) Find the least (+)ve integer so that $3|a,\ 5|a+1,\ 7|a+2$.

(iii) Obtain three consecutive integers each having a square factor.

Sol: (i) From the given conditions we have

$$a \equiv 0 \pmod 2, \qquad (1)$$
$$a+1 \equiv 0 \pmod 3 \Rightarrow a \equiv -1 \pmod 3$$
$$\Rightarrow a \equiv 2 \pmod 3 \qquad (2)$$
$$a+2 \equiv 0 \pmod 4 \Rightarrow \equiv -2 \pmod 4$$
$$\Rightarrow a \equiv 2 \pmod 4 \qquad (3)$$
$$a+3 \equiv 0 \pmod 5 \Rightarrow a \equiv -3 \pmod 5$$
$$\Rightarrow a \equiv 2 \pmod 5 \qquad (4)$$
$$a+4 \equiv 0 \pmod 6 \Rightarrow a \equiv -4 \pmod 6$$
$$\Rightarrow a \equiv 2 \pmod 6. \qquad (5)$$

So from (1), (2), (3), (4) and (5) we have

$$a \equiv 2 \pmod{[2,3,4,5,6]} \quad [[23,4,5,6] = \text{l.c.m. of } 2,\ 3,\ 4,\ 5,\ 6]$$

i.e., $a \equiv 2 \pmod{60}$

i.e., $a \equiv 62 \pmod{60}$ $(\because\ a > 2)$.

So, 62 is the smallest (+)ve integer satisfies the given conditions.

(ii) From the given conditions we have

$$a \equiv 0 \pmod 3$$

or $a \equiv 3 \pmod 3$ (1)

$$a+1 \equiv 0 \pmod 5$$

$\Rightarrow$ $a \equiv -1 \pmod 5$

$\Rightarrow$ $a \equiv 4 \pmod 5$ (2)

$$a+2 \equiv 0 \pmod 7$$

$\Rightarrow$ $a \equiv -2 \pmod 7$

$\Rightarrow$ $a \equiv 5 \pmod 7$. (3)

Apply Chinese remainder theorem.

Here $M = 3 \cdot 5 \cdot 7,\ m_1 = 3,\ m_2 = 5,\ m_3 = 7.$

3, 5, 7 are prime to each other. $b_1 = 3,\ b_2 = 4,\ b_3 = 5$

$\therefore$ $\dfrac{M}{m_1} y_1 \equiv 1 \pmod 3$

$\Rightarrow$ $35y_1 \equiv 1 \pmod 3 \Rightarrow y_1 \equiv 2 \pmod 3$ (4)

$\dfrac{M}{m_2} y_2 \equiv 1 \pmod 5$

$\Rightarrow$ $21y_2 \equiv 1 \pmod 5$

$\Rightarrow$ $y_2 \equiv 1 \pmod 5$ (5)

$\dfrac{M}{m_3} y_3 \equiv 1 \pmod 7$

$\Rightarrow$ $15y_3 \equiv 1 \pmod 7$

$\Rightarrow \qquad y_3 \equiv 1 \pmod 7$. (6)

Thus, $\quad a \equiv \frac{M}{m_1}y_1b_1 + \frac{M}{m_2}y_2b_2 + \frac{M}{m_3}y_3b_3 \pmod{105}$

i.e., $\quad a \equiv 35 \cdot 2 \cdot 3 + 21 \cdot 1 \cdot 4 + 15 \cdot 1 \cdot 5 \pmod{105}$

i.e., $\quad a \equiv 210 + 84 + 75 \pmod{105}$

i.e., $\quad a \equiv 369 \pmod{105}$

i.e., $\quad a \equiv 54 \pmod{105}$.

So least (+)ve integer $a = 54$ satisfies the given conditions.

(iii) Let $a,\ a+1,\ a+2$ be three consecutive integers such that

$$2^2|a,\ 3^2|a+1,\ 5^2|a+2.$$

Then $\quad a \equiv 0 \pmod{2^2} \Rightarrow a \equiv 4 \pmod 4$ (1)

$a + 1 \equiv 0 \pmod{3^2} \ \Rightarrow\ a \equiv 8 \pmod 9$ (2)

$a + 2 \equiv 0 \pmod{5^2} \Rightarrow a \equiv 23 \pmod{25}$. (3)

4, 9, 25 are prime to each other. We apply Chinese remainder theorem.

Here $\quad M = 4 \cdot 9 \cdot 25 = 900,\ m_1 = 4,\ m_2 = 9,\ m_3 = 25$

Let $\quad \frac{M}{m_1}y_1 \equiv 1 \pmod 4$

Then $\quad 225y_1 \equiv 1 \pmod 4$

$\Rightarrow \quad y_1 \equiv 1 \pmod 4$. (4)

Let $\quad \frac{M}{m_2}y_1 \equiv 1 \pmod 9$.

Then $\quad 100y_2 \equiv 1 \pmod 9$

$\Rightarrow \quad y_2 \equiv 1 \pmod 9$. (5)

Let $\quad \frac{M}{m_3}y_3 \equiv 1 \pmod{25}$.

Then $\quad 36y_3 \equiv 1 \pmod{25}$

$\Rightarrow \quad 36y_3 \equiv 1 - 13 \times 25 \pmod{25}$

$\Rightarrow \quad 36y_3 \equiv -324 \pmod{25}$

$\Rightarrow \quad y_3 \equiv -9 \pmod{25}$

$\Rightarrow \quad y_3 \equiv 16 \pmod{25}$. (5)

Therefore, $a \equiv \frac{M}{m_1}y_1b_1 + \frac{M}{m_2}y_2b_2 + \frac{M}{m_3}y_3b_3 \pmod{900}$.

i.e., $\quad a \equiv 225 \cdot 1 \cdot 4 + 100 \cdot 1 \cdot 8 + 36 \cdot 16 \cdot 23 \pmod{900}$

i.e., $\quad a \equiv 900 + 800 + 13248 \pmod{900}$

i.e., $\quad a \equiv 14948 \pmod{900}$

i.e., $\quad a \equiv 14948 - 16 \times 900 \pmod{900}$

i.e., $\quad a \equiv 548 \pmod{900}$.

So the least value of $a = 548$.

So, three consecutive integers are 548, 549, 550 having square factors 2^2, 3^2, 5^2 respectively.

Problem (11.85): Show that $x \equiv a \pmod{n}$

$\Rightarrow$ either $x \equiv a \pmod{2n}$

or, $x \equiv a + n \pmod{2n}$.

Sol: $x \equiv a \pmod{n}$

$\Rightarrow \quad n | x - a$

$\Rightarrow \quad x - a = nt$ for some $t \in Z$.

For, $t = 2m$, $m \in Z$, $x - a = n \cdot 2m = 2n \cdot m$

$\Rightarrow \quad 2n | x - a$

$\Rightarrow \quad x - a \equiv 0 \pmod{2n}$

$\Rightarrow \quad x \equiv a \pmod{2n}$.

Again, for $t = 2m + 1$, $m \in Z$.

$x - a = n(2m + 1) = n \cdot 2m + n$.

$\Rightarrow \quad x - a - n = 2n \cdot m$.

$\Rightarrow \quad 2n | x - a - n$

$\Rightarrow \quad x - a - n \equiv 0 \pmod{2n}$

$\Rightarrow \quad x \equiv a + n \pmod{2n}$.

Hence the results.

Expression of an integer N as a product of two factors

Let N be a given (+)ve integer.

If $\exists$ an integer x_0 such that $x_0^2 - N$ is a perfect square, say y_0^2, then $x_0^2 - N = y_0^2 \Rightarrow N = x_0^2 - y_0^2 \Rightarrow N = (x_0 - y_0)(x_0 + y_0)$, product of two factors.

Thus, to express N as the product of two factors, we have to find out the integer x_0 so that $x_0^2 - N = y_0^2$.

We apply the following method in finding x_0.

(i) We find an integer x such that $(x - 1)^2 < N < x^2$.

(ii) We then find out the values of $x^2 - N$, $(x + 1)^2 - N$, $(x + 2)^2 - N \cdots$ in succession till we get $(x + a)^2 - N =$ a perfect square.

(iii) To evaluate $(x + 1)^2$, $(x + 2)^2, \ldots$, we notice that

$(x + 1)^2 - x^2 = 2x + 1$

$(x + 2)^2 - (x + 1)^2 = 2x + 3 = (2x + 1) + 2$

$(x + 3)^2 - (x + 2)^2 = 2x + 5 = (2x + 3) + 2$ and so on.

Thus, to find $(x + 1)^2$ we add $(2x + 1)$ to x^2.

Similarly, to find $(x + 2)^2$ from $(x + 1)^2$ we addd $(2x + 1) + 2$ to $(x + 1)^2$.and so on.

Example: Let $N = 2911$.

If we choose $x = 54$, then

$(x - 1)^2 < 2911 < x^2$

i.e., $53^2 < 2911 < 54^2$.

Again, $2x + 1 = 2 \cdot 54 + 1 = 109$.

Also $x^2 - 2911 = 54^2 - 2911 = 5$.

Calculation is done in the tabular form

a	$(x+a)^2 - 2911$	Remarks
0	$54 - 2911 = 5$.	Not a perfect square.
1	$5 + 109 = 114$	Add $(2x + 1)$. Not a perfect square.
	$\frac{5}{109}$	
2	$\frac{114}{111}$	Add $(2x + 1) + 2$. Not a perfect square.

Now $114 + 111 = 225 = 15^2$, a perfect square.

Therefore, $(x + 2)^2 - N = 15^2 \Rightarrow (54 + 2)^2 - N = 15^2$

$\Rightarrow$ $N = (56)^2 - 15^2 = (56 + 15)(56 - 15) = 71.41$

i.e., $N = 2911 = 41. \cdot 71$.

Problem (11.86): Express $N = 3293$ as a product of two factors.

We choose $x = 58$

so that $(x - 1)^2 < N < x^2$ i.e., $57^2 < 3293 < 58^2$.

Now, $x^2 - N = 58^2 - 3293 = 71$

$2x + 1 = 2 \cdot 58 + 1 = 117$.

Calculation table

a	$(x+a)^2 - N$	Remarks
0	$x^2 - N = 71$	Not a perfect square.
1	$(x+1)^2 - N = \frac{71}{117} = 188$	adding $(2x + 1)$. Not a perfect square.
2.	$(x+2)^2 - N = \frac{188}{119} = 307$	adding $(2x + 1) + 2$. Not a perfect square.
3.	$(x+3)^2 - N = \frac{307}{121} = 428$	adding $(2x + 3) + 2$. Not a perfect square.
4.	$(x+4)^2 - N = \frac{428}{123} = 551$	adding $(2x + 5) + 2$. Not a perfect square.
5.	$(x+5)^2 - N = \frac{551}{125} = 676 = 26^2$	adding $(2x + 7) + 2$. A perfect square.

Thus, $(x + 5)^2 - N = 676 = 26^2$

$\Rightarrow$ $(58 + 5)^2 - N = 26^2$, i.e., $63^2 - N = 26^2$

$\Rightarrow$ $N = 63^2 - 26^2 = (63 - 26)(63 + 26) = 37 \cdot 89 =$ a product of two factors.

Problem (11.87): Find a divisor of $N = 119143$. We express N as a product of two factors.

Let $x = 346$.

Then $(x - 1)^2 < N < x^2$

$\Rightarrow$ $345^2 < N < 346^2$.

Now, $346^2 - N = 346^2 - 119143 = 119716 - 119143 = 573$.

$2x + 1 = 2 \cdot 346 + 1 = 693$.

a	$(x+a)^2 - N$	Remarks
0	$x^2 - N = 573$	Not a perfect square.
1	$(x+1)^2 - N = \frac{573}{693} = 1266$	Add $(2x+1)$. Not a perfect square.
2.	$(x+2)^2 - N = \frac{1266}{695} = 1961$	Add $(2x+1)+2$. Not a perfect square.
3.	$(x+3)^2 - N = \frac{1961}{697} = 2658$	Add $(2x+3)+2$. Not a perfect square
4.	$(x+4)^{-}N = \frac{2658}{699} = 3357$	Add $(2x+5)+2$. Not a perfect square
5.	$(x+5)^2 - N = \frac{3357}{701} = 4058$	Add $(2x+7)+2$. Not a perfect square.
6.	$(x+6)^2 - N = \frac{4058}{703} = 4761 = 69^2$	Add $(2x+9)+2$. A perfect square.

Thus, $(x+6)^2 - N = 69^2 \Rightarrow (346+6)^2 - N = 69^2$

$\Rightarrow \quad N = 352^2 - 69^2 = (352-69)\cdot(352+69) = 283 \cdot 421.$

Thus, 421 (or 283) is a divisor of $N = 119143$.

Problem (11.88): Solve (i) $8x^5 \equiv 10 \pmod{17}$

(ii) $3x^{10} \equiv 7 \pmod{17}$.

Sol: (i) $8x^5 \equiv 10 \pmod{17}$

$\Rightarrow \quad \text{ind}_2^8 + 5\text{ind}_2^x \equiv \text{ind}_2^{10} \pmod{\phi(17)}$

$\Rightarrow \quad \text{ind } 8 + 5\text{ind } x \equiv \text{ind } 10 \pmod{\phi(17)}$

$\Rightarrow \quad 5 \text{ ind } x \equiv -\text{ind } 8 + \text{ind } 10 \pmod{16}$

$\Rightarrow \quad 5 \text{ ind } x \equiv -10 + 3 \pmod{16}$ [from index table $m = 17$]

$\Rightarrow \quad 5 \text{ ind } x \equiv -7 \pmod{16}$

$\Rightarrow \quad 5 \text{ ind } x \equiv 25 \pmod{16}$.

Here $(5, 16) = 1$. It is solvable and it has a unique solution.

Therefore, ind $x \equiv 5 \pmod{16}$ has a solution only when ind $x = 5$.

i.e., when $x = 5$, since $\text{ind}_2^5 = 5$.

Thus, the solution is $x \equiv 5 \pmod{17}$.

(ii) $3x^{10} \equiv 7 \pmod{17}$

$\Rightarrow \quad \text{ind}_2^3 + 10\,\text{ind}_2^x \equiv \text{ind}_2^7 \pmod{\phi(17)}$

$\Rightarrow \quad \text{ind}_2^3 + 10\,\text{ind}_2^x \equiv \text{ind}_2^7 \pmod{16}$

$\Rightarrow \quad 10 \text{ ind}_2^x \equiv -\text{ind}_2^3 + \text{ind}_2^7 \pmod{16}$

$\Rightarrow \quad 10 \text{ ind}_2^x \equiv (-1+11) \pmod{16}$ [See index table $m = 17$]

$\Rightarrow \quad 10 \text{ ind}_2^x \equiv 10 \pmod{16}$.

Here $\quad (n, \phi(m)) = (10, 16) = 2$ and $2|10$.

Therefore, solution exists and it has two incongruent solutions.

That is, $\quad \text{ind}_2^x \equiv 1 \pmod{8}$ has solutions,

where $\quad \text{ind}_2^x = 1$ and $\text{ind}_2^x = 9$.

That is, where $x = 3$ and $x = 14$,

since from index table $\text{ind}_2^3 = 1$ and $\text{ind}_2^{14} = 9$.

Therefore, the solutions are $x \equiv 3 \pmod{17}$ and $x \equiv 14 \pmod{17}$.

Problem (11.89): From the Binet formula of the Fibonacci sequence (See Th [10.9]) show that (i) $f(2n+1) \cdot f(2n-1) - 1 = f^2(2n)$.

(ii) $f(2n+1) \cdot f(2n-1) - f(2n) \cdot f(2n+1) = 1 \; \forall n \geq 1$.

Sol: (i) From the Binet formula we have $f(n) = \dfrac{\alpha^n - \beta^n}{\alpha - \beta}$.

$$f(2n+1) = (\alpha^{2n+1} - \beta^{2n+1})/(\alpha - \beta), \; f(2n-1) = (\alpha^{2n-1} - \beta^{2n-1})/(\alpha - \beta).$$

Then, $f(2n+1) \cdot f(2n-1) - 1 = \frac{(\alpha^{2n+1} - \beta^{2n+1})(\alpha^{2n-1} - \beta^{2n-1})}{(\alpha - \beta)^2} - 1$

$$= \frac{1}{5}[\alpha^{4n} + \beta^{4n} - (\alpha\beta)^{2n-1} \cdot \alpha^2 - (\alpha\beta)^{2n-1}\beta^2 - 5] \quad [\because \alpha - \beta = \sqrt{5}]$$

$$= \frac{1}{5}[\alpha^{4n} + \beta^{4n} + \alpha^2 + \beta^2 - 5] \quad [\because \alpha\beta = -1]$$

$$= \frac{1}{5}[\alpha^{4n} + \beta^{4n} + 3 - 5] = \frac{1}{5}[\alpha^{4n} + \beta^{4n} - 2] = \frac{1}{5}[\alpha^{4n} + \beta^{4n} - 2(\alpha\beta)^{2n}]$$

$$= \left(\frac{\alpha^{2n} - \beta^{2n}}{\sqrt{5}}\right)^2 = \left(\frac{\alpha^{2n} - \beta^{2n}}{\alpha - \beta}\right)^2 = f^2(2n).$$

Hence, $f(2n+1) \cdot f(2n-1) - 1 = f^2(2n)$.

(ii) $f(2n+2) \cdot f(2n-1) - f(2n)f(2n+1)$

$$= \frac{\alpha^{2n+2} - \beta^{2n+2}}{\alpha - \beta} \cdot \frac{\alpha^{2n-1} - \beta^{2n-1}}{\alpha - \beta} - \frac{\alpha^{2n} - \beta^{2n}}{\alpha - \beta} \cdot \frac{\alpha^{2n+1} - \beta^{2n+1}}{\alpha - \beta}.$$

$$= [\alpha^{4n+1} + \beta^{4n+1} - (\alpha\beta)^{2n-1} \cdot \beta^3 - (\alpha\beta)^{2n-1} \cdot \alpha^3]/(\alpha - \beta)^2$$
$$- [\alpha^{4n+1} + \beta^{4n+1} - (\alpha\beta)^{2n} \cdot \alpha - (\alpha\beta)^{2n}\beta]/(\alpha - \beta)^2$$

$$= \frac{\alpha^{4n+1} + \beta^{4n+1} + (\beta^3 + \alpha^3)}{(\alpha - \beta)^2} - \frac{\alpha^{4n+1} + \beta^{4n+1} - (\alpha + \beta)}{(\alpha - \beta)^2}$$

$$= [(\alpha + \beta) + (\alpha^3 + \beta^3)]/(\alpha - \beta)^2 = \frac{(\alpha + \beta) + [(\alpha + \beta)^3 - 3\alpha\beta(\alpha + \beta)]}{(\alpha - \beta)^2}$$

$$= \frac{1 + (1 + 3)}{5} = \frac{5}{5} = 1.$$

Problem (11.90): Let $n = p_1^{a_1} p_2^{a_2} \cdots p_k^{a_k}, n > 1$, where p_i are primes $\forall i = 1, 2, \ldots k$. Then establish each of the following:

(i) $\displaystyle\sum_{d|n} \mu(d)\phi(d) = (2 - p_1)(2 - p_2) \cdots (2 - p_k)$

(ii) $\displaystyle\sum_{d|n} d\phi(d) = \left(\frac{p_1^{2a_1+1} + 1}{p_1 + 1}\right)\left(\frac{p_2^{2a_2+1} + 1}{p_2 + 1}\right) \cdots \left(\frac{p_k^{2a_k+1} + 1}{p_k + 1}\right)$.

(iii) $\displaystyle\sum_{d|n} \frac{\phi(d)}{d} = \left(1 + \frac{a_1(p_1 - 1)}{p_1}\right)\left(1 + \frac{a_2(p_2 - 1)}{p_2}\right) \cdots \left(1 + \frac{a_k(p_k - 1)}{p_k}\right)$.

Sol: (i) Let $P = \prod_{p_i|n}[\mu(1)\phi(1) + \mu(p_i)\phi(p_i) + \mu(p_i^2)\phi(p_i^2)\cdots + \mu(p_i^k)\phi(p_i^k)]$.

Multiplying this we obtain a sum of terms of the form:

$$\mu(p_i^{a_1})\cdot\mu(p_2^{a_2})\cdots\mu(p_k^{a_k})\cdot\phi(p_1^{a_1})\phi(p_2^{a_2})\cdots\phi(p_k^{a_k})$$
$$= \mu(p_1^{a_1}\cdot p_2^{a_2}\cdots p_k^{a_k})\phi(p_1^{a_1}p_2^{a_2}\cdots p_k^{a_k})$$

[since μ and ϕ are both multiplicative. In case of $\phi, \phi(mn) = \phi(m)\cdot\phi(n)$, where $(m,n) = 1$. Here $(p_i^{a_i}, p_j^{a_j}) = 1$]

$$= \mu(d)\phi(d),$$

where $d =$ divisor of $n = p_1^{a_1}p_2^{a_2}\cdots p_k^{a_k}$, $a_i \geq 0\ \forall\ i = 1, 2, \ldots k$.

Hence, $P = \sum_{d|n}\mu(d)\phi(d)$.

In this case, we write $\sum_{d|n}\mu(d)\phi(d) = \prod_{p_i|n}[\mu(1)\phi(1) + \mu(p_i)\phi(p_i)]$

[Since $\mu(p_i^2) = 0\ \forall\ i = 1, 2, \ldots k$, other terms in the product are zero].

$$= (\mu(1)\phi(1) + \mu(p_1)\phi(p_1))(\mu(1)\phi(1) + \mu(p_2)\phi(p_2))\cdots(\mu(1)\phi(1) + \mu(p_k)\phi(p_k))$$
$$= [(1 + (-1)(p_1 - 1))(1 + (-1)(p_2 - 1))\cdots(1 + (-1)(p_k - 1))]$$
$$= (2 - p_1)(2 - p_2)\cdots(2 - p_k).$$

(ii) $\sum d\phi(d) = \prod_{p_i|n}(1\phi(1)) + p_i\phi(p_i) + p_i^2\phi(p_i^2) + \cdots + p_i^{a_i}\phi(p_i^{a_i}), i = 1, 2, \ldots k.$

Let us consider $(1\cdot\phi(1)) + p_1\phi(p_1) + p_1^2\phi(p_1^2) + \cdots + p_1^{a_1}\phi(p_1^{a_1})$ for $i = 1$

$$= 1 + p_1(p_1 - 1) + p_1^2(p_1^2 - p_1) + p_1^3(p_1^3 - p_1^2) + \cdots + p_1^{a_1}(p_1^{a_1} - p_1^{a_1-1})$$
$$= (1 + p_1^2 + P_1^4 + p_1^6 + \cdots + p_1^{2a_1}) - p_1(1 + p_1^2 + p_1^4 + \cdots + p_1^{2a_1-2})$$
$$= \left(\frac{p_1^{2a_1+2} - 1}{p_1^2 - 1}\right) - p_1\frac{(p_1^{2a_1} - 1)}{(p_1^2 - 1)}$$
$$= \frac{p_1^{2a_1+2} - 1 - p_1^{2a_1+1} + p_1}{(p_1^2 - 1)} = \frac{p_1^{2a_1+1}(p_1 - 1) + (p_1 - 1)}{p_1^2 - 1} = \frac{(p_1 - 1)(p_1^{2a_1+1} + 1)}{p_1^2 - 1}$$
$$= \frac{p_1^{2a_1+1} + 1}{(p_1 + 1)}.$$

For $i = 2$, 2nd term of the product $= \dfrac{p_2^{2a_2+1} + 1}{(p_2 + 1)}$

and the last term of product (for $i = a_k$), $\dfrac{p_k^{2a_k+1} + 1}{(p_k + 1)}$.

Hence, $\sum_{d|n} d\phi(d) = \left(\dfrac{p_1^{2a_1+1} + 1}{p_1 + 1}\right)\left(\dfrac{p_2^{2a_2+1} + 1}{p_2 + 1}\right)\cdots\dfrac{(p_k^{2a_k+1} + 1)}{(p_k + 1)}$.

(iii) $\sum\dfrac{\phi(d)}{d} = \prod_{p_i|n}\left[\phi(1) + \dfrac{1}{p_i}\phi(p_i) + \dfrac{1}{p_i^2}\phi(p_i^2) + \cdots + \dfrac{\phi(p_i^{a_i})}{p_i^{(a_i)}}\right]$

Now for $i = 1$, the 1st term of the product

$$= \left[1 + \frac{1}{p_1}\phi(p_1) + \frac{1}{p_1^2}\phi(p_1^2) + \cdots + \frac{1}{p_1^{a_1}}\phi(p_1^{a_1})\right]$$

$$= \left[1 + \frac{(p_1-1)}{p_1} + \frac{1}{p_1^2}(p_1^2 - p_1) + \frac{1}{p_1^3}(p_1^3 - p_1^2) + \cdots + \frac{1}{p_1^{a_1}}(p_1^{a_1} - p_1^{a_1})\right]$$

$$= \left[1 + \frac{(p_1-1)}{p_1} + \frac{(p_1-1)}{p_1} + \frac{(p_1-1)}{p_1} + \cdots + \frac{(p_1-1)}{p_1}\right]$$

$$= \left[1 + \frac{a_1(p_1-1)}{p_1}\right].$$

Similarly, 2nd term $= \left[1 + \frac{a_2(p_2-1)}{p_2}\right]$ and the last term of the product $=$ $\left[1 + \frac{a_k(p_k-1)}{p_k}\right]$.

Hence, $\sum \frac{\phi(d)}{d} = \left(1 + \frac{a_1(p_1-1)}{p_1}\right)\left(1 + \frac{a_2(p_2-1)}{p_2}\right)\cdots\left(1 + \frac{a_k(p_k-1)}{p_k}\right).$

Problem (11.91): If p is prime and $k \geq 2$, then show that

$$\phi(\phi(p^k)) = p^{k-2}(\phi(p-1)^2).$$

Sol: $\phi(p^k) = p^k\left(1 - \frac{1}{p}\right) = p^{k-1}(p-1)$

Again $\phi(\phi(p^k)) = \phi(p^{k-1}(p-1)) = \phi(p^{k-1})\phi(p-1)$

$[(p^{k-1}, p-1) = 1,\ k \geq 2.$ So ϕ is multiplicative, i.e., $\phi(m,n) = \phi(m)\cdot\phi(n)]$

$$= p^{k-1}\left(1 - \frac{1}{p}\right)\phi(p-1) = p^{k-2}(p-1)\phi(p-1)$$

$= p^{k-2}\{\phi(p-1)^2\}.$ [See prob. 5.7; $\phi(n^2) = n\phi(n)$ for any (+)ve integer]

Problem (11.92): (i) Show that there are infinitely many values of n for which $\phi(n) = n|3$.

(ii) Show that there are infinitely many integers n for which $\phi(n)$ is a perfect square.

(iii) Show that there are no integer n for which $\phi(n) = n|4$.

Sol: Let $n = 2^r \cdot 3^s$, $r, s \in Z^+$.

Then $\phi(n) = n|3 \Rightarrow \phi(2^r \cdot 3^s) = \dfrac{2^r \cdot 3^s}{3}$

$\Rightarrow \quad \phi(2^r)\cdot\phi(3^s) = 2^r \cdot 3^{s-1} \Rightarrow 2^r \cdot 3^s\left(1 - \frac{1}{2}\right)\left(1 - \frac{1}{3}\right) = 2^r \cdot 3^{s-1}$

$\Rightarrow \quad 2^{r-1}\cdot 3^{s-1}\cdot 2 = 2^r \cdot 3^{s-1}$

$\Rightarrow \quad 2^r \cdot 3^{s-1} = 2^r \cdot 3^{s-1},$ (1)

which becomes an identity. So $\exists$ infinite number of values of r and s satisfying (1).

Hence, n is infinite.

(ii) Let $n = 2^{2k+1}, k = 1, 2, 3, \ldots$

Then $\phi(n) = \phi(2^{2k+1}) = 2^{2k+1}\left(1 - \frac{1}{2}\right) = 2^{2k}$,

which is a perfect square for infinite values of k.

(iii) If possible, let $\exists$ an integer $n = 2^r \cdot 3^s$ such that $\phi(n) = \frac{n}{4}$.

Then $\phi(2^r \cdot 3^s) = \frac{2^r 3^s}{4}$, r, s are (+)ve integers

$$\Rightarrow \quad 2^r \cdot 3^s \left(1 - \frac{1}{2}\right)\left(1 - \frac{1}{3}\right) = 2^{r-2} \cdot 3^s$$

$$\Rightarrow \quad 2^{r-1} 3^{s-1} \cdot 2 = 2^{r-2} \cdot 3^s.$$

$$\Rightarrow \quad 2^r \cdot 3^{s-1} = 2^{r-2} \cdot 3^s.$$

$\Rightarrow \quad 2^2 = 3$, which is impossible.

In more general, let $n = p_1^n \cdot p_2^s$, p_1, p_2 are primes and $r, s \in Z^+$.

Then proceeding as above we obtain $p_1^2 = p_2$; square of any prime number is a prime number which is not true.

Problem (11.93): For a (+)ve integer n, prove that $\sum_{d|n} \frac{\mu^2(d)}{\phi(d)} = \frac{n}{\phi(n)}$.

Sol: Let $n = p_1^{k_1} p_2^{k_2} \cdots p_m^{k_m}$, then $d|n$ for $d = p_i, p_i p_j, p_i p_j p_k, \ldots, p_1 p_2 \cdots p_m$, where p_i are primes, $i = 1, 2, \ldots m$. For other values of $d = p_i^r$, $\mu(p_i^r) = 0$ for $r \geq 2$. Here μ, ϕ are multiplicative.

$$\text{So} \quad \sum \frac{\mu^2(d)}{\phi(d)} = \prod_{i=1}^{m} \left[\frac{\mu^2(1)}{\phi(1)} + \frac{\mu^2(p_i)}{\phi(p_i)}\right]$$

$$= \left[\left(\frac{\mu^2(1)}{\phi(1)} + \frac{\mu^2(p_i)}{\phi(p_i)}\right)\left(\frac{\mu^2(1)}{\phi(1)} + \frac{\mu^2(p_2)}{\phi(p_2)}\right)\left(\frac{\mu^2(1)}{\phi(1)} + \frac{\mu^2(p_3)}{\phi(p_3)}\right) \cdot \right.$$

$$\left. + \cdots + \left(\frac{\mu^2(1)}{\phi(1)} + \frac{\mu^2(p_m)}{\phi(p_m m)}\right)\right]$$

$$= \left[\left(1 + \frac{1}{\phi(p_1)}\right)\left(1 + \frac{1}{\phi(p_2)}\right)\left(1 + \frac{1}{\phi(p_3)}\right) \cdots \left(1 + \frac{1}{\phi(p_m)}\right)\right]$$

$$= \frac{(1 + \phi(p_1))}{\phi(p_1)} \cdot \frac{(1 + \phi(p_2))}{\phi(p_2)} \frac{(1 + \phi(p_3))}{\phi(p_3)} \cdots \left(\frac{1 + \phi(p_)}{\phi(p_m)}\right)$$

$$[\because \mu^2(p_i) = \mu(p_i) \cdot \mu(p_i) = (-1) \cdot (-1) = 1 \; \forall \; i = 1, 2, \ldots, m]$$

$$= \frac{1 + (p_1 - 1)}{\phi(p_1)} \cdot \frac{1 + (p_2 - 1)}{\phi(p_2)} \cdot \frac{1 + (p_3 - 1)}{\phi(p_3)} \cdots \frac{1 + (p_m - 1)}{\phi(p_m)}$$

$$= \frac{p_1 \cdot p_2 \cdot p_3 \cdots p_m}{\phi(p_1)\phi(p_2) \cdots \phi(p_m)} = \frac{(p_1 p_2 p_3 \cdots p_m) \cdot p_1^{k_1 - 1} \cdot p_2^{k_2 - 1} \cdots p_m^{k_m - 1}}{p_1^{k_1 - 1} \cdot (p_1 - 1) p_2^{k_2 - 1}(p_2 - 1) \cdots p_m^{k_m - 1}(p_m - 1)}$$

$$= \frac{p_1^{k_1} p_2^{k_2} \cdots p_m^{k_m}}{p_1^{k_1}\left(1 - \frac{1}{p_1}\right) p_2^{k_2}\left(1 - \frac{1}{p_2}\right) \cdots p_m^{k_m}\left(1 - \frac{1}{p_m}\right)}$$

$$= \frac{1}{\Pi_{i=1}^{k_m}\left(1 - \frac{1}{p_i}\right)} = \frac{1}{\frac{\phi(n)}{n}} = \frac{n}{\phi(n)} \quad \text{[by (3) of Th. 5.4]}$$

Problem (11.94): Show that, for any (+)ve integer n, $\sum_{d|m}(-1)^{\frac{n}{d}}\phi(d) = \sum_{d|2^{k-1}m}\phi(d) - \sum_{d|m}\phi(2^k d)$, if $n = 2^k m$, where m is any odd integer.

Hence, show that $\sum_{d|n}(-1)^{\frac{n}{d}}\phi(d) = \begin{cases} 0, & \text{if } n \text{ is even} \\ -n, & \text{if } n \text{ is odd.}\end{cases}$

Verify the results for $n = 24$ and $n = 15$.

Sol: For $n = 2^k m,\ k = 1, 2, 3, \ldots$

$d = 1, 2, 2^2, \ldots, 2^{k-1}, 2^k,\ m, 2^{k-1}m, 2^k m$ so that $d|n$.

$$\text{Then } \sum_{d|n}(-1)^{\frac{n}{d}}\phi(d) = (-1)^{\frac{2^k m}{1}}\phi(1) + (-1)^{\frac{2^k m}{2}}\phi(2) + (-1)^{\frac{2^k m}{2^2}}\phi(2^2) + \cdots$$

$$+ (-1)^{\frac{2^k m}{2^{k-1}}}\phi(2^{k-1}) + (-1)^{\frac{2^k m}{2^k}}\phi(2^k) + (-1)^{\frac{2^k m}{m}}\phi(m) + (-1)^{\frac{2^k m}{2^{k-1}m}}\phi(2^{k-1}m) + (-1)^{\frac{2^k m}{2^k m}}\phi(2^k m)$$

$$= \phi(1) + \phi(2) + \phi(2^2) + \cdots + \phi(2^{k-1}) - \phi(2^k) + \phi(m) + \phi(2^{k-1}m) - \phi(2^k m)$$

$$= \{\phi(1) + \phi(2) + \phi(2^2) + \cdots + \phi(2^{k-1})\} + \phi(m) - \{\phi(2^k) + \phi(2^k m)\} + \phi(2^{k-1}m). \quad (1)$$

Again $\sum_{d|2^{k-1}m}\phi(d)$, where $d = 1, 2, 2^2, \ldots, 2^{k-1}, m, 2^{k-1}m$

$$= \phi(1) + \phi(2) + \phi(2^2) + \cdots + \phi(2^{k-1}) + \phi(m) + \phi(2^{k-1}m) \quad (2)$$

and $\sum_{d|m}\phi(2^k d)$, where $d = 1,\ m$

$$= \phi(2^k) + \phi(2^k m) \quad (3)$$

So from (1) by (2) and (3) it follows that

$$\sum_{d|n}(-1)^{\frac{n}{d}}\phi(d) = \sum_{d|2^{k-1}m}\phi(d) - \sum_{d|m}\phi(2^k d).$$

Let n be an even integer. Then two cases may arise:

case (i) $n = 2^k$, case (ii) $n = 2^k p$ where p is an odd prime.

For case (i), $d = 1, 2, 2^2, \ldots 2^{k-1}, 2^k$ so that

$$\sum_{d|n}(-1)^{\frac{n}{d}}\phi(d) = 1 + \phi(2) + \phi(2^2) + \phi(2^3) + \cdots + (2^{k-1}) - \phi(2^k)$$

$$= 1 + \left(1 + 2^2 \cdot \frac{1}{2} + 2^3 \cdot \frac{1}{2} + \cdots + 2^{k-1} \cdot \frac{1}{2}\right) - 2^{k-1}.$$

$$= 1 + (1 + 2 + 2^2 + \cdots + 2^{k-2}) - 2^{k-1}$$

$$= 1\frac{2^{k-1} - 1}{2 - 1} - 2^{k-1} = 1 + 2^{k-1} - 1 - 2^{k-1} = 0.$$

Case (ii) Let $n = 2^k p$, p is an odd prime.

Then, $d = 1, 2, 2^2, \ldots, 2^k, p, 2p, 2^2p, \ldots 2^k p.$

Therefore, $$\sum_{d|n}(-1)^{\frac{n}{d}}\phi(d) = \{\phi(1) + \phi(2) + \phi(2^2) + \phi(2^3) + \cdots + \phi(2^{k-1}) - \phi(2^k)\}$$
$$+ \{\phi(p) + \phi(2p) + \phi(2^2p) + \cdots + \phi(2^{k-1}p) - \phi(2^kp)\}$$
$$= 0 + \phi(p)\{1 + \phi(2) + \phi(2^2) + \cdots + \phi(2^{k-1})\} - \phi(2^k)\phi(p) \quad \text{[by case (i)]}$$

[$\because$ $(2^k, p) = 1\ \forall\ k = 1, 2, \ldots$ and $\phi(2^kp) = \phi(2^k)\phi(p)$, since p is an odd prime]

$$= 0 + \phi(p)[\{1 + 1 + 2 + 2^2 + \cdots + 2^{k-2}\} - 2^{k-1}]$$
$$= 0 + \phi(p)\cdot 0 \text{ [by case (i)]}$$
$$= 0.$$

Hence, $$\sum_{d|n}(-1)^{\frac{n}{d}}\phi(d) = 0 \text{ when } n \text{ is even.}$$

When n is an odd integer. Two cases arise :

Case (i) $n = m$, m is an odd prime.

Case (ii) $n = p_1^{k_1}\ p_2^{k_2} \cdots p_r^{k_r}$, $k_i \geq 0$, p_i are odd primes when n is composite of primes.

For (i): $n = m$, $\Rightarrow$ $d = 1, m$.

$$\therefore \sum_{d|n}(-1)^{\frac{n}{d}}\phi(d) = -\phi(1) - \phi(m) = -\phi(1) - (m-1) = -m = -n.$$

For (ii): $$\sum_{d|n}(-1)^{\frac{n}{d}}\phi(d) = -\sum_{d|n}\phi(d) \quad [\because\ (-1)^{\frac{n}{d}} = -1\ \forall\ d]$$
$$= -\left[\prod_{i=1}^{r}\{1 + \phi(p_i) + \phi(p_i^2) + \cdots + \phi(p_i^{k_i})\}\right]$$
$$= -[\Pi\{1 + (p_i - 1) + (p_i^2 - p_i) + (p_i^3 - p_i^2) + \cdots + (p_i^{k_i} - p_i^{k_i - 1})\}]$$
$$= -\Pi p_i^{k_i}$$
$$= -(p_1^{k_1}\cdot p_2^{k_2}\cdots p_r^{kr}) = -n.$$

Hence, where n is an odd integer,

$$\sum_{d|n}(-1)^{\frac{n}{d}}\phi(d) = -n \text{ [by cases (i) \& (ii)].}$$

Verifications of the above results for $n = 24$ and $n = 15$.

For $n = 24 = 2^3\cdot 3$ $\quad d = 1, 2, 2^2, 2^3, 3, 2\cdot 3, 2^2\cdot 3,\ 2^3\cdot 3.$

$$\sum_{d|n}(-1)^{\frac{n}{d}}\phi(d) = \phi(1) + \phi(2) + \phi(2^2) - \phi(2^3) + \phi(3) + \phi(2\cdot 3) + \phi(2^2\cdot 3) - \phi(2^3\cdot 3)$$
$$= 1 + 1 + 2^2\cdot\frac{1}{2} - 2^3\cdot\frac{1}{2} + (3-1) + \phi(2)\cdot\phi(3) + \phi(2^2)\cdot\phi(3)$$
$$- \phi(2^3)\cdot\phi(3)$$
$$= 1 + 1 + 2 - 2^2 + 2 + 1\cdot(3-1) + 2(3-1) - 2^2(3-1)$$
$$= 4 - 4 + 4 + 4 - 8 = 0.$$

Again, for $n = 15$ (composite number) $= 3 \cdot 5$, $\quad d = 1, 3, 5, 3 \cdot 5$

$$\sum_{d|n}(-1)^{\frac{n}{d}}\phi(d) = (-1)^{15}\phi(1) + (-1)^{5}\phi(3) + (-1)^{3}\phi(5) + (-1)^{1}\phi(3 \cdot 5)$$

$$= -\phi(1) - \phi(3) - \phi(5) - \phi(3) \cdot \phi(5)$$

$$= -1 - 2 - 4 - 2 \cdot 4 = -7 - 8 = -15.$$

Problem (11.95): (1) For a given even perfect number $n = 2^{k-1}(2^k - 1)$, show that $\phi(n) = n - 2^{2k-2}$.

Sol: (1) $n = 2^{k-1}(2^k - 1)$, Here $2^k - 1$ is prime for $k > 1$.

$(2^{k-1},\ 2^k - 1) = 1.$

So $\phi(n) = \phi(2^{k-1}) \cdot \phi(2^k - 1)$

$$= 2^{k-1} \cdot \frac{1}{2}\{(2^k - 1) - 1\} \quad [\text{for prime } p, \phi(p) = p - 1]$$

$$= 2^{k-2}(2^k - 2)$$

$$= 2^{k-2} \cdot 2^k - 2^{k-1} = 2^{2k-2} - 2^{k-1} = 2^{k-1}(2^{k-1} - 1) \qquad \text{(i)}$$

$$= 2^{2k-2} + 2^{2k-1} - 2^{2k-1} - 2^{k-1}$$

$$= (2^{2k-1} - 2^{k-1}) - 2^{2k-1}\left(1 - \frac{1}{2}\right) = n - 2^{2k-2}.$$

(2) Given that n is a perfect even number, show that $n = 1 + 2 + 3 + \cdots + (2^k - 1)$ and also show that $\phi(n) = 2^{k-1}(2^{k-1} - 1)$.

Sol: Let $n = 2^{k-1}(2^k - 1)$.

Now $1 + 2 + 3 + \cdots + (2^k - 1) =$ sum of the 1st $(2^k - 1)$ natural numbers

$$= \frac{(2^k - 1)(2^k - 1 + 1)}{2} = (2^k - 1) \cdot 2^{k-1} = 2^{k-1}(2^k - 1) = n.$$

Again, from (i) of prob (1) of (11.95), $\phi(n) = 2^{k-2} - 2^{k-1} = 2^{k-1}(2^{k-1} - 1)$.

Problem (11.96): Show that the integer $n = 2^{10}(2^{11} - 1)$ is a perfect number.

Sol: n is a perfect number if $\sigma(n) = 2n$, [by Def. 10.1], where $\sigma(n)$ denotes the sum of the (+)ve divisors of n.

For example, $\sigma(6) = 1 + 2 + 3 + 6 = 12 = 2 \cdot 6$, so 6 is a perfect number.

Here $\sigma(n) = \sigma(2^{10}) \cdot \sigma(2^{11} - 1)$

$$= (1 + 2 + 2^2 + \cdots + 2^{10})(2^{11} - 1 + 1) \quad [\because 2^{11} - 1 \text{ is prime}]$$

$$= \frac{2^{11} - 1}{2 - 1} \cdot 2^{11} = (2^{11} - 1)2^{10} \cdot 2$$

$$= 2 \cdot 2^{10}(2^{11} - 1) = 2n.$$

Thus, $\sigma(n) = 2n$. Hence, n is a perfect number.

Problem (11.97): If $n_1, n_2, \ldots, n_r$ are distinct even perfect number, show that $\phi(n_1 n_2 \cdots n_r) = 2^{r-1}\phi(n_1)\phi(n_2)\cdots\phi(n_r)$.

Sol: Let $n_1 = 2^{k_1-1}(2^{k_1} - 1)$, $n_2 = 2^{k_2-1}(2^{k_2} - 1)$, $\ldots$, $n_r = 2^{k_r-1}(2^{k_r} - 1)$.

From (i) of problem (1) of (11.95) we have

$$\phi(n_1) = 2^{k_1-1}(2^{k_1-1} - 1),$$

$\phi(n_2) = 2^{k_2-1}(2^{k_2-1}-1), \ldots, \phi(n_r) = 2^{k_r-1}(2^{k_r-1}-1).$

Then, $\phi(n_1)\phi(n_2)\cdots\phi(n_r) = 2^{\sum(k_i-1)}\Pi(2^{k_i-1}-1),\ i = 1, 2, \ldots r.$

$= 2^{\sum k_i - r}\Pi(2^{k_i-1}-1).$

$\therefore \quad 2^{r-1}\phi(n_1)\phi(n_2)\cdots\phi(n_r) = 2^{\sum k_i - 1}\Pi(2^{k_i-1}-1).$ (i)

Again $n_1 \cdot n_2 \cdots n_r = 2^{\sum k_i - r} \cdot (2^{k_1}-1)(2^{k_2}-1)\cdots(2^{k_r}-1)$

$\therefore \quad \phi(n_1 \cdot n_2 \cdots n_r) = \phi(2^{\sum k_i - r}) \cdot \phi(2^{k_1}-1) \cdot \phi(2^{k_2}-1)\cdots\phi(2^{k_r}-1)$

$= 2^{\sum k_i - r} \cdot \frac{1}{2}(2^{k_1}-2)(2^{k_2}-2)\cdots(2^{k_r}-2)$ [$\because\ 2^k - 1$ is prime]

$= 2^{\sum k_i - r - 1} \cdot 2^r(2^{k_1-1}-1)(2^{k_2-1}-1)\cdots(2^{k_r-1}-1)$

$= 2^{\sum k_i - 1}\Pi(2^{k_i-1}-1),\ i = 1, 2, \cdots r.$ (ii)

From (i) & (ii), $\phi(n_1 n_2 \cdots n_r) = 2^{r-1}\phi(n_1)\phi(n_2)\cdots\phi(n_r).$

Problem (11.98): Show that the following system of equations $x \equiv 5 \pmod 6$ and $x \equiv 7 \pmod{15}$ has no solution.

Sol: Here $(6, 15) = 3$.

$b - a = 7 - 5 = 2. \quad 3 \nmid 2$

So by Theorem [3.14], the system has no solution.

Problem (11.99): Show that the system of equations $x \equiv 5 \pmod 6$ and $x = 8 \pmod{15}$ has a solution and find it.

Sol: Here $(6, 15) = 3 (= d)$, $b - a = 8 - 5 = 3$.

$\therefore \quad d | b - a$ as $3|3$.

Therefore, the system has a solution [by Th [3.14]]

Now $x \equiv 5 \pmod 6$

$\Rightarrow \quad x = 5 + 6y$, where y is given by $5 + 6y \equiv 8 \pmod{15}$ (i)

or, $6y \equiv 3 \pmod{15}$.

which has a solution $y = 3(= y_1) < \frac{q}{d} = \frac{15}{3} = 5$ (See Th [3.14]).

The general values of y are given by

$y = y_1 + \frac{q}{d}t = 3 + 5t,\ t \in Z.$

Then $x = 5 + 6(3 + 5t)$, by (i)

$= 23 + 30t$

$\Rightarrow \quad x \equiv 23 \pmod{30}$.

For $t = 0$, $x = 23$, satisfies the given system of equations.

Problem (11.100): When $n = 14, 206$ and 957, show that $\sigma(n) = \sigma(n + 1)$.

Sol: (i) For $n = 14,\ d = 1, 2, 7, 14$ so that $d|n$.

Then $\quad \sigma(n) = 1 + 2 + 7 + 14 = 24$

$(n + 1) = 15,\ d = 1, 3, 5, 15$ so that $d|(n + 1)$.

$\therefore \quad \sigma(n + 1) = 1 + 3 + 5 + 15 = 24.$

Therefore, $\sigma(n) = \sigma(n+1)$.

(ii) $n = 206,\ d = 1, 2, 103, 206$ so that $d|n$.

$\therefore \quad \sigma(n) = 1 + 2 + 103 + 206 = 312$

$(n+1) = 207,\ d = 1, 3, 3^2, 23, 69, 207$ so that $d|n$.

$\therefore \quad \sigma(n+1) = 1 + 3 + 9 + 23 + 69 + 207 = 312.$

Hence $\quad \sigma(n) = \sigma(n+1)$.

(iii) For $n = 957$, then $d = 1, 3, 11, 29, 33, 87, 319, 957$ so that $d|n$.

$\therefore \quad \sigma(957) = 1 + 3 + 11 + 29 + 33 + 87 + 319 + 957 = 1440$

$(n+1) = 958$, then $d = 1, 2, 479, 958$.

$\therefore \quad \sigma(958) = 1 + 2 + 479 + 958 = 1440.$

$\therefore \quad \sigma(n) = \sigma(n+1)$.

Problem (11.101): Prove that (i) $\tau(n)$ is an odd integer, if and only if n is a perfect square.

(ii) $\sigma(n)$ is an odd integer, if and only if n is a perfect square or twice a perfect square.

Sol: (i) Let $n = p_1^{k_1} p_2^{k_2} \cdots p_r^{k_r}$, where p_i are odd prime for $i = 1, 2, \ldots, r$.

Now, $\quad 1 + p + p^2 + \cdots + p^k$ is odd if and only if k is even.

That is, $\dfrac{(p^{k+1} - 1)}{p - 1}$ is odd, iff k is even $\quad (1)$

Let $\quad k_1 = 2m_1,\ k_2 = 2m_2, \ldots, k_r = 2m_r$ so that k_i are even $\forall i = 1, 2 \cdots r$.

Consequently, $\tau(n) = (k_1 + 1)(k_2 + 1) \cdots (k_r + 1) =$ an odd integer

[$\because$ the product of odd integers is an odd integer]

and subsequently $n = p_1^{2m_1} p_2^{2m_2} \cdots p_r^{2m_r} = (p_1^{m_1} p_2^{m_2} \cdots p_r^{m_r})^2 =$ a perfect square. (2)

It immediately follows from (1) and (2) that

$$\sigma(n) = \left(\frac{p_1^{k_1+1} - 1}{p_1 - 1}\right) \cdot \left(\frac{p_2^{k_2+1} - 1}{p_2 - 1}\right) \cdots \left(\frac{p_r^{k_r+1} - 1}{p_r - 1}\right) = \text{an odd integer,}$$

while n is a perfect square.

Again, if $n = 2 \cdot (p_1^{m_1} p_2^{m_2} \cdots p_r^{m_r})^2 = 2 \cdot p_1^{2m_1} \cdot p_2^{2m_2} \cdots p_r^{2m_r} = 2^1 \cdot p_1^{k_1} \cdot p_2^{k_2} \cdots p_r^{k_r}$.

Then $\quad \sigma(n) = \left(\dfrac{2^{1+1} - 1}{2 - 1}\right) \cdot \left(\dfrac{p_1^{k_1+1} - 1}{p_1 - 1}\right) \cdots \left(\dfrac{p_r^{k_r+1} - 1}{p_r - 1}\right)$

$= 3 \times$ odd integers $=$ an odd integer.

[Note in this case, $\tau(n)$ becomes an even integer]

Problem (11.102): Prove that the integer $53^{103} + 103^{53}$ is divisible by 39 and the integer $111^{333} + 333^{111}$ is divisible by 7.

Sol: $53 \equiv 14 \pmod{39}$

$53^2 \equiv 196 \pmod{39}$

$\Rightarrow \quad 53^2 \equiv 1 \pmod{39}$

and $\quad 53^3 \equiv 14 \pmod{39}$.

Again, $\quad (53^2)^{50} \equiv 1 \pmod{39} \Rightarrow 53^{100} \equiv 1 \pmod{39}$.

Therefore, $53^{100} \cdot 53^3 \equiv 1 \cdot 14 \pmod{39}$

$\Rightarrow \quad 53^{103} \equiv 14 \pmod{39}$. (1)

Now, $\quad 103 \equiv 64 \pmod{39}$

$\Rightarrow \quad 103 \equiv -14 \pmod{39}$

$\Rightarrow \quad (103)^2 \equiv 196 \pmod{39}$

$\Rightarrow \quad (103)^2 \equiv 1 \pmod{39}$

$\Rightarrow \quad (103)^3 \equiv -14 \pmod{39}$.

Therefore, $(103^2)^{25} \cdot (103)^3 \equiv -14 \pmod{39}$

$\Rightarrow \quad 103^{53} \equiv -14 \pmod{39}$. (2)

(1) + (2) gives $53^{103} + 103^{53} \equiv 0 \pmod{39}$.

Hence the result.

For the next case, $111 \equiv 6 \pmod 7$

$\Rightarrow \quad 111 \equiv -1 \pmod 7$

$\Rightarrow \quad (111)^2 \equiv 1 \pmod 7$

$\Rightarrow \quad \{(111)^2\}^{166} \equiv 1 \pmod 7$

$\Rightarrow \quad 111^{332} \equiv 1 \pmod 7$.

Therefore, $111^{332} \cdot 111 \equiv -1 \pmod 7$

$\Rightarrow \quad 111^{333} \equiv -1 \pmod 7$. (3)

Again, $\quad 333 \equiv -3 \pmod 7$ i.e., $333 \equiv 2^2 \pmod 7$.

$(333)^2 \equiv 9 \pmod 7 \Rightarrow (333)^2 \equiv 2 \pmod 7$.

$\Rightarrow \quad (333^2)^{55} \equiv 2^{55} \pmod 7$

$\Rightarrow \quad (333)^{110} \equiv 2^{55} \pmod 7$.

Therefore, $(333)^{111} \equiv 2^{57} \pmod 7$.

Now, $\quad 2^3 \equiv 1 \pmod 7 \Rightarrow (2^3)^{19} \equiv 1 \pmod 7$

$\Rightarrow \quad 2^{57} \equiv 1 \pmod 7$.

So $\quad (333)^{111} \equiv 1 \pmod 7$. (4)

(3) + (4) gives $111^{333} + 333^{111} \equiv 0 \pmod 7$.

Hence the result.

Corollary: Show that $7^9 + 9^7$ is divisible by 8.

Problem (11.103): Show that the highest power h_p of a prime p occuring in $\dfrac{2n!}{n!n!}$ is $h_p = \sum_{n>0} \left[\dfrac{2n}{p^r}\right] - 2\sum_{r>0} \left[\dfrac{n}{p^r}\right]$, where $\left[\dfrac{n}{p^r}\right]$ represents the integers divisible by p^r.

Sol: Go through the examples

(i) For $n = 6$, $\dfrac{2n!}{n!n!} = \dfrac{12!}{6!6!}$.

Let us consider prime 5.

$$12! = 12 \cdot 11 \cdot 10 \cdot 9 \cdot 8 \cdot 7 \cdot 6 \cdot 5 \cdot 4 \cdot 3 \cdot 2 \cdot 1$$
$$= 12 \cdot 11 \cdot (2 \cdot 5) \cdot 9 \cdot 8 \cdot 7 \cdot 6 \cdot 5 \cdot 4 \cdot 3 \cdot 2 \cdot 1$$
$$h_5 = h'_p = \text{highest power of 5 in } 12! = 2$$

5 also divides 6! and h''_5 = highest power of 5 in $6! = 1$

and consequently, h_p = highest power in $\dfrac{12!}{6!6!} = h'_5 - h''_5 = 2 - 2 \times 1 = 0$.

Let $p = 7$. 7 divides 12! but 7 does not divide 6!

$$h'_p = \left[\frac{12}{7}\right] = 1 < 7.$$

$$h''_p = 0 \quad [\because\ 5 < 7].$$

Consequently, h_p = highest power of 7 in $\dfrac{12!}{6!6!} = 1 - 2 \times 0 = 1$.

(ii) We consider $\dfrac{500!}{250!250!}$ for $n = 250$, and let $p = 13$.

$$h'_p = \text{highest power of } p = 13 \text{ in } 500! = \left[\frac{500}{13}\right] + \left[\frac{500}{13^2}\right].$$

$$= 38 + 2 = 40 \quad \left[\because\ \left[\frac{500}{13^2}\right] = \left[\frac{38}{13}\right] = 2\ (12 < 13)\right]$$

$$h''_p = \text{highest power of } p = 13 \text{ dividing } 250!$$
$$= \left[\frac{250}{13}\right] + \left[\frac{250}{13^2}\right]$$
$$= 19 + 1 = 20 \quad \left[\because\ \left[\frac{250}{13^2}\right] = \left[\frac{19}{13}\right] = 1,\ [6 < 13]\right]$$

which occurs twice in $(n!n!)$.

Therefore, h_p = highest power of 13 in $\dfrac{500!}{250!250!} = 40 - 2 \cdot 20 = 0$.

We now proceed the general case of the problem.

Let p (a prime) divide $\dfrac{2n!}{n!n!}$.

Then it necessarily divides $2n!$.

But p may or may not divide $n!$ [as shown in Ex(i) for a prime $p = 7$]. If p does not divide $n!$, then $p > n$ and as such $\left[\dfrac{n}{p^r}\right] = 0\ \forall\ r > 0$.

Hence, it follows that

$$h_p = \text{highest power of } p \text{ in } \frac{2n!}{n!n!}$$
$$= h'_p \text{ (highest powers of } p) \text{ in } 2n! - 2 \times h''_p \text{ (highest power of } p) \text{ in } n!$$
$$= \sum_{r>0}\left[\frac{2n}{p^r}\right] - 2\sum_{r>0}\left[\frac{n}{p^r}\right] \quad [\because\ h''_p \text{ occurs twice in } n!n!]$$

Index of Special Symbols

Index